THE

OFFICIAL

URBAN & WILDERNESS

EMERGENCY SURVIVAL

GUIDE

A Handbook Needed by Every American

to Combat Today's Terrorist Threat

Robert W. Pelton

FOUR COMPLETE URBAN & WILDERNESS EMERGENCY SURVIVAL GUIDES IN ONE:

Section 1: The Official Urban & Wilderness Emergency Survival Manual

Section 2: The Official Urban & Wilderness Emergency Medical Survival Manual

Section 3: The Official Urban & Wilderness Edible Plant Survival Manual

Section 4: The Official Urban & Wilderness Medicinal Plant Survival Manual

ISBN 0-7414-0946-1

Published by:

519 West Lancaster Avenue
Haverford, PA 19041-1413
Info@buybooksontheweb.com
www.buybooksontheweb.com
Toll-free (877) BUY BOOK
Local Phone (610) 520-2500
Fax (610) 519-0261

Printed in the United States of America

Printed on Recycled Paper

Published February, 2003

First Edition

ENDORSEMENTS

ABOUT THE AUTHOR

Robert W. Pelton is one of the nation's most prolific freelance writers and authors. He has published hundreds of feature articles in a variety of magazines and numerous books in his 30 year writing career. Pelton lectures widely, has appeared on many television shows, has been a guest on numerous popular radio talk shows, and has even hosted his own.

Mr. Pelton is considered to be one of the most respected survival experts in the country. He has published a series of four outstanding pocket survival manuals which were the result of 25 years of intense data collecting and firsthand experience, both in and outside the military. These unique survival guides have been widely proclaimed by many leading authorities to be the best and most practical of their kind in the world.

Pelton has been much in demand as a speaker to diverse groups all over the United States on a great variety of survival subjects – pertaining to both urban and wilderness survival techniques. Included are such topics as buying and stocking a backpack, building shelters in the wilderness, long-term food storage for the home, edible and medicinal plants, and so on. Tom R. Murray of the Council of Conservative Citizens, after hearing him speak, offers this: "Mr. Pelton puts together rare combinations of intellectual energies as a writer and speaker that will captivate all levels of an audience while preparing and teaching the listeners. In today's complicated world I feel that no one can involve an audience and deliver an important message more than Mr. Robert Pelton."

Mr. Pelton may be contacted for convention speaking engagements, speaking before church groups, scouting and other groups. Additonally, he is available for book signings and can be reached at:

Survival Specialists
P.O. Box 12619
Knoxville, Tennessee 12619-0619

Fax: 865-633-8398
e-mail: survival@rwpelton.com

Dedicated

To

Vic Harris – a long time best friend going all the way back to the Vietnam era. Vic is one of the nations foremost survival experts and trainers – a man who has taught me much over the years and one who has willingly and unselfishly shared his knowledge and expertise.

And to

Barbara Cram – no doubt America's premier female survival authority and my personal friend of many years standing. This fine (and beautiful) lady is also one of the country's most formidable karate experts and most exciting teachers of the martial arts.

SECTION 1

THE

OFFICIAL

URBAN & WILDERNESS

EMERGENCY

SURVIVAL MANUAL

CONTENTS

1

Introducing The Official Urban & Wilderness Emergency Survival Guide

What should you and your family do if you found yourselves lost while hiking in the mountains, woods or desert? What should you and your family do if while enjoying a weekend camping excursion in the wilds you were suddenly trapped by a severe earthquake or a blizzard? What should you and your family do if a tornado, hurricane, flood or a political upheaval forced you to flee from your home?

What should you do when your clothing gets wet and the temperature is still dropping? What must you do if snow is covering the ground and nothing can be found to make a suitable shelter? Did you know about the ordinary household item that can readily be used to purify water and make it safe for drinking? And food – what can you do when you run out of supplies and face starvation? **The Official Urban & Wilderness Emergency Survival Guide** provides answers to all of the above questions and many more. As survival expert and widely respected survival authority Vic Harris states: *"This manual is absolutely the best I have ever seen. It's full of no-nonsense life saving information. I can't recommend it highly enough."* Yes, **The Official Urban & Wilderness Emergency Survival Guide** is without a doubt the finest book of its kind in the world today.

No one is properly prepared unless they own a useable survival manual. That is why everyone needs **The Official Urban & Wilderness Emergency Survival Guide**. Every American should keep one in their home as well as a copy in each of their motor vehicles. Unfortunately, most of the other purported survival manuals on the market today can't really be taken seriously! They are terribly impractical! They're loaded with excess verbiage spread over hundreds of unnecessary pages. Imagine trying to read the fine print found in so many of them – especially after the sun goes down. Sections dealing with edible and medicinal plants even go so far as to give the plants their Latin names! This is certainly meaningless to a person whose only interest is in trying to stay alive!

Imagine trying to quickly find instructions for starting a fire on a rainy and bitter cold night! You'll freeze first! And if you don't, you'll end up catching pneumonia. Imagine thumbing through these useless volumes in a futile effort to find directions for making an emergency shelter to protect you and your family from the elements. This would be an impossibility! Yes, these books are called survival manuals! Yet, none are much good when you need them most – in a real life survival situation! Nevertheless, one good use does come to mind! The incredible number of pages could be used to start a campfire! Here's what Tom Dodge, outdoors writer for **Heartland USA** magazine has to say about **The Official Urban & Wilderness Emergency Survival Guide**: *"This is certainly the perfect survival manual. It is an indispensable survival tool."*

The first and foremost rule of wilderness survival? Never panic when in the wilds! Always remember that it is normal to experience feelings of fear in certain situations. It's a powerful defense mechanism against the unknown. Fear raises the level of a person's senses. It attunes the mind and body to potential dangers or hazards. Fear under control is wonderful. Out of control fear leads to panic. And panic must be avoided!

The second most important point to remember in any survival situation: *Don't give up!* Remember – nature *always* provides the necessities required for survival – shelter, food, water and fuel (wood, etc.). The essential knowledge as well as the equipment needed to survive in any outdoor emergency is easily found in **THE OFFICIAL URBAN & WILDERNESS EMERGENCY SURVIVAL GUIDE**.

Yes, all of the above things and more are meticulously covered. Bruce Hopkins of Best Prices Storable Foods offers this: *"Pelton's book is truly a great work. It's sure to become a survival standard. Everyone should own at least two copies."* It's ideal for fathers and mothers, members of the military, backpackers, scouts, hikers, campers, hunters, fishermen and all others. It's a most practical, fully illustrated quick reference manual. Yes, **THE OFFICIAL URBAN & WILDERNESS EMERGENCY SURVIVAL GUIDE** is designed to help keep you alive when you are confronted with any dire emergency situation.

Building a Fire

1. Building a fire should be the number one priority of anyone caught out in the wilderness.

2. A person's chances of survival are increased or decreased according to their ability to build a fire anytime, anyplace.

3. A good fire in an emergency situation is important for a number of reasons:

    a. To provide warmth and for keeping dry.

    b. To provide light.

    c. For cooking meals.

    d. For heating water to make beverages (coffee, bouillon, hot chocolate, tea, etc.).

    e. To purify water by boiling.

THINGS NEEDED TO START A FIRE

1. Matches (waterproof/windproof matches are the best) but a magnifying glass will work on sunny days.

2. A good fire starter. *See table below:*

FIRE STARTER TABLE

STARTER TYPE	EASE OF LIGHTING	AMOUNT OF SMOKE	BURN TIME	HEAT INTENSITY
Trioxane Military	Very easy when either wet or dry	Almost none	10+ minutes	Extremely high
Hexamine Military	Quite difficult when wet	Very little	Approximately 6 minutes	Quite high
Tinder Quick U.S.A.F	Difficult with flint	Very little	1 to 2 minutes	Quite low
Sure Start Commercial	Easy when wet or dry	Quite a lot	15+ minutes	High
Magnesium Shavings or Powder	Extremely easy with flint or matches	Almost none to speak of	30 seconds to 1 minute	Extremely high – 5500 degrees

3. Finely shredded dry tinder to provide a low point of combustion. The tinder must be fluffed to allow oxygen to readily flow through.

4. Tinder saturated with Chapstick, vaseline, insect repellent, etc., burns longer and hotter.

5. A few kinds of tinder would Include bird's nests, twine or rope strands, wood chips and slivers cut from dead limbs, finely shredded dry bark and dry leaves.

6. Other tinder includes tiny brittle tree branches, dry moss, dried animal dung (deer, etc.), wood dust produced by bugs under dead tree bark and pine needles.

7. Dry pieces of kindling wood small enough to ignite from the small flame of the tinder.

WHERE TO BUILD A FIRE

1. Build a fire in a well protected area such as against a fallen tree trunk if windy.

2. Build under a tree if it's raining or snowing.

3. Build a fire in a cave if one can be found.

4. When the ground is covered with snow, or when it's wet, build your fire up off the ground.

5. Do this on a green log platform or on a platform made with rocks.

PREPARATIONS FOR BUILDING A FIRE

1. Construct a heat reflector/wind break by stacking a row of green logs behind your fire in order togenerate the most warmth from it.

2. A heat reflector may be more intricate, if desired, and if time permits.

3. Remember – a small fire with a good heat reflector is better than a large fire without one.

4. Heat reflectors concentrate the heat in the direction desired and are also windbreaks which allow your fire to burn more slowly, last longer, and use less fuel.

5. Pick up small logs, wood chunks, and look for fallen tree trunks. These are always a good source of dry wood — even in wet weather.

6. Gather up a good supply of tinder and kindling for your fire.

7. The ideal set-up for your lean-to and a heat reflector. A reflector should be constructed of green logs to best reflect the heat from your fire into your shelter.

8. When there are no trees to be found, gather dried animal dung, look for pieces of coal, etc.

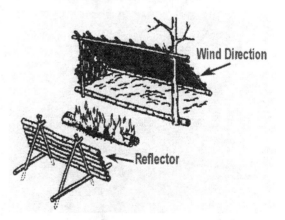

Wind Direction

Reflector

BUILDING YOUR FIRE

1. Put a small piece of fire starter (Trioxane, Hexamine, etc.) on the ground and add tinder in a small neat pile.

2. Build a small tepee structure with pieces of dry, brittle branches taken from dead trees.

3. A "Tepee" fire produces a more concentrated heat and is an excellent fire for cooking.

4. Another method of building a fire is the "Log Cabin Pile." This fire produces lots of heat and light.

5. It is also good for drying wet firewood and it provides a good supply of coals for cooking.

6. The "Pyramid Fire" is also a good choice as a heat producer, dryer of wet firewood, and a coal producer for cooking.

7. **Other fire building methods are the:**

KEYHOLE FIRE

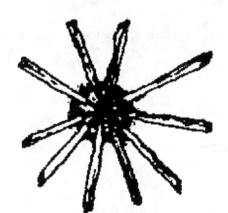

STAR FIRE

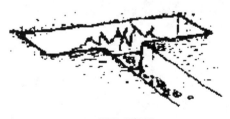

"T"-FIRE

LONG FIRE

8. Start your fire after everything is ready and lightly blow on it to increase the burn rate.

9. Add more tinder as needed and then put on more kindling.

10. Kindling can be dry wood strips, large chunks of bark, pine cones, pine knots and small tree branches.

11. Place two to five inch diameter logs, one to two feet long, on the fire. Split logs catch fire more easily and burn faster than do whole logs.

12. **NOTE**: Keep in mind that a circle of small fires gives off more heat than does one big fire.

KEEPING YOUR FIRE BURNING

1. Add green logs to your fire rather than dry ones as this makes your fire burn longer.
2. Use a heat reflector or wind break as this will keep the embers from blowing in the wind and will stop an unwanted fire from starting in the woods.
3. A windbreak also slows the amount of air going through your fire and allows it to burn longer.
4. Always remember — it's a lot easier to keep a fire going, than it is to build a new one.

BUILDING A FIRE WITHOUT MATCHES

1. Prepare a small amount of extremely dry tinder and carefully shelter this from the rain and wind.
2. A magnifying glass can be used on sunny days to concentrate the sun's rays and ignite the tinder.
3. This glass can be a lens taken from binoculars, a telescopic sight or camera, or ordinary eyeglasses.
4. Use your knife and scrape minuscule shavings from a magnesium bar into a small pile the size of a quarter.
5. Surround the magnesium shavings with tinder and kindling.
6. Strike the flint on the back of the magnesium bar with your knife blade and the sparks will ignite the magnesium shavings.
7. This in turn ignites the tinder which ultimately sets fire to the kindling.

THE FLINT AND STEEL METHOD

1. Gather together your dry tinder and arrange some kindling around it.
2. Hold your flint close to the tinder and strike downward with your knife blade.
3. Continue striking the flint so the sparks hit the tinder and it begins to smolder.
4. Blow gently or fan the air near the tinder until a flame is evident and add more sticks of kindling. Add wood once the fire has caught.

3

How and What To Cook

A FIRE FOR COOKING

1. Build a good fire and let your large firewood logs and pieces burn down to a uniform bed of coals. Coals give off the best heat for cooking.

2. A narrow trench can be dug for a fire and can be used to support your pot or pan.

A FEW GOOD COOKING METHODS

1. Construct a simple cooking structure and place two green logs or some large stones on opposite sides of your fire. This will support your cooking pot or pan.

A piece of sheet metal, if available, can be used to make an effective cooking surface.

2. A "Hobo Stove" made with a large tin can is excellent for cooking in the wilderness. Can be used for frying, making coffee, and a multitude of other things.

3. A long, trimmed tree branch can be propped over your fire and held in place by a shorter forked stick. This will easily hold a pot for cooking.

4. It's also a good way to roast meat. Just slide the meat on the stick. Push the meat down until it's over the fire. Baste to crust the edges and retain its juices.

5. Fish can be cooked over an open fire by using an improvised crane.

6. They can be roasted on an improvised grill made with green sticks (tree branches).

7. They can be wrapped in leaves and clay and then slowly baked.

AN OLD INDIAN STYLE UNDERGROUND STOVE

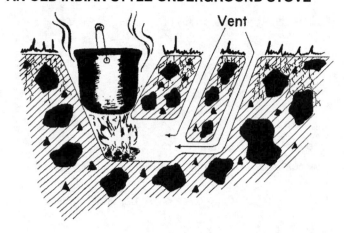

Vent

1. Note the vents on the upwind side. They provide a draft for the fire under the cooking pot.

2. This cooking method has some important advantages in survival situations as it negates the effect of high winds on the fire while cooking.

3. This is an excellent fire to build when evasion is an important factor as it reduces the amount of smoke given off.

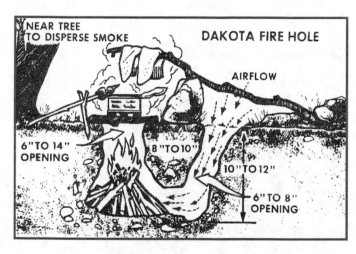

NEAR TREE TO DISPERSE SMOKE

DAKOTA FIRE HOLE

AIRFLOW

6" TO 14" OPENING

8" TO 10"

10" TO 12"

6" TO 8" OPENING

A DAKOTA FIRE HOLE

One of the best fires to build when there are high winds or when evasion is a priority.

A SIMPLE PIT FIRE

1. Scoop out a small hole in the ground.

2. Cut two forks and place them in the ground at opposite ends of the cooking pit.

3. Slide your meat or fish on a long slender tree branch.

4. Lay this branch across the pit and in each of the pit forks.

5. Rotate fish or game occasionally until finished broiling.

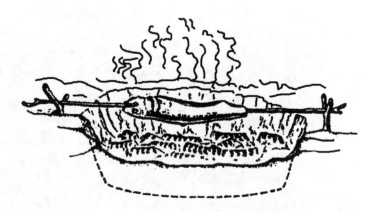

BOILING YOUR FOOD

1. Tough meat and other foods requiring long cooking times, can first be boiled to prepare them for later baking, roasting or frying.

2. Boiling is one of the best cooking methods because it conserves the food's natural juices.

3. This "stock" is excellent nourishment and it can be used to make soup.

BAKING YOUR FOOD

1. Baking is merely cooking in a closed space (an oven) over steady, moderate heat.

2. An oven may be a pit lined with stones (to hold more heat) under your fire, a closed vessel, or a leaf and/or a clay wrapping.

3. To bake in a pit, first fill the pit with hot coals and set your covered pot containing the food and water in the pit.

4. Put a layer of hot coals over the pot and cover this with a thin layer of dirt.

5. Pit cooking is especially good because it gives off no flame at night, and it protects your food from flies and other pests.

STEAMING YOUR FOOD

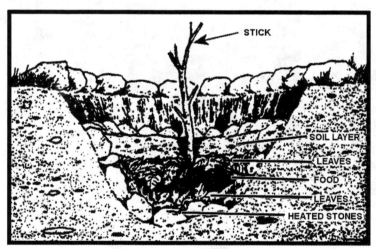

A STEAMING PIT

1. Steaming is a good way to prepare foods that require little cooking such as shellfish.

2. Steaming can be done with no pots and pans. Place your food in a pit with heated stones covered with layers of leaves.

3. Pile more leaves over your food and then push a stick down through the leaves to the food pocket.

4. Pack a dirt layer on top of the leaves and around the stick.

5. Pull out the stick and pour some water down the hole that's left.

6. Steaming is slow, but it works well.

SMOKING YOUR MEAT

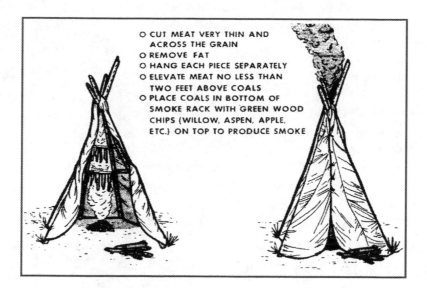

O CUT MEAT VERY THIN AND
ACROSS THE GRAIN
O REMOVE FAT
O HANG EACH PIECE SEPARATELY
O ELEVATE MEAT NO LESS THAN
TWO FEET ABOVE COALS
O PLACE COALS IN BOTTOM OF
SMOKE RACK WITH GREEN WOOD
CHIPS (WILLOW, ASPEN, APPLE,
ETC.) ON TOP TO PRODUCE SMOKE

1. A tepee structure makes an excellent smokehouse when the top flaps are closed.

2. Hang your meat up high.

3. Use a wooden grate for drying the meat and build a slow smoldering fire under it.

4. Smoking removes moisture and preserves the meat.

5. Use salt, if available, as this improves the flavor of the meat and it promotes faster drying.

6. Don't use pitch wood such as pine or fir as they produce unsightly soot and give meat an undesirable flavor.

ANOTHER WAY OF SMOKING MEAT

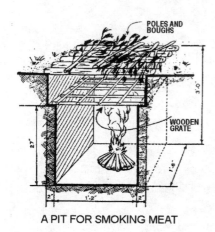

A PIT FOR SMOKING MEAT

1. Dig a large hole in the ground, build a small fire on the bottom, and add plenty of green wood for smoke.

2. Make a wooden grate about 3/4 from the bottom of the hole and cover the pit with leaves, boughs, etc.

3. Your meat will be edible for about a week after one night of heavy smoking.

4. Two nights of smoking will keep the meat edible for two to four weeks or more.

SUN DRYING MEAT

1. String thinly sliced strips of meat between two trees or poles where they will dry in the sun.
2. Fish can be dried in the same manner. Hang the strips as shown here.
3. Fish can also be dried by spreading them on hot rocks in direct sunlight.

COOKING EDIBLE PLANTS

1. Parching is a desirable way to prepare most grains, nuts and seeds.
2. Parching is done by crushing the food and placing it in a metal container or on a flat, heated stone.
3. Allow the food to become thoroughly scorched.
4. After scorching, seeds and grains can be ground into a flour or meal.

COOKING NUTS

1. Most nuts can be eaten raw.
2. Chestnuts are good baked, steamed or roasted.
3. To improve the taste of some nuts, try soaking, parboiling, roasting, baking, boiling, cooking or leeching.
4. Acorns and other bad tasting nuts can be made palatable if they are leeched.
   a. Leeching is crushing the nuts, putting them in a strainer, and pouring boiling water over them.

COOKING POT HERBS

1. Boil leaves, stems and buds until tender.
2. Several changes of water will help to eliminate any bitterness.

TUBERS AND ROOTS

1. Tubers and roots can be boiled, baked or roasted.

COOKING FRUIT

1. Tough and thick-skinned fruit should always be baked or roasted.
2. Juicy fruits should be boiled.
3. Many wild growing fruits can be eaten raw.

COOKING MEAT FROM LARGE GAME

1. All wild game must be thoroughly cooked to kill intestinal parasites.
2. Boil the meat of any animal larger than a house cat and then broil or roast the meat.
3. Cook the meat quickly when broiling as the meat of wild game will get tough over a slow fire.
4. If the meat is exceptionally tough, stew it with vegetables.
5. When baking or broiling meat, allow some of the fat to run over the meat.

COOKING MEAT FROM SMALL GAME

1. All small game and scavenger birds must be thoroughly cooked to kill internal parasites.
2. Small animals and birds can be cooked whole or when cut into pieces.
3. The entrails and sex glands must always be removed prior to cooking.

4. It's best to boil small game as there is little waste this way.

5. A bird can be wrapped in clay and then baked. The clay wrapping removes the feathers when it is later broken from the carcass.

6. Add taste to the bird by stuffing it with greens (dandelions, etc.), roots (onions, etc.), berries and grains.

COOKING REPTILES AND AMPHIBIANS

1. Frogs, small snakes and lizards (salamanders, etc.) can be roasted over a fire on a stick.

2. All frogs and snakes should be skinned before roasting.

3. Eels and large snakes are best if they are boiled prior to roasting.

4. Boil a turtle until the shell falls off, cut up the meat and add tubers (onions, etc.) and plenty of greens to make soup.

COOKING CRUSTACEANS

1. Crabs, prawns, shrimp, crayfish, etc., must be cooked to kill disease-producing organisms.

2. Crustaceans should always be cooked immediately after their capture by dropping them in boiling water.

COOKING MOLLUSKS

1. All shell fish can be steamed, baked in their shell, or boiled.

2. Add tubers (onions, etc.) and greens to make a great tasting soup.

COOKING INSECTS

1. Grasshoppers, large grubs, locusts, ants, termites and most other insects are edible.

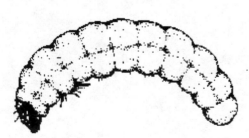

2. They are relatively easy to catch, provide nourishment in an emergency, and are quite tasty when properly prepared.

3. Insects can be used to provide stock for soup and add protein to stew.

4. Cook grasshoppers before eating, in order to kill the parasites in their bodies. They are delicious crisp.

COOKING EGGS

1. Eggs from all birds are edible at all stages of embryo development. Eggs can be eaten raw or cooked in a variety of ways such as fried or boiled.

SIMPLIFIED BREAD BAKING

1. Bread can be made with a simple mixture of flour and water. Knead your dough thoroughly.

2. Remove the bark from a green tree branch. Chew the end of the branch to see if it tastes bitter as it will adversely affect the taste of your bread.

3. If not bitter, then twist the dough around the stick and place it over the fire.

4. Bread can also be made by spreading the dough into thin sheets on a hot rock.

5. A small amount of leaven (dough set aside to deliberately sour) can be added to the bread dough. The adding of leaven will vastly improve the finished bread.

4

Any Shelter is Better Than No Shelter

BUILDING YOUR SHELTER

1. Once your fire has been started and it's burning well, it's time to begin building your shelter.
2. A shelter should ideally be constructed where water is close at hand, where materials are readily available and where there is fuel for your fire.
3. Your shelter should be in a place safe from hazards such as floods, avalanches, poison ivy and other such plants, and pests such as fire ants, mosquitos, etc.
4. The kind of shelter you build will depend upon how much time you have.
5. Factors also include how much daylight is left and the weather conditions (rain, snow, hot, cold, etc.).
6. Also important are the tools you have on hand. Do you have a survival knife? A wire saw? A hatchet?
7. What can your surroundings (nature) supply for your shelter (Leaves? Pine needles? Tree limbs?).
8. What your physical condition is at the time. Are you fatigued? Do you have injuries? Are you dehydrated?
9. And lastly, and most important, is how ingenious (creative) you happen to be.

IMPORTANT THINGS TO CONSIDER ABOUT YOUR SHELTER

1. Is your shelter to be built near the fire?
2. Is your shelter roomy enough to sleep in comfortably?
3. Is there enough room to store your gear?
4. Is the shelter strong enough to hold up?

SOME TYPES OF SHELTERS

PONCHO SHELTERS

1. The simplest kind of shelter can be made by pulling your poncho over your sleeping bag.
2. Easy-to-set-up shelters can be made by attaching your poncho to trees, tree branches, or poles.

IMPROVISED SHELTERS

1. During the summer, you need protection from mosquitos and other pesky insects. During the winter, you can't stay in the open for long unless you're on the move.
2. Tents or other regular shelters are sometimes not available in an emergency situation. If they aren't, a temporary improvised shelter must be built.
3. Suitable natural shelters may be available. Natural shelters would include caves, rock ledges or overhangs and crevices.
4. The shelter you build will depend on the tools you have and available materials. Some sort of shelter can be built in any season by properly using available materials:

    a. In open terrain, a shelter can be built using ponchos, canvas, snow blocks, etc.

    b. Snow houses, snow caves, snow holes, or snow trenches can be built in the winter.

    c. In the woods, a lean-to is usually preferred to other kinds of shelter.

5. Whatever the weather conditions, nature provides the materials to prepare a shelter. Your comfort, however, greatly depends on your initiative and skill at improvising.

A SIMPLE ONE-PERSON SHELTER

1. Spread your poncho on the ground with the hood side up. Close hood by pulling and tying the drawstrings.

2. Raise the poncho in the middle of its shortest side to form a ridge.

3. Stake out the corners first, then the sides.

4. Sod or branches (fir, if available) can be used to seal one side of your shelter from the wind and help retain heat.

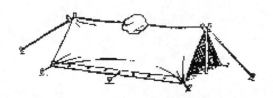

A SIMPLE TWO-PERSON SHELTER

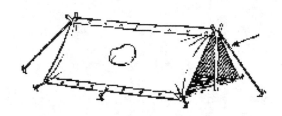

1. Spread two ponchos on the ground with the hood sides facing up. Close both hoods by pulling and tying the drawstrings.

2. The ponchos long sides are then snapped together and raised where they are joined to form a ridge.

3. Stake out the corners first, then the sides.

4. If a third poncho is available, snap it onto the other ponchos to form a waterproof ground cloth.

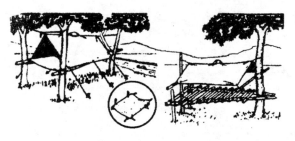

ELEVATED SHELTERS

1. In wet climates, an enclosed elevated shelter is a necessity to protect you from insects and dampness.

A VARIETY OF TEMPORARY SHELTERS

1. You may be lucky and come across a natural shelter suitable for at least some weather conditions.

2. This shelter would take few, if any improvements, other than draping a poncho over it to keep out the rain.

3. Here's a simple shelter to put together. Simply drape a poncho over the top to keep out rain, snow and wind.

BOUGH SHELTER

4. One of the simplest of all temporary shelters is called the "Bough Shelter." The branches on the underside can be trimmed away and used for insulation on the ground.

5. Bough shelters have disadvantages. They don't reflect the heat of a fire and they don't keep rain out.

6. Another temporary shelter to put together in a hurry is called a "Log Shelter." It's quick and easy.

7. Place two poles on a large log and cover the frame with foliage or a poncho if one is available.

8. This shelter, obviously not permanent, doesn't ward off rain when foliage is used for the roof.

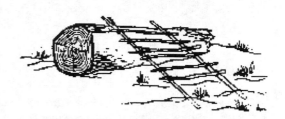

TEMPERATE OR MODERATE CLIMATE SHELTERS

NO POLE A-FRAME

ONE POLE A-FRAME

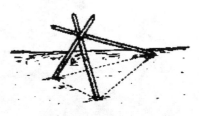

These are all excellent shelters. They're easy to build. They can be quickly completed. Each protects well from wind and rain.

THREE-POLE A-FRAME

METHOD OF PITCHING LEAN-TO

A SIMPLE LEAN-TO

1. A simple but quite suitable lean-to can be constructed between two trees.
2. Cover with brush, leaves, etc., and drape a poncho or a tarp over it to keep out the elements.

ANOTHER LEAN-TO

5. Here's another method of constructing a sturdy lean-to made of trees and tree limbs.
6. Use two trees or two forked poles to support your lean-to crosspiece. Or build two A-frames to hold the crosspiece (pole).
7. Lay a large log at the rear of the lean-to and lay stringers (poles) about one foot apart or less from the front crosspiece to the rear log.
8. Available roof covers -- ponchos, boughs, etc., -- are to be placed on top of the roof stringers.
9. Make a double lean-to by building two single lean-tos face-to-face with a fire between them.

SAPLING SHELTER

1. The framework for a decent shelter can be made by tying willow or other kinds of saplings together.
2. This shelter frame can be covered with a poncho or a tarp and is big enough for one man and his gear.
3. Be sure to place the open end of this shelter at right angles to the prevailing winds.
4. If in cold weather and snow is on the ground, pack the edges of the cover with some of the snow. This will stop the wind from blowing in.

20

A TEPEE OR WIGWAM

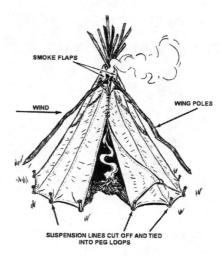

1. A tepee can be constructed in a wooded area by tying a number of poles (tree branches) together at the top.
2. Stand the poles up and spread them at the bottom to form a large circle.
3. This framework is then covered with ponchos or other suitable material.

COLD WEATHER SHELTERS

CONSTRUCTING A SIMPLE BUT DURABLE SNOW SHELTER

1. This is known as a "thermal A-frame". First construct a frame for your shelter from tree branches or whatever else is available.

2. Cover this with one or more ponchos, a tarp, or plastic sheeting. This makes a suitable shelter for rainy weather.

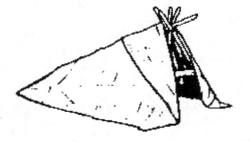

3. Now blanket the entire outside and the floor with brush and small tree limbs (preferably fir). This gives you an ideal moderate weather shelter.

4. This identical shelter can be easily converted to a cold weather shelter with no modifications. Simply cover with lots of snow for better insulation.

THREE STEP MOLDED DOME SHELTER

1. Gather up a large pile of brush. Lay a tarp or poncho liner over the brush. Cover all this with a thick layer of snow. Pack down nicely.

2. When done applying lots of snow, pull out all the brush and the poncho.

3. Make a block of snow to cover the entrance.

4. Here's how your molded dome snow shelter looks in a cutaway view.

CONSTRUCTING A SNOW CAVE

1. A snow cave can be dug out on the lee side of a steep ridge or riverbank. This is where drifted snow collects, and is packed by the wind. Dig an entrance tunnel to your snow cave 18" wide and chest high.

2. Dig out a rectangular portion of snow crossways to the entrance.

3. Shovel out the entrance about two feet inward and about one foot down. Dig upwards in all directions leaving the sleeping floor flat.

4. The roof must be arched for strength. The arched roof will make water forming on the inside follow the curvature of the roof and sides.

5. When finished excavating the inside of your snow cave, the wall thickness should be no less than 20 inches.

6. Now cut some snow blocks and carefully place across the top of the entrance.

7. Pack the cracks between these entrance blocks with snow.

THE COMPLETED SNOW CAVE SHELTER

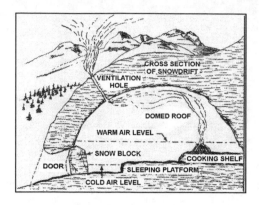

1. Your snow cave must be high enough to provide comfortable sitting space.
2. The sleeping area must be at a higher level than the highest point of the tunnel entrance where there is warmer air.
3. Heat the snow cave with a candle or small stove.
4. Allow no fire to burn while you are sleeping.
5. Cover the entrance with your backpack, a poncho, or a block of snow.
6. When not in use, use your poncho, or tree boughs for insulation on your snow cave floor.

AN IGLOO AS A SHELTER

1. An igloo makes a good shelter in snow areas when other materials aren't available.

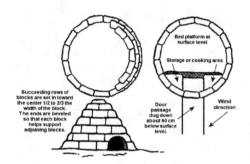

2. Cut the snow into blocks and cover the cracks between the blocks with loose snow.
3. A passageway can be built around the entrance about two yards long, one yard wide and one yard high.
4. The passageway provides more warmth inside your igloo and it's a place to keep a backpack and other gear.
5. Your igloo can be heated with a candle or a one-burner stove.

A TREE PIT SHELTER

1. Tree pit shelters offer excellent protection from the elements.
2. Choose a large fir tree with thick lower branches and surrounded with deep snow.
3. Dig out a large pit around the tree and line the walls and floor with branches.

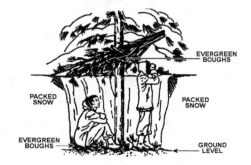

A SNOW WALL

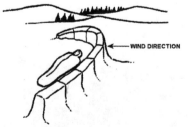

1. A snow wall will provide protection against freezing winds in open terrain with snow and ice.
2. Blocks of snow and ice are all that's needed to build a good wind break.

A SNOW TRENCH

1. A snow trench is one method of quickly making a shelter by simply burrowing into an existing snow drift or by digging out a trench in the snow with the prevailing wind against the side.
2. Make a roof with a poncho or snow blocks or use boughs for a roof covering as well as for a bed.

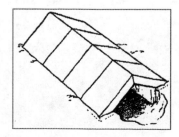

IMPORTANT POINTERS ON SNOW SHELTERS

1. As a general rule of thumb, a snow cave is too warm unless you can see your breath when inside. If too warm, the snow will melt and drip.
2. The best snow cave will have one-third of its living space below the ground level.
3. The inside of a snow cave is never really warm. When it's five degrees above zero outside, it's not more than ten degrees above zero on the inside.
4. Snow cave shelters are a requirement when in cold climate. They are particularly important when the wind chill factor is below -20 degrees.

COLD WEATHER DANGER

1. Suffocation by carbon monoxide is a grave danger during cold weather in a snow shelter.
2. The desire to get warm and stay warm often overrules common sense when a person is freezing.
3. Always depend on your clothing to keep you warm — not a fire in a snow shelter!
4. Fires and heaters are for cooking only when in a temporary shelter.
5. A fire burning for as little as 30 minutes is dangerous in a poorly ventilated shelter. A fire produces a hazardous amount of odorless carbon monoxide fumes.
6. Ventilation can be provided by leaving the top open as an exhaust for fumes leaving the shelter and another opening close to the ground for bringing in fresh air.
7. This could be a partially open door flap.

A FLOOR BED IN YOUR SHELTER

WHY A FLOOR BED?

1. A floor bed in a shelter is of the utmost importance in protecting you from the cold and damp ground.
2. It also provides something comfortable on which to sit or lie down to sleep.

TO FIX A FLOOR BED IN YOUR SHELTER

1. All kinds of plant leaves, ferns, cattail leaves, grasses, etc., can be used as well as boughs from cedar, spruce and other kinds of evergreens.

MAKING A BOUGH BED

1. Select a young fir tree with plenty of easy to reach boughs and cut them off as you walk around the tree.
2. Break apart the small branches and pick out those that are as thick as pencil lead.
3. Cut or find some 4" to 6" thick logs and place them as a border to hold the fir boughs in place.
4. Gather up the boughs in a pile and start placing them in rows against the log frame.
5. Fill in the entire shelter floor area with boughs about 8" to 10" deep.
6. Add additional boughs as necessary for comfort, as many layers mean better insulation from the ground.
7. When lying down at night, give special attention to protecting special areas of your body where heat loss is the greatest such as the head, neck and shoulders.
8. Wearing a sweatshirt with a hood is highly recommended. So is wearing a scarf and a cap.

5

You Can't Do Without Water and Stay Alive for Long

WATER -- YOUR GREATEST NEED!

1. Your chances of living without water are zero. All the food you find means nothing if you don't have water to drink. This is even more true when you sweat abundantly.

2. The body needs at least two quarts of water a day in cold weather. Less than this reduces your ability to find food and do the other things necessary for survival.

THINGS YOU SHOULD KNOW

1. Physical activities should be limited in order to conserve water.

2. Higher altitudes cause a lack of oxygen, which in turn requires more water.

3. Illness or injury create more need for water.

4. You should maintain your body fluid level by drinking as much water as possible.

SOURCES OF WATER

1. Surface water such as streams, rivers, lakes and springs.

2. Precipitation including rain, dew, snow, sleet and hail.

3. Subsurface water such as cisterns and wells.

SIGNS OF WATER

1. Swarming insects indicate that water is close at hand.

2. Early morning or late afternoon bird flights may indicate the direction where water can be found.

3. The "V" of a game trail usually points to water.

4. Look for lush grass in large clumps or signs of other abundant vegetation.

WATER LOCATIONS: WHERE TO LOOK

1. When no surface water is to be found, ground water can often be located in areas as noted above.

2. Limestone caverns have more and larger springs and seepages than any other kind of rock.

3. Lava is porous and a good place to find seeping ground water. Look for springs along the walls of valleys crossing the lava floor.

4. Look for seepage where a dry canyon cuts through a layer of porous sandstone.

5. In areas with lots of granite, look for hillsides with green grass. Dig a small ditch at the base of the greenest area and wait for the water to seep in.

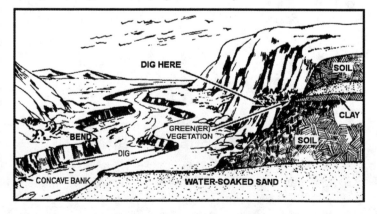

DEHYDRATION CAN BE DEADLY

1. Dehydration is when your body loses water through sweating.

2. Water loss must be replaced by drinking more water. Failure to do this results in a loss of energy and efficiency.

3. Drink plenty of water, especially at mealtimes. If you drink water only during meals, you tend to dehydrate between meals.

4. As more water is lost from your body, you slow down, you lose your appetite, you feel uncomfortably thirsty, you get sleepy and your temperature rises.

5. By the time you dehydrate 5% of your body weight, you will feel nauseated. By the time you lose 6% to 10% of your body weight, the danger signs increase ten-fold.

6. These danger signs include dizziness, headaches, breathing difficulties, tingling arms and legs, dry mouth, skin turns bluish, indistinct speech and inability to walk.

7. Losing as much as 10% of your body weight through dehydration causes no permanent harm, but you must drink enough water to gain it all back.

8. When the temperature is 85 degrees or cooler, you can survive a 25% dehydration. When the air temperature is in the 90s or higher, 15% dehydration could be fatal.

PREVENTING DEHYDRATION

1. Drinking water when dehydrated will return you to normal if taken in time.

2. Drinking water in sufficient amounts to quench your thirst won't prevent dehydration. It simply slows down the process.

3. There is absolutely no substitute for water to prevent dehydration. Only water will keep your body operating at normal efficiency.

USE WATER INTELLIGENTLY!

1. Drink sparingly when water is scarce. Sip small amounts of water at a time when extremely thirsty—but drink your fill.

2. Never drink urine. The waste materials in it will make you sick.

3. Purify all of your water if possible! If you can't purify it, try to get it from a cold, clear, clean and fast running source.

DANGERS OF DRINKING WATER FROM A RIVER, LAKE, POND, ETC.

1. Don't drink water from a river, stream, lake or pond, no matter how overpowering your thirst may seem.

2. Such water usually teems with waterborne disease organisms. These are one of the worst hazards to survival.

3. One disease commonly contracted by drinking untreated water is dysentery which causes severe diarrhea, bloody stool, and high fever.

MAKING WATER SAFE TO DRINK

1. All water should be boiled for at least one minute before it will be safe to drink.

2. Water purification tablets can be safely used in lieu of boiling your water.

3. Water can be aerated to make it taste better by pouring it from one container to another.

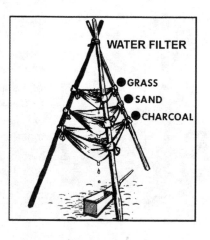

WATER FILTER
● GRASS
● SAND
● CHARCOAL

POLLUTED, STAGNANT & MUDDY WATER

1. You may drink water from a stagnant or muddy pool, puddle or pond, even if it has an unpleasant smell.

2. Muddy water can be made clear by allowing it to stand for 12 or more hours. Or you can repeatedly pour the muddy water through a sand-filled piece of cloth.

3. Charcoal pieces from your fire can be added to boiled polluted water to rid it of obnoxious odors.

WHAT ABOUT SNOW AND ICE?

1. Never eat snow and ice during the winter as a source of water because it can lower your body temperature, induce dehydration and injure your lips and mouth.

2. Put ice or snow in a waterproof container and place it between layers of your clothing, not next to your skin. Use your body heat to melt the snow.

PARACHUTE MATERIAL

SNOW

BUILD A SNOW AND ICE MELTING MACHINE

1. Construct a simple snow and ice melting machine.

MORE WATER SOURCES

1. Plants that collect water in their hollow segments. Water produced by live plants require no special purification treatment.

LEANING TREE METHOD

1. Getting water from a leaning tree is quite uncomplicated. A cloth wrapped around the trunk absorbs rain water and allows it to drop into a container.

GETTING WATER FROM A VINE

1. Not every vine will be a source of drinkable water. Try any vine you can find.

2. Always cut the top of the vine first. Cut a deep notch in it as high as you can reach.

3. Don't drink if a milky sap appears when you cut into the bark. If the juice is clear and like water, then cut the vine close to the ground.

4. Pour some of the juice in your hand and check the taste to see if it's drinkable.

5. Then let the water drip into your mouth or in a container (can, jar, canteen, etc.). Never allow the vine to touch your lips.

IMPROVISED MOISTURE COLLECTORS

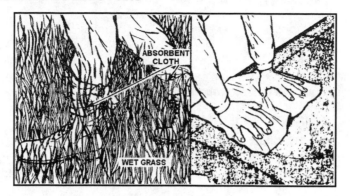

THE BARREL CACTUS AS A WATER SOURCE

1. Cut off the top of this tough, spine-studded rind. Smash the pulp inside the cactus and catch the liquid in a container.
2. One three-foot high barrel cactus will yield about one quart of drinkable juice.

ROOTS OF DESERT PLANTS

1. Many plants in desert areas have roots close to the surface.
2. Pry these roots from the ground, cut into 24 to 36 inch lengths, remove the bark and suck out the water.

SIGNS OF WATER IN ARID DESERT AREAS

1. Watch carefully for nature's plant indicators of water in the desert. These grow only where ground water is close to the surface.
2. Find one or more of these plants and start digging: willows, elderberry, cattails, greasewoods, rushes and salt grass.

PURIFYING WATER WITH ORDINARY HOUSEHOLD BLEACH

1. To use for water storage: plastic gallon and half gallon milk and juice containers, two and three liter soft drink bottles and empty gallon bleach containers.
2. To sterilize storage containers, use one cup of bleach for each gallon of water in your sink. Keep bottles and caps in the sterilizing water for three (3) minutes.
3. To sterilize tap water for long term storage after sterilizing your bottles and caps:

Tap water	Bleach	Tap water	Bleach
1 gallon	8 drops	3 liters	6 drops
1/2 gallon	4 drops	2 liters	4 drops
1 quart	2 drops	1 liter	2 drops
1 pint	1 drop		

4. To sterilize tap water for long term storage without sterilizing your bottles and caps:

Tap water	Bleach		Tap water	Bleach
1 gallon	16 drops		3 liters	12 drops
1/2 gallon	8 drops		2 liters	8 drops
1 quart	4 drops		1 liter	4 drops
1 pint	2 drops			

5. **Note:** The bleach used for water purification is the common household variety used for every day laundry. It isn't harmful if used properly.

OBTAINING DRINKING WATER ON A COAST

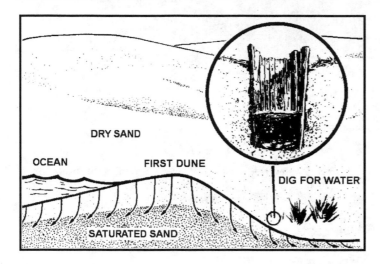

DRY SAND

OCEAN   FIRST DUNE

DIG FOR WATER

SATURATED SAND

1. Water can be located in the sand dunes above a beach and sometimes on the beach itself.

2. Look in the hollows between sand dunes for visible water and dig where the sand appears to be moist.

3. Try digging a "beach well" as shown above.

4. **WARNING:** Never drink sea water as the salt concentration is so high that body fluids have to be drawn to eliminate it. Your kidneys will eventually stop working.

OTHER METHODS OF OBTAINING DRINKING WATER

WATER TRANSPIRATION BAG

WATER TRANSPIRATION BAG

1. Use a clear plastic bag as shown. The collected water will have a taste like the plant smells.

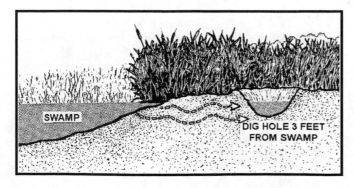

SWAMP

DIG HOLE 3 FEET FROM SWAMP

HERE'S HOW A SEEPAGE BASIN WORKS

1. The water is filtered through the sand.

BUILDING A GOOD SOLAR STILL

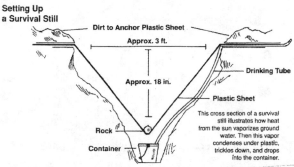

Setting Up
a Survival Still

Dirt to Anchor Plastic Sheet

Approx. 3 ft.

Approx. 18 in.

Drinking Tube

Plastic Sheet

This cross section of a survival still illustrates how heat from the sun vaporizes ground water. Then this vapor condenses under plastic, trickles down, and drops into the container.

Rock

Container

1. A solar still should be built in a low area where there is little or no shade.

2. The still hole should be bowl shaped and have a slightly rounded top edge.

3. Dig a small hole within the still to hold a pail or other kind of water container. The plastic sucking tube is put in this water container and taped to hold in place.

4. Cover the walls of the solar still hole with leaves and plants.

5. Place the plastic sheet over the hole, but leave enough slack to make a downward slope. Seal the plastic sheet all around the top with dirt and rocks.

6. Put a rock in the center of the plastic and over the container. This rock should pull the plastic taunt but it shouldn't touch the container or the sides of the hole.

HOW AND WHY A SOLAR STILL WORKS DURING THE DAY

1. The sun's heat raises the air and soil temperature under the plastic. The air becomes saturated and unable to hold any more water vapor.

2. The vapor condenses in small drops of water on the plastic's underside. These drops of water run down the sloping underside of the plastic and drip into your bucket.

3. **NOTE**: Don't expect to have instant drinking water. You should have from a pint to a quart in 24 hours.

HOW YOUR STILL CONTINUES TO WORK AFTER SUNDOWN

1. After the sun goes down, the plastic cools while the soil temperature remains relatively high. Water vapor continues to condense on the plastic undersurface.

2. A water still will produce approximately half as much water from 5:00 p.m. to 8:00 a.m. as it will during sunny daylight hours.

MATERIALS YOU'LL NEED TO SET UP A SURVIVAL WATER STILL

1. One 6' x 6' (or larger) sheet of clear plastic. A painting drop cloth will do nicely.

2. A one gallon bucket or whatever else may be available.

3. 5'- 0" (Five feet) of plastic tubing. This tubing allows you to drink without pulling your bucket out of the hole.

YOUR WATER STILL ALSO A GOOD FOOD SOURCE?

1. Yes, the water bucket under the plastic sheeting attracts small animals and snakes.

2. Snakes and small animals may crawl down the plastic sheet in search of water.

3. They are then trapped since they are unable to crawl out on the slippery plastic sheet.

6

If You Run Out of Food — Edible Plants

HARVESTED VEGETABLE AND GRAIN FIELDS

5. Look for old pea patches as well as turnip, corn and potato fields. Dig into the hills of a potato field for potatoes that were overlooked when harvesting.

6. Look for fields with vegetable stalks still in the ground after harvesting. Included are beets, rutabagas, radishes, carrots and turnips.

7. Clean and peel beets, rutabagas and turnips. Eat raw or cooked. Radishes can be eaten raw.

8. Find abandoned corn fields and look for discarded ears of corn. Eat corn kernels cooked or raw.

9. Parch ears of corn in hot ashes or over a fire and grind into flour or meal. A handful mixed with water makes a highly nutritious food called *pinole*.

POINTERS ABOUT WILD PLANTS AS FOOD

1. Plants eaten by birds and animals are generally safe to eat, yet there are some specific plants eaten by animals that are poisonous to humans.

2. For example, poison ivy is eaten by horses. Squirrels eat every kind of mushroom. Baneberry is eaten by bears. And seeds containing strychnine are eaten by birds.

3. A word of caution: There are few plants around of which every part of the plant is edible.

BE CAREFUL WHEN CHOOSING WHAT PLANTS YOU EAT

1. Mushroom and fungi should not be chosen as a food possibility. Many are agonizingly deadly.

2. Avoid plants with umbrella-shaped clusters of flowers.

3. Avoid plants with beans and peas.

4. Avoid plants with things resembling melons, cucumbers or parsnips.

5. Avoid plants having a carrot-like foliage.

6. As a general rule, all bulbs should be avoided.

7. Don't eat any plant resembling dill or parsley.

8. Avoid white and yellow berries as they are almost always poisonous.

9. Be careful with red berries as approximately half of them are poisonous.

10. Don't eat the seeds and nuts of fruit as some contain strychnine.

11. Plants with shiny leaves should be considered poisonous.

12. A milky sap usually indicates a poisonous plant with very few exceptions (a dandelion is one exception).

13. Plants having yellow, red, orange, dark or soapy tasting sap should be avoided.

14. Avoid plants with sap that quickly turns dark when exposed to the air.

15. Plants that irritate the skin (for example, poison ivy) should not be eaten.

16. Never eat grasses or grass seeds that have turned black.

A FEW POINTERS ON SAFE PLANTS TO EAT

1. Black and blue berries are almost always edible.

2. Aggregated berries (raspberry, blackberry, thimbleberry) are usually safe to eat.

3. Single fruits on a stem are generally considered safe to eat.

4. Leaves from violets, wild strawberries and rose hips are all edible and are excellent sources of vitamin C.

5. Vitamin C is also obtained by making a tea from the needles of pine, hemlock and spruce trees.

6. Almost all nuts are suitable to eat.

EDIBILITY TESTS FOR PLANTS

1. Never apply the wild plant test to more than one plant at a time. If a person gets sick or suffers other ill effects, it will be obvious which plant caused the problem.

2. Touch the plant's juice to your inner forearm or tongue tip. Tasting of a poisonous plant won't kill you.

3. If there are no ill effects such as a rash, numbing sensation, etc., proceed with the rest of the steps.

4. Boil the plant or plant part twice for five minutes in two changes of water. The poison properties in many plants are either water soluble or heat destroys them.

5. Place one teaspoon of the boiled plant in your mouth and chew but don't swallow.

6. If unpleasant effects occur — bitter, burning, or nauseating taste — immediately spit out the material and discard that plant as a source of food.

7. If no unpleasant effects are noted, swallow the teaspoon of plant material and wait for eight hours.

8. If no unpleasant effects are noticed after eight hours — diarrhea, cramps, nausea — take two more teaspoons of plant material and wait another eight hours.

9. If no unpleasant effects are noticed at the end of this second eight hour period, the plant can be considered edible.

EDIBLE ROOTS AND ROOTSTALKS

1. These starch storing foods include tubers, root stalks and bulbs. All tubers are found underground.

2. Some edible roots grow to seven feet long. Rootstalks are several inches thick, short, and pointed.

3. Roots and tubers have more food value than do the leaves. Cook tubers only by roasting or boiling.

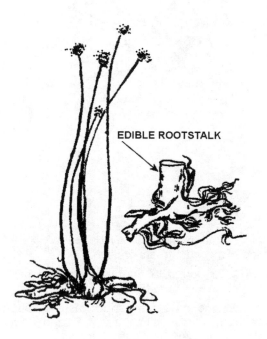

EDIBLE ROOTSTALK

BULRUSH

1. A tall plant usually found in wet swampy areas.

2. White stem and roots can be eaten raw or cooked.

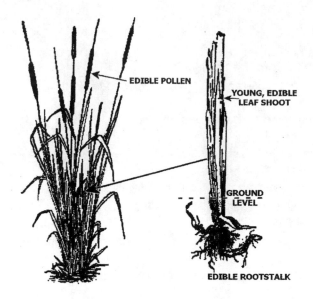

EDIBLE POLLEN

YOUNG, EDIBLE LEAF SHOOT

GROUND LEVEL

EDIBLE ROOTSTALK

CATTAIL

1. Found along lakes, ponds and rivers.

2. Grows six to 15 feet high with long, thin, pale green leaves and rootstalks to one inch thick.

3. Cattail roots contain 46% starch and 11% sugar.

4. Peel off the outer covering, grate the inner white part and eat raw or boiled.

5. Mix the yellow pollen from the flowers with water and then steam as bread cakes. The young shoots are delicious when boiled as you would asparagus.

DANDELION

1. Leaves and roots can be eaten raw but taste better if boiled or fried. Makes a great omelet if the leaves are blended with eggs and fried.

2. Clean, split and cut the dandelion roots into small pieces and roast. Then grind between stones and brew to make a coffee-like drink.

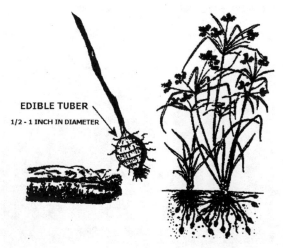

EDIBLE TUBER

1/2 - 1 INCH IN DIAMETER

NUT GRASS

1. Can be found in moist sandy places along the edge of a pond, stream or ditch.

2. Thick underground tubers grow from one-half to one inch thick and taste sweet and nutty.

3. Boil, peel and grind between stones into flour. Ground nut grass tubers can be used as a coffee substitute.

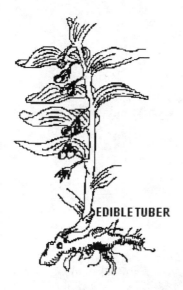

SOLOMON'S SEAL

1. Tubers grow on small plants and taste much like parsnips when roasted or boiled.

EDIBLE TUBER

WATER CHESTNUT

1. Water chestnuts are found floating in large areas on rivers, lakes and ponds.

2. Nuts from one to two inches around, with spines resembling a horned steer, are found under the water. The nuts inner seed can be eaten when roasted or broiled.

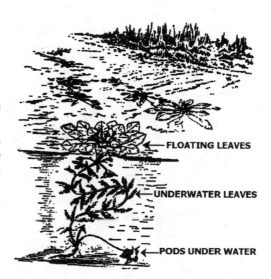

FLOATING LEAVES

UNDERWATER LEAVES

PODS UNDER WATER

WHITE FLOWERS

EDIBLE STARCHY ROOTSTOCKS

WATER PLANTAIN

1. A white flowering plant with a long stalk and smooth, heart-shaped leaves. It's usually found around lakes, ponds and streams. Also abundant in marshy areas.

2. Most often partly submerged under water.

3. Thick, bulb-like, acrid tasting, root stalks grow below the ground. The acrid taste disappears when this plant is dried. Cook the way you would cook potatoes.

WILD POTATO

1. The wild potato is an edible tuber found all over America. Eat only after cooking for it's poisonous when consumed raw.

EDIBLE TUBER

WILLOW

1. Easily identified by flowers or fruit clusters that develop into roughly one inch caterpillar-like spikes.

2. An extremely rich source of vitamin C.

EDIBLE BULBS

All bulbs contain a lot of starch. They all, with the exception of the wild onion, taste better cooked.

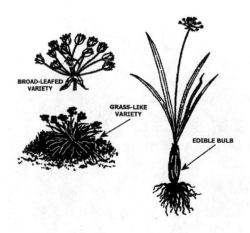

BROAD-LEAFED VARIETY

GRASS-LIKE VARIETY

EDIBLE BULB

WILD ONION

1. This is the most common of all wild bulbs and is found all over the United States. All varieties of wild onion can be detected by their onion smell.

2. All onion bulbs are edible. The leaves can be cooked and eaten (only before flowerstalks appear) as greens or eaten raw in salads.

36

EDIBLE SHOOTS AND STEMS

1. Shoots and stems grow much like asparagus and the young ones make excellent food.
2. Most are better if parboiled for 10 minutes, drained off, and then reboiled until tender.

WILD GOURD OR LUFFA SPONGE

1. Grows like a watermelon, cucumber or cantaloupe with vines three to eight inches across.
2. You might find this growing wild in old gardens and clearings with smooth, seedy and cylindrical fruit. It should be boiled and eaten when half ripe.
3. Shoots, flowers and new leaves should be cooked before eating. Roast the seeds and eat them like peanuts.

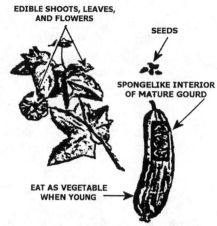

EDIBLE SHOOTS, LEAVES, AND FLOWERS

SEEDS

SPONGELIKE INTERIOR OF MATURE GOURD

EAT AS VEGETABLE WHEN YOUNG

EDIBLE FERNS

Ferns are abundant all over the United States. They are found in wooded areas, along streams, gullies, etc.

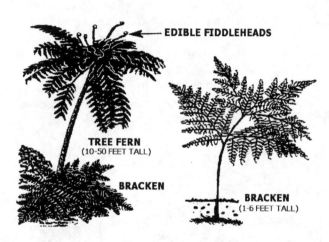

EDIBLE FIDDLEHEADS

TREE FERN
(10-50 FEET TALL)

BRACKEN

BRACKEN
(1-6 FEET TALL)

BRACKEN FERN

Select the young stalks not more than six to eight inches high and break off low on the plant. Wash and then boil in salted water or steam until tender.

EDIBLE FLOWERS

1. There are lots of edible flowers including the Bachelor's Button, Carnation, Clover, Daisy, Dandelion, Day-Lily, Forget-Me-Not and Gardenia.

2. Other edible flowers include the Scented Geranium, Honeysuckle, Lilac, Marigold, Nasturtium, Pansy, Redbud, Rose and Violet.

3. The stamen and the style in the flower's center should be removed before the flower is eaten. This part of the flower contains the pollen which sometimes causes serious allergic reactions.

EDIBLE LEAVES

Plants with edible leaves are the most numerous of all plant foods. Most can be eaten cooked or raw but overcooking them will destroy most of the vitamins.

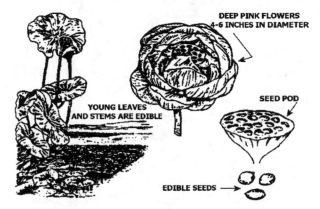

YOUNG LEAVES AND STEMS ARE EDIBLE

DEEP PINK FLOWERS 4-6 INCHES IN DIAMETER

SEED POD

EDIBLE SEEDS

LOTUS LILY

1. Grows in lakes, ponds and slow streams where the plant stands five to six feet above the water's surface.

2. The rootstalks grow 50 feet long with bubble-like enlargements and has white, yellow or pink flowers and leaves measuring three to five feet across.

3. The seeds are edible when ripe after removing the bitter embryo and then boiling or roasting.

4. The rootstalks can be boiled and eaten like potatoes while the stems and leaves can be boiled and eaten after removing the rough outer layer.

PRICKLY PEAR

1. This plant has a water-filled, one inch diameter stem covered with short, needle-like spines.

2. Slice off the top of the egg-shaped fruit, peel back the outer layer, and eat it all.

3. Cut off the sharp spines, slice the stem lengthwise into thin strips, and boil or eat raw.

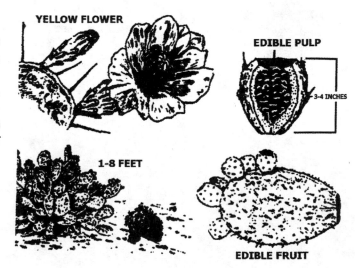

YELLOW FLOWER

EDIBLE PULP

3-4 INCHES

1-8 FEET

EDIBLE FRUIT

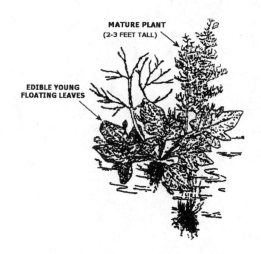

MATURE PLANT
(2-3 FEET TALL)

EDIBLE YOUNG
FLOATING LEAVES

WATER LETTUCE

1. Usually found as a floating plant covering large areas in ponds, lakes and backwaters. Boil before eating.

WILD CHICORY

1. Wild chicory grows as a weed along the sides of roads and in fields with abundant flowers and stems two to four feet long.

2. Has a long carrot-like root which can be ground and used as a substitute for coffee.

3. Leaves clustered close to the ground look like dandelion leaves but are rougher and thicker. They are tender and can be eaten raw.

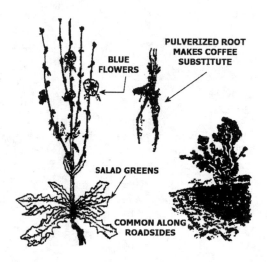

BLUE FLOWERS

PULVERIZED ROOT MAKES COFFEE SUBSTITUTE

SALAD GREENS

COMMON ALONG ROADSIDES

EDIBLE NUTS

Nuts are among the most nutritious of all plant foods. They contain much valuable protein.

EDIBLE ACORNS

ACORNS

1. Acorn trees grow to a height of 60 feet.

2. Acorns aren't edible raw because of their extremely bitter taste. Boil them for two hours, pour out the water, and soak for three to four hours in cold water.

3. Grind nuts to a paste with stones, mix with water to make a mush and cook. Or spread the acorn mush on stones and let dry to make a flour.

BEECHNUT

1. Beechnut trees grow wild in moist areas.
2. Mature beechnuts fall out of their husks.
3. Roast and then pulverize the nut kernel.
4. Boil the powder for a coffee substitute.

HUSK SURROUNDING NUT

EDIBLE NUT

HAZELNUT

1. Grows on six to twelve-foot bushes in dense thickets along stream banks as well as in open places.
2. Hazelnuts ripen in the fall and can be eaten fresh or when roasted. Their great food value is derived from the heavy oil content.

HAZELNUTS

VARIETIES OF EDIBLE NUTS

EDIBLE SEEDS AND GRAINS

1. The seeds of many plants – buckwheat, beans, peas, etc. -- contain oil rich in protein. Grains used for cereal and other grasses are also rich in plant protein.
2. They can be ground between stones, mixed with water and made into porridge. Grains such as corn can be parched and preserved for future use.

WILD RICE

1. Rice grains can be roasted and beaten into flour for making bread or they can be simply boiled and eaten.

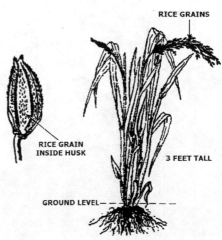

RICE GRAINS

RICE GRAIN INSIDE HUSK

3 FEET TALL

GROUND LEVEL

40

EDIBLE FRUIT

Wild fruit is plentiful. Included are crab apples, cherries, plums, apples, raspberries and others.

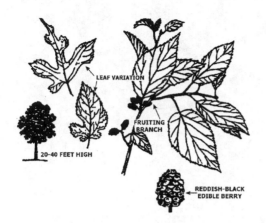

MULBERRY

1. Found in forested areas, alongside roads, and in abandoned fields, mulberry vines grow 20 to 60 feet tall and have one to two inch fruit resembling the blackberry.

2. Each berry is about as thick as the index finger and varies in color from black to red.

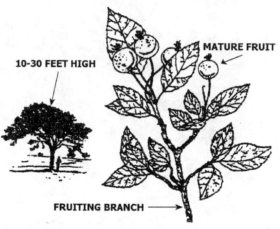

WILD CRAB APPLE

1. Look for this tree in woodlands, fields, and the edge of woods. Crab apples can be eaten whole or cut into small slices and dried in the sun.

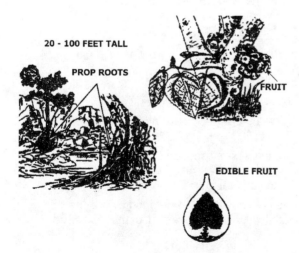

WILD FIGS

1. A few kinds of wild fig trees are primarily found in the desert areas of the United States.

2. Fig trees are evergreen with large leathery leaves. The fruit resembling a pear is red, green or black, almost hairless, and soft when ripe enough to eat.

WILD GRAPEVINE

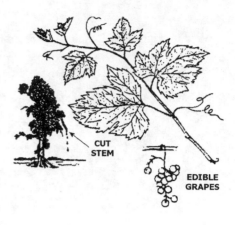

1. The grapes hang in large bunches and are rich in energy giving sugar. Water can be extracted from the vine.

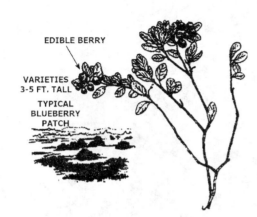

WILD HUCKLEBERRIES, BLUEBERRIES AND WHORTLEBERRIES

1. Large patches of these tasty berries thrive in the late summer.

EDIBLE BARK

1. The inner bark of many trees -- aspen, willow, cottonwood, birch and pine -- may be cooked or eaten raw.

2. Pine bark and bark from other cone bearing trees is particularly rich in vitamin C.

3. Scrape away the outer bark in order to strip the inner bark from the tree trunk. Pulverize the inner bark into flour and then eat it cooked, dried or fresh.

EDIBLE SEAWEED

1. Properly prepared seaweed is a natural source for iodine and vitamin C.

2. Wash thick leathery seaweed and then soften it by boiling. Get fresh seaweed attached to rocks or that which is seen to be floating free.

3. Tender bunches of seaweed can be dried over a fire or in the sun until crisp. Crush and use this for soup flavoring.

FRESH WATER ALGAE

1. This type of seaweed is commonly found growing in pools and ponds during the spring and summer. It forms round, green, jelly-like globs about the size of marbles.

2. Dry this algae and use in soup.

42

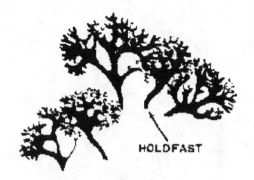

IRISH MOSS

1. Commonly found below the high tide line on the beach. Always boil Irish moss before eating.

HOLDFAST

KELP

1. Look for short, cylindrical stem and thin wavy fronds measuring from one to several feet long. Boil before eating and mix with vegetables or add to soup.

LAVER

1. Usually red, dark purple or purplish-brown with satiny sheen. Found at low tide on beaches. Use as relish or mix with crushed grain, make into pancakes, and fry.

STALKS 1-5 FEET LONG

SUGAR WRACK

1. The stalks of this edible brown seaweed plant are sweet to the taste.

HOLDFAST

FUNGI

1. There are thousands of varieties of mushrooms or fungi.

2. Mushrooms are not really worth picking and eating. They have no nutritional value.

Why even bother learning to identify edible mushrooms? Why go to the trouble of preparing mushrooms to eat? And better yet, why even bother eating mushrooms in the first place? No logical answers can be given to the above questions – at least from the standpoint of survival!

The United States military takes this approach to the identifying, collecting and eating of mushrooms. The official Army survival manual, FM 21-76 gives this warning:

"DO NOT EAT MUSHROOMS IN A SURVIVAL SITUATION! THE ONLY WAY TO TELL IF A MUSHROOM IS EDIBLE IS BY POSITIVE IDENTIFICATION. THERE IS NO ROOM FOR EXPERIMENTATION. SYMPTOMS OF THE MOST DANGEROUS MUSHROOMS AFFECTING THE CENTRAL NERVOUS SYSTEM MAY SHOW UP AFTER SEVERAL DAYS HAVE PASSED WHEN IT IS TOO LATE TO REVERSE THEIR EFFECTS."

Need more be said?

7

Obtaining Animals for Food

1. All animals are edible but they are the most difficult to obtain of all survival foods. Hunting and trapping animals isn't easy even for the experienced.
2. Those who have had no hunting experience should hide near a place where animals pass. This includes a trail, a watering hole or a feeding ground.
3. This "still hunt" hiding place must always be downwind to prevent the animals from picking up your scent and being scared away.
4. Remain motionless and wait until the animal comes into the range of your weapon or walks into your trap.

STALKING AN ANIMAL

1. Stalk downwind of the animal so no scent is picked up. Move noiselessly and slowly when the animal is looking the other way or feeding.
2. Freeze and remain motionless when the animal looks your way.

HUNTING GAME

1. The secret to successful hunting is simply to see your quarry before it sees you. Carefully watch for signs that reveal the presence of game.
2. When coming upon a clearing, lake or a ridge, slow down and look first at distant, then closer ground.
3. Hide when finding signs of game at a water hole. Wait until an animal approaches. Be patient for this may take hours.
4. Hunt early in the morning or at dusk for best results. Look for these clues: animal tracks, a game run or trail, trampled down underbrush and animal droppings.

BIRDS

1. Birds can see and hear exceptionally well but they lack a good sense of smell.
2. Birds fear man less while nesting and are easier to catch in the spring and summer while nesting in trees, marshes, bushes, and on cliffs, etc.
3. Watch for older birds and you will be able to locate eggs.

SHOOTING GAME FOR SURVIVAL

1. An animal usually stops if you whistle sharply. This gives you time to stand and get off one shot. Aim for the neck, lung or head when shooting large game.
2. If animal is wounded and runs, slowly and deliberately follow the trail of blood. If severely wounded, it will lie down, weaken and be unable to rise.

AFTER THE KILL

1. After killing a large animal such as a deer, gut and bleed it immediately.
2. Cut the musk glands from between the hind legs and at the joints of the hind legs. Be careful to not burst the bladder while removing it.

TRAPPING SMALL GAME

YOU MUST KNOW YOUR SMALL GAME

1. To successfully trap small game, you must decide the kind of animal you want to trap. You must know how the animal will react. You must know what kind of bait to use.

2. Squirrels, rabbits, rats and mice are all relatively easy to trap. They follow a regular pattern of habits and confine themselves to limited areas of activity.

3. Rabbits run in a circle when they are scared. One way to catch a rabbit or other small game is with a wire snare, placed on a game trail.

HANGING SNARES

1. Fasten a slip noose to the end of a bent sapling. Open it enough to slip over animal's head but not wide enough for the animal's body to slip through.

2. Secure the trigger so it holds the sapling as shown above. Make this noose large enough so a slight jerk will free the trigger.

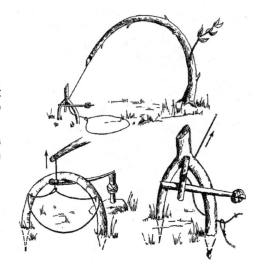

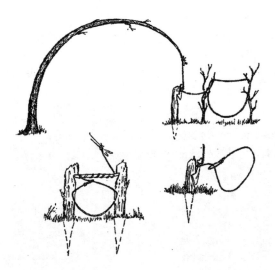

FIXED SNARE

1. This is one of the simplest, yet most effective snares for catching rabbits. Fasten the loop to a log or a forked stick and set this near a limb or bush as shown.

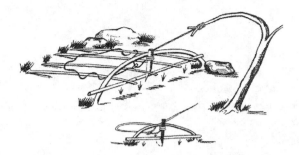

TREADLE SPRING SNARE

1. The above snare is effective for small animals and birds. Cover the treadle with grass and leaves.

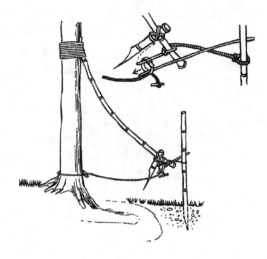

SPRING AND SPEAR TRAP

1. Trap small animals by using a tree branch spring and a spear snare.

2. An animal hits the cord or wire secured to the trigger mechanism. The trigger if released and the spear is driven into the animal by the force of the spring.

A SQUIRREL POLE

1. A "squirrel pole" can be constructed with a simple loop snare.

2. Make the noose opening slightly larger than the animal's head. Squirrels require the width of three fingers while rabbits require a fist-size width.

DEAD FALL FOR SMALLER GAME

1. Build a deadfall as shown above. Construct a figure four trigger and tie your bait on this trigger.

2. Let a heavy rock or log rest on the baited trigger at a steep angle. When the small game goes after the bait, the weight will drop on its head.

DEAD FALL FOR LARGER SMALL GAME

1. Medium to large animals can be killed in a dead fall like the one shown above.
2. Build this kind of dead fall beside a stream, on a ridge and across a game trail. Be sure the upper log (fall log) slides smoothly between the upright guideposts.
3. Be sure to place the bait far enough in from the bottom log. This will allow time enough for the top log to drop before the animal can withdraw its head.

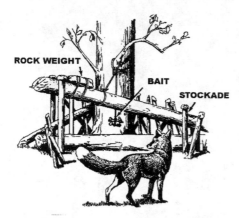

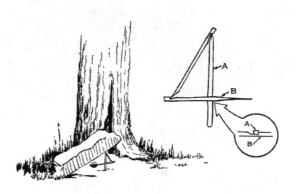

TRAPPING HINTS

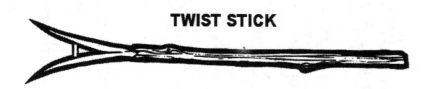

TWIST STICK

1. A forked stick is an excellent tool to use for trapping small game.
2. A forked stick is safer to use than your hands when going after small game in an animal's den, or in the hollow part of a tree.
3. Insert the forked stick into the animal's den until you feel something soft. Then twist, wrapping the animal's loose skin around the fork.
4. Pull the animal from its den, keeping the forked stick taunt. Be prepared to kill the animal as it could at this point be dangerous.

NOOSE STICK

1. Animals living in burrows should be smoked out of their dens. You can use a noose stick secured to a long pole to snare your quarry as it emerges from its hole.

2. A noose snare can be placed at the opening of the animal's den.
3. This snare can also be placed on a trail leading to a watering hole, feeding or bedding area, and where fresh animal droppings or tracks are found.

THE ART OF FUNNELING

1. After setting a trap in an animal's runway, erect a barrier on both sides constructed of sticks, dry branches and leaves.

2. Shape all of these things to make a large "V" to funnel the animal into the trap.

3. The animal won't try jumping over or walking on the barrier. Instead it will travel parallel to the barrier and go straight to your trap.

4. After developing the barriers, spread animal blood or bladder contents around the area to eliminate the human scent.

5. When this isn't possible, build a fire and smoke the area thoroughly. Animals won't suspect anything after an area has been smoked.

49

8

Catching Fish for Food

FISH AND FISHING

1. Keep on the lookout for streams, rivers, lakes and ponds as they are abundant food sources.

2. They support more aquatic life in a small area than the land does animals and it's often easier to acquire.

3. Count on finding such life as fish, snails, crabs, frogs and turtles in and around most inland waters.

4. Fish can be caught with the most primitive fishing gear but you must know when, where and how to fish.

TRY FISHING WHEN THE FISH ARE FEEDING

1. As a general rule, look for fish to feed just before dawn and just after dusk. Also try your luck immediately before a storm.

2. Try fishing at night during a full moon, or when the moon is waning. And try when minnows are seen jumping.

RECOMMENDED PLACES TO FISH

1. The place to start fishing depends on the time of day and the kind of water. In fast streams during the hottest part of the day – try deep pools below the riffles.

2. In fast moving streams in early morning or late afternoon – let your bait float over the riffles. Aim for bush overhangs, underwater logs and undercut river banks.

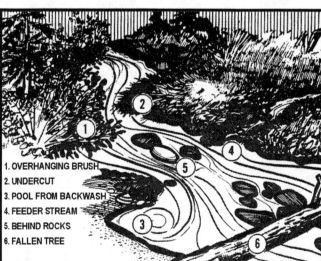

1. OVERHANGING BRUSH
2. UNDERCUT
3. POOL FROM BACKWASH
4. FEEDER STREAM
5. BEHIND ROCKS
6. FALLEN TREE

3. On lakes in the heat of the afternoon – fish deep as fish seek the coolness of deeper water.

4. Fish are more likely in the summer to feed in shallow water and at the edges of a lake. Try in the early morning or evening.

5. Fishing in a lake in the spring or fall will be better in shallow water and on the lake's edge because the fish will be seeking warmer water or they'll be bedding.

BAIT FOR FISHING

1. Fish bite better on bait taken from their native water. Look in water close to shore for fish eggs, crabs, minnows, etc. Look on the bank for worms and insects.

2. Look in the stomach of a fish you catch to see what it's been eating. Try to match this for your bait.

3. Use the eyes and intestines of a fish you catch if other bait sources prove to be unproductive.

4. Cover your hook completely if using worms. Pass your hook through the body of a minnow -- under the backbone to the rear of its dorsal fin.

5. You can make fake bait with small pieces of brightly colored cloth, feathers, metal pieces, etc. These should be made to imitate worms, insects and minnows.

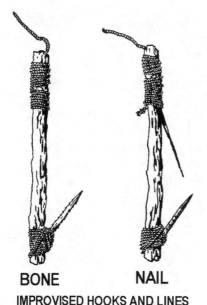

BONE   **NAIL**

IMPROVISED HOOKS AND LINES

IMPROVISED FISH HOOKS AND LINES

1. Fish hooks and line should be carried in every back pack or fanny pack. If you have none, improvise some with pieces of bone, safety pins, wood pieces or nails.

2. Fishing line can be improvised by using the inner bark of a tree. Knot the ends of two strands of inner bark and secure them to a solid base.

3. Hold a strand in each hand and twist clockwise crossing one above the other. Add strands as needed to increase the length of the line.

METHODS OF CATCHING FISH

SET LINES

1. Set lines provide a practical method of fishing if you are near a lake, pond or stream for a period of time. Tie a number of hooks on the same line and bait them all.

2. Tie your line to a low hanging branch that will bend when a fish is hooked. Check your line periodically to remove any fish and to rebait the hooks.

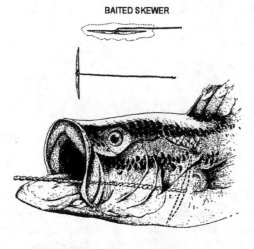

BAITED SKEWER

THE GORGE OR SKEWER HOOK

1. The gorge or skewer hook is excellent with a set line.

2. Cover the skewer with a chunk of bait. After a fish swallows the bait, the skewer opens and lodges itself in the stomach.

FISHING WITH YOUR HANDS

1. Hand fishing can be done in small streams with undercut banks and with large flat rocks near the edge, where fish can hide.

2. Put a cotton glove or a sock over your hand or wrap a piece of gauze bandage around your hand to prevent fish from slipping away.

3. Slowly reach under the rock or bank and feel around until you find a fish. Grasp the fish firmly and bring it out of the water.

SPEARING FISH

1. Spearing is a difficult way to catch fish. It's best to try this in a small stream where fish are plentiful. To make a fishing spear, tie a knife on the end of a pole.
2. Other improvised fishing spears: sharpen the end of a slender tree limb, lash two or more long thorns on a stick, or whittle a spear point from a piece of bone.
3. Position yourself on a rock over a fish run. Wait quietly for one or more fish to swim by.

USE AN IMPROVISED NET

1. A net can be used when the fish are too small to spear or catch with bait on a hook. Start by finding a sapling and bending it into a circular frame.
2. Stitch or tie a shirt or other material to the frame. Scoop with this in water whenever you see fish.

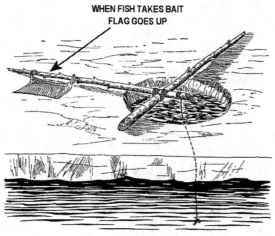

WHEN FISH TAKES BAIT
FLAG GOES UP

AUTOMATIC FISHERMAN

FISHING THROUGH ICE

1. Winter fishing is done through a hole in the ice covering a pond, stream, lake or river.
2. When not fishing, keep the hole in the ice open by covering with brush and heaping snow over the brush.
3. Fish generally gather in deep pools. Cut holes in the ice over the deepest part of the lake, pond or river. Place a fishing rig (as shown above) over several holes.
4. When a fish bites, the flag will jump to the upright position. Pull in the fish, remove it from your line and immediately put new bait on your hook.

A WORD OF CAUTION ABOUT EATING FISH

Never eat any fish you catch with an unpleasant smell, flabby skin, sunken eyes, pale slimy gills, or flesh that remains dented when pressed.

FROGS AS FOOD

1. Hunt frogs after the sun goes down. They can be easily located by listening for their croaking. Frogs can be clubbed to death with a tree limb.
2. Most people cook and eat only frog legs, not the whole frog. They are delicious when fried or roasted.

52

MOLLUSKS AS FOOD

1. Mollusks include such things as snails, mussels, clams, etc. Be sure any mollusks you find are fresh.

2. Never eat mollusks raw as their parasites can be a serious health hazard. Boil them before eating.

CRUSTACEANS AS FOOD

1. Crustaceans include such things as crabs, crayfish, shrimp, lobster, prawns, etc. Always eat crustaceans after boiling to kill harmful parasites.

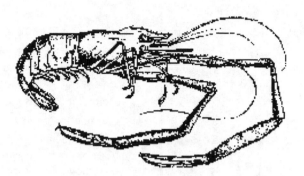

9

Preparing Animals, Fish and Fowl for Cooking

SKINNING AND DRESSING LARGER ANIMALS

1. Clean and dress the carcass as soon as possible after the animal is killed. Hang it from a convenient tree limb with its head facing down toward the ground.

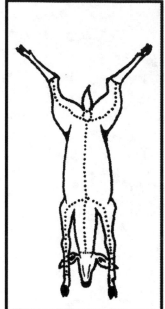

2. Cut the animals's throat and let the blood drain into a container. Boil the blood thoroughly as it's a valuable source of salt and food.

3. Make a ring cut at the knee and elbow joints. Make a "Y" cut down the front of each of the hind legs and make a cut down the belly as far as the throat.

4. From the belly make a cut down each foreleg and make a clean circular cut around the sex organs.

5. Working from the knee downward, remove the skin. The skin of a freshly killed carcass will pull off like a well-fitted glove.

6. Cut open the belly and pin the flesh back with wooden skewers. Remove the entrails and clear the entire mass with a firm circular cut to remove the sex organs.

7. Don't throw any part of the animal away as everything is either edible or useable in some other way.

8. Save and use the fat around the intestines. The brain and tongue can be eaten. The reproductive organs and entrails are good for baiting traps and fish lines.

9. Check the heart, liver and kidneys for spots before cooking. If animal is diseased, it may be dangerous to handle the meat and prepare it for cooking.

10. Be sure to save and use the animal's skin to make a warm bed cover or a piece of clothing. Cure the skin by removing all the flesh and fat.

11. Stretch the hide (skin) on a frame.

12. Soak the hide in a tannic acid solution, the stronger the better. The more times this is done, the better the finished product.

13. Tannic acid is obtained by stripping the inner bark from oak trees and soaking in water. It is also found in mimosa, hemlock and chestnut trees as well as in tea.

SKINNING AND BUTCHERING SMALL ANIMALS

REMOVING THE ANIMAL'S SKIN

1. Open the abdominal cavity and remove and discard the intestines taking care not to rupture them.

2. Save the organs -- kidneys, heart and liver. If spotted, throw them away.

3. Save all meaty parts of the skull -- brains, tongue and eyes. Wash the meat and then cook.

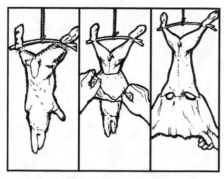

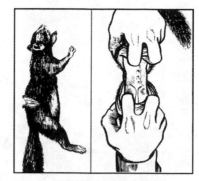

ONE-CUT SKINNING GLOVE SKINNING

RATS AND MICE

1. Mice and rats are an excellent source of palatable meat. Skin, gut and boil these rodents for at least ten minutes before eating.

2. They are especially good in a stew, or cooked with a batch of dandelion leaves. Always include the liver.

RABBITS

1. To skin, make an incision behind the head and peel back the skin. Or bite out a piece of the skin to allow your fingers to be inserted to pull it back.

2. To clean a rabbit, simply make an incision down its belly. Spread open and shake vigorously until the intestines fall out. Scrape and wash out whatever remains.

OTHER EDIBLE ANIMALS

1. All mammals are edible, no matter what they are.

2. Skin and gut dogs, cats, badgers, porcupines and others before cooking. Make them into a stew with bunches of edible leaves.

FISH

1. Scale the fish and then gut it by slitting open its stomach and scraping it clean. Cut out the large blood vessels next to the backbone.

2. Cut off the head of your fish. Leave the head on if the fish is to be broiled on a spit.

3. Catfish are simply skinned before cooking since they have no scales.

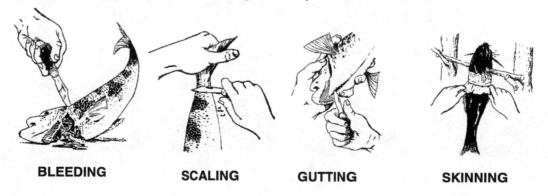

BLEEDING   SCALING   GUTTING   SKINNING

FOWL

FEATHERING

CUTTING AND GUTTING

1. Scald and pluck the feathers from the fowl. Save the feathers for bedding.

2. Cut its neck off close to the body. Clean out the insides through this neck cavity.

3. Wash out the fowl with fresh clean water. Save the liver, heart, gizzard and neck for making soup or stew.

4. Scavenger birds like buzzards and vultures can also be eaten. These scavengers must be boiled for at least 20 minutes to kill the parasites.

REPTILES

1. Snakes and lizards are edible. Simply cut off the head and skin the snake before cooking and eating.

2. Lizards are found everywhere and their meat can be broiled or fried before eating.

10

Hygiene is Important Even in the Wilderness

AN IMPROVISED LATRINE

1. Adopt strict measures for the disposal of human waste and garbage. The latrine to the left is simple yet effective.

2. Immediately set up a latrine downwind of your site if you are going to be in an area for a length of time. One is sufficient for quite a number of people.

ROPE OR WIRE

3. This latrine can be made windproof with tree branches, ponchos or log walls. In the winter months, snow blocks made into walls will act as a wind break.

CLEANLINESS

1. Keeping your body clean in a survival situation is essential to good health.

2. The face, armpits, hands and crotch should be washed daily. If soap isn't available, substitute white ashes, sand or loamy soil.

3. If water isn't available, "bathe" by rubbing down with cornstarch. The cornstarch removes perspiration and excess oil from your skin.

4. Check your body carefully and regularly for ticks, fleas and lice, etc. Pick any of the above off your body and crush (including eggs if possible).

5. Wash and fumigate your clothing and equipment with smoke to get rid of any ticks, fleas and lice, etc.

MOUTH AND TEETH

1. Brush your teeth every day. If you have no toothbrush, rub your teeth with a piece of cloth wound around your finger.

2. A small green twig can be frayed by chewing on one end. This makes a decent substitute for a toothbrush.

3. If you have no toothpaste, brush your teeth with soap, salt or baking soda.

4. Rubbing your teeth with a clean finger is another way of cleaning them. Gums should also be stimulated by rubbing them with a clean finger.

5. Gargling with a salt and water solution helps prevent sore throats and it helps clean the teeth and gums.

CARING FOR YOUR FEET

1. Wash, dry and massage your feet each day. Check regularly for blisters and sores. Tape or band aids will help prevent further problems.

HAIR CARE

1. Keep face shaven and hair trimmed to help prevent parasites and bacterial growth.

CLOTHING CARE

1. Wash your clothing when dirty – underclothing and socks are especially important to launder.

2. If laundering is impossible, then shake out your clothing and hang in the sun every day. This will destroy mildew and bacteria.

A SUMMARY OF RULES TO AVOID ILLNESS

1. All water obtained from natural sources should be purified before drinking. Boil it for at least one minute or use water purification tablets.

2. Always wash your hands and carefully clean under your fingernails before preparing food.

3. All cooking and eating utensils should be cleaned and sterilized with boiling water after each use.

4. Clean your mouth and teeth thoroughly at least once each day.

5. Insect bites can be prevented by proper use of insect repellant, equipment (head nets, etc.) and clothing.

6. Keep flies and other vermin off of and away from your food and drink.

7. Dry your wet clothing as soon as possible.

8. Try to get at least seven to eight hours of sleep each night.

11

Survival Medical Tips and Techniques

WHEN BREATHING STOPS

1. If person appears to be unconscious, shake him and shout in an effort to revive.
2. If no response, tilt the victim's head with their chin pointing up. Immediately look, listen and feel for air.
3. If the victim isn't breathing, give four quick breaths into his or her mouth. If there is no exchange of air, reposition the victim's head and try again.
4. If still no exchange of air, turn the victim on his or her side and give four sharp blows between the shoulder blades.
5. Turn victim on their back and give four sharp abdominal thrusts. Check mouth for obstruction. If still no breathing, start rate of one breath every five seconds.

TO CONTROL BLEEDING

1. Use a sterile pad or clean cloth to apply direct pressure to wound. Elevate injured limb higher than heart. If bleeding doesn't stop, apply tighter dressing.
2. Bandage wounds which don't require a tourniquet only tightly enough to check the bleeding. Loosen when bleeding has been controlled.
3. Use a tourniquet only after every other method has been attempted.
4. After 20 minutes, gradually loosen tourniquet. If bleeding has ceased, remove tourniquet. If bleeding continues, reapply tourniquet.

SCALP OR TEMPLE

LOWER FACE

NECK

SHOULDER OR UPPER ARM

UPPER ARM AND ELBOW

LOWER ARM

HAND

THIGH

THIGH

LOWER LEG

FOOT

If blood is spurting from wound (artery), press at the point or site where main artery supplying the wounded area lies near skin surface or over bone as shown. This pressure shuts off or slows down the flow of blood from the heart to the wound until a pressure dressing can be unwrapped and applied. You will know you have located the artery when you feel a pulse.

DIGITAL PRESSURE (Pressure with fingers, thumb, or hands)

CONTROL BLEEDING BY USING PRESSURE POINTS

TO CONTROL BLEEDING IN FREEZING WEATHER

1. Apply a tourniquet immediately when blood is spurting from a wound in freezing weather.
2. Once applied, the tourniquet must be left on, despite the probable loss of a limb due to freezing, since no replacement for lost blood will be available.

FRACTURES, SPRAINS AND DISLOCATIONS

1. Treat wounds, stop bleeding and apply clean, dry dressings. Apply a splint immobilizing joints above and below the injury.

2. Apply cold initially to control swelling and after 48 hours, apply heat. Check circulation periodically.

SHOCK (TREAT FOR ALL INJURIES)

1. First treat the injury and then cover the person to prevent the loss of any body heat.

2. Lower extremities should be elevated if there are no head injuries, if there are no abdominal injuries, and if injury won't be aggravated.

3. Don't give any liquids to an unconscious person.

GARLIC OIL AND CAYENNE PEPPER

1. Every Russian soldier is required to carry garlic oil on his person. It not only kills bacteria, but induces fast healing.

2. Cayenne pepper is another unusually good medical aid. This common household item quickly stops profuse bleeding and it arrests heart attacks!

TREATING COMMON SURVIVAL ILLNESSES AND INJURIES

BERIBERI

1. Beriberi symptoms: twitching legs, cramping muscles, a loss of appetite and paralysis are caused by a lack of vitamin B.

2. The cure is quite simple: eating green foods or drinking bark tea which is made by boiling the outer layer of bark for five minutes and then drinking the liquid.

BURNS AND SCALDS

FIRST DEGREE BURNS

1. First degree burns have mild swelling and pain and the surface of the skin turns red.

2. Treat first degree burns by putting burned area in cold water, if at all possible, or put cool, wet cloths on the burns.

3. Use dry sterile pads to blot and change these as needed to prevent infection. Do not use material with loose fibers to cover a burn.

4. Do not try to clean a burn by removing clothing or dirt from the area as these things are sterile.

5. Drink water with salt, if available. This should be 1/4 teaspoon of salt per quart of water.

6. Do not break burn blisters and don't put greasy medication on a burn.

SECOND DEGREE BURNS

1. Second degree burns have much swelling, intense pain and blisters. The victim's skin is bright red and blotchy and looks wet.

2. To treat, immerse burned area in cold water. Put cool wet sterile cloths on the burn. Change these as needed to prevent infection.

3. Drink water with salt, if available. This should be 1/4 teaspoon of salt per quart of water.

4. Elevate burned arms or legs above level of the victim's heart.

5. Watch continuously for signs of infection.

CONSTIPATION AND DIARRHEA

1. Constipation can be prevented by eating fruit regularly, drinking plenty of water, and frequent exercise. Don't take laxatives.

DIARRHEA OR DYSENTERY

1. Diarrhea or dysentery can be treated by drinking tea leaves in water, drinking plenty of water, or going on a liquid diet.
2. Try eating a paste made of water, chalk and charcoal or charred bone. This acts as a regulator.

EYE PROBLEMS

FOREIGN OBJECTS IN YOUR EYES

1. Wash eyes out with clean water.
2. Make gentle swipes over affected area with an improvised cotton swab. This may remove stubborn particles.
3. If unsuccessful, apply eye ointment if available and wear a patch over eye for 24-hours.

SUN AND SNOW BLINDNESS

1. Sun blindness is caused by the reflection or glare caused by the sun on water, sand, etc.
2. Snow blindness is caused by the reflection or glare caused by the sun on snow and ice.
3. Sun and snow blindness can occur even during cloudy weather as well as when its sunny.
4. The first symptom of sun and snow blindness can be noted with a change in a person's vision. Variations in the level of the ground can no longer be detected.
5. A burning sensation is noted in the eyes as well as a scratchy, gritty or sandy feeling under the eyelids.
6. Also to be experienced are redness and watering of the eyes, headaches, and the eyes hurting when exposed to even a weak light.
7. Complete darkness is the best medicine. Wear a dry, sterile blindfold or bandage for 18 hours.
8. Sun and snow blindness can be prevented by wearing sun glasses at all times.
9. If no sunglasses are available, wear a pair of improvised leather, rubber, bark, etc., goggles (see right) with narrow slits cut in them.
10. Blacken your cheeks and nose to help reduce the sun's glare.

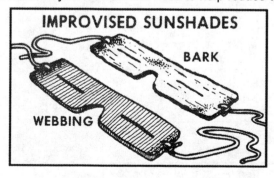

IMPROVISED SUNSHADES

BARK

WEBBING

HEAT DISORDERS

HEAT CRAMPS

1. Heat cramps are brought about by an excessive loss of salt in the body.
2. Symptoms are cramps in the abdomen, arms and legs, pale and wet skin, and extreme thirst and dizziness.
3. The treatment is to get plenty of rest and drink 1/2 pint of salt water to start.
4. Then drink a quart of salt water (1/4 teaspoon of salt stirred in) over a 30 minute period or simply swallow a packet of salt and wash it down with water.

HEAT EXHAUSTION

1. Heat exhaustion is caused by dehydration and a loss of body salt due to extreme physical activity.
2. The symptoms are weakness, dizzy, faint, heavy sweating, pale, moist cool skin and headache or nausea.
3. The treatment is to rest in the shade and drink a quart of water with 1/4 teaspoon of salt stirred in.

HEAT STROKE

1. The symptoms of heat stroke are a strong, fast pulse, hot and dry skin, and no sweating.
2. The treatment is to handle gently, cool down as rapidly as possible, and saturate clothing with water.
3. Do not give or take any kind of stimulant and avoid overcooling.

HYPOTHERMIA

HYPOTHERMIA TABLE			
Temperature		**Stage**	**Symptoms**
98°F	37°C	Mild	Cold at first, then stops shivering. Rigid muscles
95°F	34°C	Moderate	Poor coordination, impaired speech - slow and slurring. Memory loss. Convulsions.
88°F	32°C	Severe	Unintelligible pulse. Pupils dilated. Shallow breathing. Unconscious. Glassy stare.
82°F	28°C	Extreme	Heart rhythm may change. Heart stops beating.

2. Hypothermia takes place when the temperature of the body's inner core (98.6 degrees F) drops below 98 degrees F (37 degrees C).

1. Hypothermia is the result of the body losing heat faster than it can produce it. Such a condition should be treated immediately.
2. Hypothermia takes place when the temperature of the body's inner core (98.6 degrees F) drops below 98 degrees F (37 degrees C).

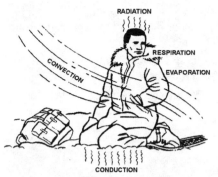

HOW HEAT IS LOST FROM THE BODY

PREVENTING HYPOTHERMIA

1. Avoid drinking cold water if hypothermia conditions exist. Drink high energy liquids such as hot chocolate, coffee and tea.

2. Eat properly and often and eat simple sugar food whenever possible.

3. Keep active but take frequent breaks to rest. Overexertion should be avoided.

4. Keep warm and stay dry. Protect yourself (face, ears and head) from the wind, wet and cold.

TREATMENT OF HYPOTHERMIA

1. Warm as rapidly as possible. Prevent further heat loss from the body (including the head). Get victim out of the wind and into the best available shelter.

2. Remove clothes, put on dry clothing and get victim in a sleeping bag. Place insulation (blankets, boughs, poncho, etc.) between victim and the ground.

3. Add warmth (hot water bottle, rocks heated over a fire and wrapped in cloth, etc.) to the sides of the victim's chest, neck and groin area.

4. If conscious, feed the victim warm, nutritious sugar-sweetened drinks and food high in carbohydrates.

5. WARNING: The victim shouldn't be given any alcohol to drink nor should he or she be massaged.

POISON

SNAKE BITE TREATMENT -- CURRENT METHOD

1. Snake bite victim must avoid physical exertion.

2. No incision over the bite is to be made.

3. Keep bite area as immobile as possible.

4. Use no ice pack on the bite and no immersion of the bite in cold water.

5. Keep bite area below the level of the heart.

6. Do not use a tourniquet on the arm or leg.

7. Use a vacuum pump suction device that comes with a small snake bite kit (if one is available).

8. Get the snake bite victim medical attention as soon as possible (this will be impossible in most cases).

SNAKE BITE TREATMENT — AN ALTERNATE METHOD

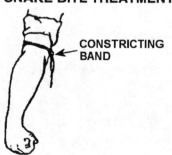

CONSTRICTING BAND

1. This is the better method to use when no medical facility is close at hand.

2. Don't allow snake bite victim to move about and exert himself physically.

3. Apply a constricting band or string two to four inches from the snake bite between 15 and 30 minutes from the time of the bite.

4. The band should be between the bite and the heart and tight enough to cut off the flow of venom. Loosen every 10 to 15 minutes for a period of 90 second.

5. Use fire from a match or a piece of firewood to sterilize a knife or a razor blade.

6. Make a single slice 1/4 inch long and 1/4 inch deep through each fang mark along the line of the muscle.

7. This should be done within 30 minutes of the bite, the sooner the better.

8. Suction should be applied by mouth if no vacuum pump suction device (in all snake bite kits) is available.

9. No person with open sores in their mouth should apply suction as the poison can get in their system.

NON-POISONOUS SNAKE BITES

1. Simply clean and bandage snake bite wound.

INGESTED POISONS

1. Lie down, stay quiet and drink lots of water.

2. Induce vomiting unless victim is unconscious, burns start to appear on the lips, and perhaps the throat, or poison is determined to be petroleum based.

SURFACE POISONS

1. Wash with large amounts of soap and water and keep from scratching by covering body parts.

PROTEIN DEFICIENCY

1. Protein deficiency causes appetite loss, muscle waste, vomiting, diarrhea, irritability and fluid retention.

2. Protein deficiency can be cured by eating lots of grains and nuts, meat and eggs and insects.

SCURVY MAY BE A PROBLEM

1. Scurvy causes cuts and wounds not to heal, bleeding gums, joint swelling and teeth loosening.

2. Scurvy is caused by a serious lack of vitamin C and can be cured by eating raw greens, fruit and drinking evergreen tea.

3. Boil evergreen needles for five minutes to make this tea. Then throw out the needles and drink the liquid.

SKIN PROTECTION PROBLEMS

SKIN PROTECTION

1. Suntan oil is a valuable asset to have in any survival situation.
2. When in cold weather, use at least once a week on exposed skin.
3. When in hot weather, use daily on exposed skin.

FROSTBITE (FREEZING OF YOUR BODY TISSUE)

1. Frostbite is the result of tissue freezing from exposure to temperatures below 32 degrees F.
2. Frostbite can be superficial (only skin tissue is affected) or deep (frozen to the bone).
3. The degree of frostbite injury depends upon the wind chill factor, how long the person was exposed to the elements, and how much proper protection was available.
4. If exposure was short, the frostbite will be superficial. If long exposure, the frostbite will be deep.

SYMPTOMS OF SUPERFICIAL FROSTBITE

1. A stinging, tingling sensation on the affected skin area, aching, stiffness, cramping, numbness and skin blistering and peeling.
2. Skin will turn bright red and then change to a waxy white or pale gray.

SYMPTOMS OF DEEP FROSTBITE

1. Severe stiffness, numbness, and blue skin.

FROSTBITE TREATMENT

1. The frostbite victim must first be put in a warm, sheltered place.
2. The victim's frostbitten parts must be rewarmed. Rewarm the face and ears by covering with your hands.
3. Rewarm the victim using your body heat. Place victim's hands under your clothing and against your body.
4. Take off victim's shoes and socks and place his feet under your clothing and against your abdomen. After feet are warmed, put on dry socks.
5. Button or zip the victim's clothing to prevent further loss of body heat.

THINGS NOT TO DO FOR FROSTBITE

1. Don't give any alcohol to the frostbite victim as alcohol increases the loss of body heat.
2. Don't allow the victim to smoke as tobacco causes the blood vessels to narrow.
3. Don't break or open blisters found on the arms and legs, don't rub frostbitten areas with snow, and don't soak the hands or feet in cold water.
4. Don't rewarm the frostbitten hands, feet, etc., by massaging them and don't expose frostbitten parts to an open fire.
5. Do not neglect frostbite. To do so is to invite gangrene.

12

Day or Night — You Don't Have to Get Lost

DETERMINING DIRECTION

1. Using a compass is the most common way of finding the correct direction.
2. Make certain your compass isn't affected by iron fragments in the ground, nearby vehicles, or metal objects carried in your backpack or your pockets.
3. Keep in mind that the magnetic field gets weaker as you get closer to the magnetic poles.

BE AWARE OF THINGS AROUND YOU

1. Check the bark on poplar and birch trees. The bark will always be darkest on the north side and whitest on the south side.
2. Carefully observe evergreen trees growing away from others. The south side will always be the bushiest.

USING THE SUN TO TELL DIRECTION

1. There are only two ways to determine direction by using the sun:

    a. Northern hemisphere: At noon shadows fall to the north of trees and other objects.

    b. Southern hemisphere: At noon shadows fall to the south of trees and other objects.

USING YOUR WATCH AND THE SUN TO TELL DIRECTION

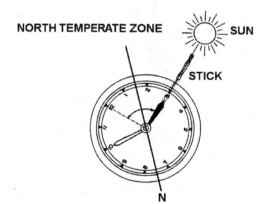

NORTH TEMPERATE ZONE · SUN · STICK · N

1. An ordinary watch can be used to find the approximate true north.
2. Northern latitudes: Point the hour hand toward the sun. An imaginary line halfway between the hour hand and 12 noon will point south.
3. For daylight saving time, the north-south line will fall between the hour hand and one o'clock.

4. Southern latitudes (south temperate zone): Point 12 noon toward the sun. An imaginary line halfway between the hour hand and 12 noon will point north.
5. For daylight saving time, the north-south line lies midway between the hour hand and one o'clock.
6. Try always to remember: The sun is in the eastern part of the sky before noon and the western part of the sky in the afternoon.

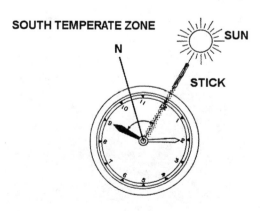

SOUTH TEMPERATE ZONE · N · SUN · STICK

DETERMINING DIRECTION ON CLOUDY DAYS

1. Place a stick at the center of your watch and hold it so its shadow falls along the hour hand.
2. One half distance between the shadow and 12 noon in north.

DETERMINING DIRECTION BY THE SUN DURING THE DAY

1. If you have no compass, the method outlined below can be used any time the sun is bright enough for a stick in the ground to cast a shadow.
2. Find a stick about three feet long and follow these simple steps:

- **Step 1:** Push the stick straight up and down into the ground. Note the shadow.
- **Step 2:** Mark the tip of the shadow with a stick or a stone. Wait 10 to 15 minutes until the tip of the shadow moves a few inches.
- **Step 3:** Mark the new position of tip of the second shadow.

- **Step 4:** Draw a straight line from the first tip to the second tip. Extend this line one foot past the second tip marking.

- **Step 5:** Stand with the toe of your left foot at the first rock. The toe of your right foot should be placed at the end of the line you drew.
- **Step 6:** You are now facing north. Find other directions by recalling their relationship to north.

3. **Basic rule to remember:** The sun rises in the east and sets in the west. The shadow tip explained above moves just the opposite. Therefore, the initial shadow tip is always in the west direction. The second mark is always in the east direction. This holds true anywhere you may be at any given time.

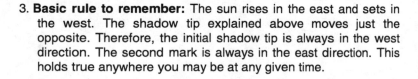

FINDING DIRECTION AT NIGHT

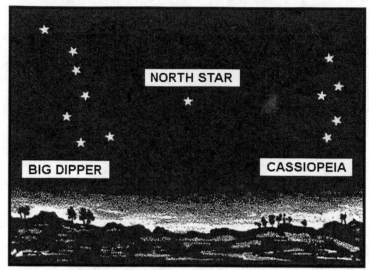

1. Your direction can be determined at night even if you have no compass.

2. Locate the Big Dipper. The two stars found at the end of the dipper bowl are called "pointers." The North Star is on a straight line from these "pointers."

3. Now find the constellation Cassiopeia. This bright star grouping is shaped like a lopsided "M", or "W" when it's low in the sky.

4. The North Star is straight out and to the left of the center star and about the same distance as it is from the Big Dipper. The North Star can be used to find a true north direction. In latitudes under 70 degrees when traveling north, the North Star makes a reliable steering mark.

   a. Its bearing is usually only one degree from true north.

   b. It's never more than 2 ½ degrees away from true north.

5. In latitudes higher than 70 degrees, the North Star is too high up to accurately indicate direction.

13

Insects, Plants, Snakes and Other Hazards

TAKE PRECAUTIONS AGAINST INSECTS

AVOIDING MOSQUITO BITES

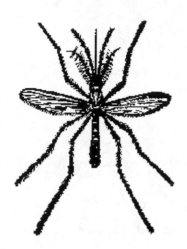

1. Set up camp on high ground away from swamps.

2. Always sleep under mosquito netting. If netting isn't available, use some other kind of material.

3. Keep mud smeared on your arms, face and any other uncovered part of your body. Do this especially before going to bed.

4. Wear all clothing, especially after dark. Tuck your pants into the tops of your shoes. Wear a mosquito head net and gloves.

5. Use mosquito repellant. Repellant applied on your skin will remain effective for a few hours. Repellant sprayed on your clothing will remain effective for weeks.

6. Smoke will reduce your exposure to mosquitos. Use this as a last resort.

FLIES

1. Flies, like mosquitoes, vary in size, have varying breeding habits, and vary in the discomfort or danger they can cause.

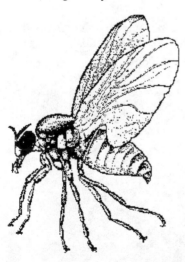

2. The protection used against mosquitoes is also effective against flies.

Black Flies

1. These flies are most bothersome and they sometimes transmit filarial worms.

2. Some people react severely to the fly bite. The danger is most often infection from scratching.

Deer and Horse Flies

1. Usually light colored with a stout body.

2. Seen during the day in areas where hoofed animals abound.

Punkie Flies or No-see-ums (Gnats)

1. These tiny flies are a lot less bothersome than black flies.

2. Punkies inflict an itching bite and some carry filarial worms.

3. If these gnats are abundant its best to move on to another place. They are seldom encountered more than a half mile from their breeding place.

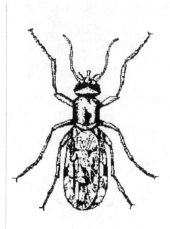

Sand Flies

1. Sand flies are blood sucking insects suspected of carrying a number of diseases.

2. Sand flies are so small they can pass through ordinary netting.

Eye Gnats

1. These obnoxious gnats like to hover around the eyes. They sometimes carry dangerous eye infections.

Screw Worm Flies

1. Found predominantly in the more tropical areas of the United States and are active during the day.

2. Screwworm flies deposit their eggs in the nostrils while a person is napping. This is especially true if the nasal passages are irritated with colds, sores, etc.

3. The screwworm larvae in the nasal tissues cause severe pain and swelling.

Bot Flies

1. Bot flies are dangerous because of their larvae. The larvae bore into the skin and cause a painful swelling that looks like a boil.

2. Frequent application of tobacco will eventually kill the larvae. They can then be squeezed out of the skin.

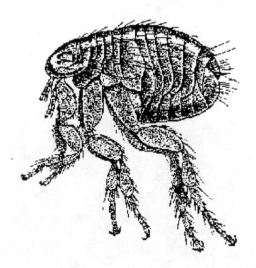

Fleas

1. Fleas are wingless and can be extremely dangerous as they can transmit plague to humans after feeding on plague-ridden rodents.

2. To protect against fleas use a flea repellant powder and wear tight fitting leggings or boots.

3. If you kill a rodent for food, don't handle until it's cold. Any fleas will leave a cold body.

TICKS

1. There are two kinds of ticks – the hard or wood tick and the soft tick.

2. These oval, flat pests are carriers of tick relapsing fever, tick typhus and Rocky Mountain spotted fever, an infection that can be fatal.

3. Symptoms of Rocky Mountain spotted fever include a rash, chills, fever and severe pain in the arms and legs.

4. Never squash a tick on your skin and never pull a tick off your skin once it has imbedded its head.

5. Coat an imbedded tick with alcohol, gasoline, kerosene or iodine or even spit. The tick will free itself and be easy to remove.

6. Hold a lighted cigarette close to the tick's body. This will also cause the tick to withdraw its head.

7. If a tick's head still remains imbedded in the skin, hold the point of a knife blade over a flame. Then touch the tick with the hot point of the blade.

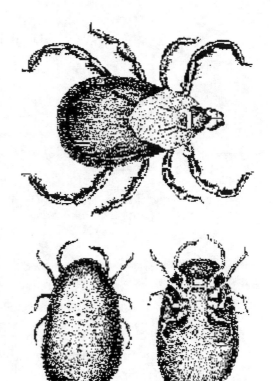

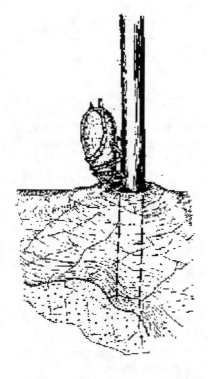

CHIGGERS, MITES AND LICE

1. Chiggers, mites and lice all bore into the skin and cause unbelievable itching and discomfort. They can rankle a person entirely out of proportion to their size.

2. Some people become sick from chigger bites.

Mites

1. Mites cause a variety of skin diseases including scabies. Scratching mite bites often causes infections.

Lice

1. If bitten by lice, try not to scratch the bites as this will spread the louse feces into the bites.

2. Boil your clothing to rid it of lice. Clothes can also be hung in direct sunlight for a few hours. Be sure to expose the seams.

3. Frequently inspect the hairy parts of your body for lice. When found, expose your body to direct sunlight. Then bathe carefully with soap.

a. If no soap is available, use sand or the sediment from the bottom of a stream.

SPIDERS

BLACK WIDOW SPIDER

1. Spiders in general aren't dangerous. Even a tarantula bite isn't fatal or even serious. However, take pains to avoid the black widow and the brown recluse spider.

2. The disastrous black widow or hourglass spider is always dark with white, yellow or red spots.

3. The black widow's bite causes severe pain, much swelling, severe abdominal cramps, and possible death.

BROWN RECLUSE SPIDER

4. The abdominal cramps are often mistaken for appendicitis or acute indigestion. This suffering comes and goes for a couple of days.

SCORPIONS

1. Scorpion stings are seldom fatal, although they are extremely painful. Only stings from the larger species are dangerous and can result in death for the bitten person.

2. Scorpions are a real danger since they hide in the dark recesses of shoes and socks, clothing and bedding. Always shake everything out before wearing or going to bed.

3. If stung by a scorpion, use cold compresses and/or mud. Coconut meat, if available, can be locally applied to the bite.

CATERPILLARS AND CENTIPEDES

1. Large caterpillars and centipedes can inflict very painful bites. Some species cause severe itching and inflammation when merely brushed against.

2. Caterpillars can also cause painful blistering of the skin.

3. Centipedes rarely bite except when cornered and unable to escape. They aren't dangerous except

when hiding in clothing about to be worn.

WASPS, BEES AND HORNETS

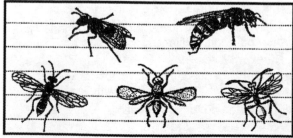

WASPS AND HORNETS

1. Getting stung by a swarm of wasps, bees or hornets can be fatal.

2. If attacked by these dangerous entities, plunge into some dense brush. The twigs and leaves will help beat off the insects as they snap back into position.

72

3. If you get stung, scrape off the barbed stingers with a knife. This will stop the poison from going into the bite wound.

4. Climbing hempweed, found near streams, swamps and seashores is a good antidote for these stings.

BEE

TAKE PRECAUTIONS AGAINST POISONOUS PLANTS

POISON SUMAC

1. The danger from poisonous plants is not a serious hazard. But under certain conditions, poisonous plants can be extremely dangerous.

2. There are two general kinds of poisonous plants: those poisonous to the touch and those poisonous to eat.

3. There are three most important poisonous plants: poison ivy, poison oak and poison sumac. All of these plants have small, round, white or gray-green fruit.

4. Symptoms of plant poisoning are: itching, swelling, reddening of the skin and blisters.

5. The best treatment after having touched the plant is to thoroughly wash the area with a strong soap.

6. The best treatment if symptoms have begun to show is to apply a paste of wood ashes and water to each infected area of the body.

7. It is important to remember that sweating and overheating greatly increase the danger of becoming contaminated by poisonous plants.

8. Also remember that the juices of poisonous plants are especially dangerous in the vicinity of the eyes.

9. Lastly, it is highly dangerous to use the wood of any contact poisonous plant for firewood.

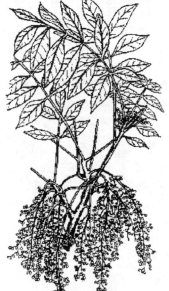

POISON IVY

POISON OAK

73

MUSHROOMS

1. Dig any gilled mushroom completely out of the ground and throw away any having a cup or vulva at the base. Avoid young gilled mushrooms with a button-like appearance.

2. The button-like gilled mushroom can be told from an edible puff ball because of it's short stem. Short stems aren't characteristic of the puff ball mushroom.

3. Avoid mushrooms with the underside of the cup covered with tiny reddish spores.

4. Avoid gilled mushrooms with scaly white gills at the base, and membrane-like cups.

5. Avoid gilled mushrooms with white or pale milky juice.

IDENTIFYING CUPS

REDDISH-WHITE CAP WITH WHITE FLECKS

6. Avoid gilled mushrooms found in the woods with a smooth reddish top and white gills radiating out from the stem like spokes.

7. Avoid yellowish orange and yellow mushrooms growing on an old stump. They are poisonous if they have a surface that glows in the dark.

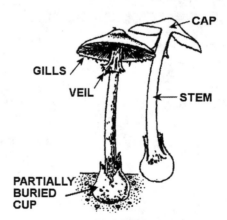

CAP

GILLS

VEIL

STEM

PARTIALLY BURIED CUP

DEATH ANGEL WITH GILLS, VEIL, STEM AND CUP

8. The above mushrooms are also poisonous if they have solid, crowded stems, convex overlapping cups, and broad gills running randomly running down the stem.

9. Avoid mushrooms when they are overripe, spoiled, water-soaked or are found to have maggots.

10. Be familiar with the Amanite family of poisonous mushrooms. The most deadly is the widespread Death Angel.

11. The Death Angel produces tox-albumin poison, or the kind of poison found in rattlesnakes. The amount of this mushroom needed to kill a human being is small.

TAKE PRECAUTIONS AGAINST POISONOUS SNAKES AND OTHER HAZARDS

Some Facts About Snakes

1. It is needless to fear snakes because only a small percentage of them are poisonous.

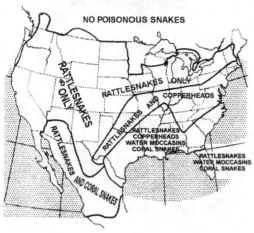

DISTRIBUTION OF POISONOUS SNAKES OF NORTH AMERICA

2. Aggressive behavior in snakes is the exception rather than the rule. They usually try to get out of your way if given the opportunity.

3. Snakes are active both day and night during the warm summer months. They are inactive in colder weather and some are known to hibernate.

4. In desert and semi-desert regions, snakes are active in the early morning, stay hidden in the shade during the day, and are active again after the sun goes down.

5. The distance at which a snake can bite has been grossly exaggerated. Striking distance is rarely more than half the snake's length.

6. Snakes can't outrun a man, nor can most snakes leap clear of the ground.

Suggestions to Help Avoid Snake Bites

1. Walk carefully and watch your step. Watch where you place your hands when climbing or when picking something up off the ground.

2. Learn the habits and living spaces of poisonous snakes in your area. Never tease or pick up a snake and avoid any sudden movements around a snake.

3. Wear leather boots or loose fitting clothing around your legs.

4. Know exactly what to do if you are bitten.

Categories of Snakes

RATTLE SNAKE

WATER MOCCASIN (COTTON MOUTH)

COPPERHEAD

1. Poisonous long-fanged venomous snakes are rattlesnakes, copperheads and cottonmouth moccasins.

2. The bite of one of these snakes is incredibly painful. There is initial local swelling which quickly increases as the venom spreads throughout the tissue.

3. A poisonous short-fanged venomous snake is the coral snake. The venom of this short-fanged snake is the most deadly among poisonous snakes.

4. Because of the short fangs, people wearing even light clothing have little to fear regarding being bitten.

CORAL SNAKE

SNAKE DESCRIPTIONS

COPPERHEADS

1. The average copperhead reaches a length of two and one half feet. Some grow to between four and five.
2. Copperheads are pale brown with darker cross bands. The head is usually copper red while its belly is light and mottled.

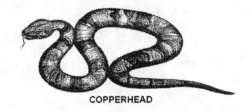

COPPERHEAD

Copperhead Habits

1. Generally found in the woods of the north and on dry ground anywhere in fields and woods of the south.
2. Copperheads are timid and usually stay hidden. They try to escape when discovered and rarely bite.
3. Copperheads vibrate their tail when cornered and make a distinct buzzing sound in vegetation.
4. Copperhead venom is rather weak yet this snake's bite can cause incredible swelling, pain and suffering.
5. The copperhead is also known as an upland moccasin, death adder, pilot snake and chunk head.

WATER MOCCASIN

1. The average length of a water moccasin is three to four feet long while some grow to six feet in length.
2. Water moccasins are dull brown or olive and have indistinct bands. The belly is yellow with darker blotches. Young water moccasins are brilliantly colored.

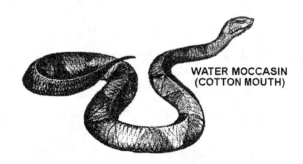

WATER MOCCASIN
(COTTON MOUTH)

Water Moccasin Habits

1. The water moccasin lives close to the water, is an excellent swimmer, exceedingly timid and quickly retreats when disturbed.
2. These snakes are found sunning on logs or trees leaning out over the water in swamps, sluggish streams and rivers and bayous.
3. Water moccasin venom is extremely poisonous and the bite of a large snake is often fatal.
4. The water moccasin is also known as a cotton mouth, trap jaw and gapper.

RATTLESNAKE

1. There are 27 different species and their sizes vary with the region they inhabit. Western diamondbacks grow up to eight feet long.

RATTLE SNAKE

2. Rattlesnake colors vary from gray to black and some of these snakes have stunningly beautiful markings.
3. The rattle on the end of the tail and its chilling sound is the best way to identify a rattlesnake.
4. The bite of a small rattler isn't likely to kill anyone. It will be excruciatingly painful (like a

horrible toothache all over your body) and will make you deathly ill.

5. The bite of a large rattlesnake from 3 to 8 feet long could be fatal, and if it turns out not to be, you may end up wishing it was.

Rattlesnake Habits

1. Rattlesnakes prefer rocky ledges or open sandy places to sun themselves.

2. These snakes don't always give a warning rattle before they strike. They may strike first and rattle afterwards if taken by surprise.

3. Rattlesnakes usually try to get away without a fight. They are basically shy as are all other snakes.

TAKE PRECAUTIONS AGAINST LIZARDS

GILA MONSTER

1. The gila monster in one of only two poisonous lizards in the United States. The other is the beaded lizard of the Southwest.

2. Both of these lizards are found only in desert areas. They are of little danger to man because of their sluggishness.

14

When You Don't Wish to be Seen

NATURAL CONCEALMENT

SHADOWS

1. Take advantage of all means of concealment provided by nature.
2. Be careful when using natural camouflage on your clothing. Foliage fades and wilts. Change it regularly.
3. Take advantage of shade and shadows but remember that they shift with the sun.
4. Unnecessary movement should be avoided.

5. Always avoid the skyline when moving.

AVOID SKYLINE

AVOIDING SHINE

6. Never leave anything exposed that may reflect sunlight (glasses, watch, pen, etc.).

NATURAL CONCEALMENT

7. Always try to break up the outline of your body.

OBSERVATION

8. Always observe an area or animals from a concealment prone position

9. Blend with the environment by using local vegetation, grass and berry stains, charcoal and dirt for concealment.

CAMOUFLAGE APPLICATION PATTERNS

1. Face camouflage: use dark colors of high spots and lighter colors on remaining exposed areas of your face.

2. Ear camouflage: use two colors to help break the outline of the inside and the back of your ears.

3. Head and neck camouflage: use a scarf, some local vegetation (leaves, branches, etc., or netting).

4. Hand camouflage: use color between your fingers and on the back of your wrists.

5. Body camouflage: use dark and light colored vegetation to break up the "V" of your crotch and armpits.

6. Smells may give you away despite all of the evasive precautions you put into practice. Avoid using scented soap, shampoo, shaving cream and aftershave lotion.

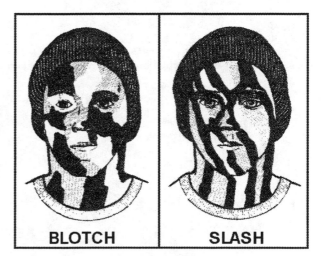

BLOTCH   **SLASH**

WHEN REST OR SLEEP IS NEEDED

NATURAL SHELTERS

1. Select your temporary sleep shelter carefully. Safe concealment is needed with minimum or no preparation. Try to find a natural concealmen spot.

2. A good temporary concealment is one with observable approaches where possible, decent escape routes, and good overhead concealment.

GROUND MOVEMENT

3. A safe concealment site is one in which you don't corner yourself. Entrances and exits should be hidden in brush, along ridges and in ditches and rocks.

4. Select a site least likely to be noticed such as in rugged terrain or a drainage ditch, etc.

TRAVEL RESTRICTION

5. Select a least obvious location or one that blends in and looks like everything else.

6. Remember this acronym:

- **B** — **B**liss
- **L** — **L**ow Silhouette
- **I** — **I**rregular Shape
- **S** — **S**mall
- **S** — **S**ecluded

MOVEMENT ON THE GROUND

1. Any movement at the wrong time will expose your location. If you must travel from your shelter, be sure your movement is masked by natural ground cover as shown here.

2. If you must travel from your shelter, try to do this during bad weather or during periods of limited light.

3. The important thing to try is to break up the recognizable lines of the human shape.

BREAKING UP OUTLINE

CONCEALMENT WHILE HUNTING, ETC.

1. A good technique for concealing or at least limiting evidence of your movement is to try never to disturb any vegetation above the level of your knees.

2. Another technique is to avoid breaking grass stems, branches lying on the ground and disturbing leaves.

3. Yet another is to make a path in the vegetation ahead by parting it with a stick and then pushing it back into its original position.

4. Don't grab and then push away small trees or brush. This may scruff the bark off at eye level.

5. Be careful to avoid overturning rocks and sticks, making scuff marks on logs, sticks and bark, and mangling low grass and bushes that normally spring back.

80

TRAVEL CONCEALMENT

6. Try not to make unnecessary noise by breaking sticks on the ground. Wrapping your feet in cloth will help muffle this.

7. When making tell-tale tracks in soft soil, try to hide them by walking in the shadow of trees and bushes, snow drifts or fallen logs.

8. Always try and take advantage of solid walking surfaces which leave no evidence of movement or travel such as logs, rocks, etc.

9. Travel during rainy times is always good as your tracks will probably be filled in and covered over.

REMOVING EVIDENCE OF TRAVEL

1. Tracks can be brushed or patted out to make them disappear or look old.

2. Hide or bury anything you must discard since any litter may expose you.

GETTING THROUGH OBSTACLES

REMOVING TRAVEL EVIDENCE

1. Go through a rail or log fence by going under or between the lower rails. If this isn't practical, climb over the top and present as low a silhouette as possible.

RAIL FENCES

CROSSING ROADS

1. Roads should be crossed only at places offering cover such as bushes, trees, a bend in the road or shadows.

ROAD CROSSING

RAILROAD TRACKS

1. Align your body parallel to the tracks, lay face down and move across the track using a semi-pushup movement. Repeat this technique for the second track.

RAILROAD TRACKS

15

Properly Using Your Sleeping Bag

WHEN USING YOUR SLEEPING BAG

1. Wear whatever clothing is necessary to stay comfortable when the temperature goes down.
2. In lower temperatures, wear long underwear, dry socks, head protection and a sleep shirt.
3. Avoid sweating by wearing the fewest clothes necessary to stay warm.
4. Your sleeping bag must be kept dry if you are to keep warm. A bag primarily becomes wet from outside moisture and from sweating when in the bag.

SUGGESTIONS FOR BEING COMFORTABLE IN A SLEEPING BAG

1. Don't cover your head and breathe into your sleeping bag. Moisture will collect and wet the bag or ice crystals will form in the bag during cold spells.
2. Eat shortly before getting into your sleeping bag as this will help you to stay warm.
3. Always relieve yourself immediately before getting into your sleeping bag for the night.

KEEPING MOISTURE OUT OF YOUR SLEEPING BAG

1. Keep outside moisture from getting in and wetting your sleeping bag by putting a protective barrier underneath.
2. This barrier could be a bough bed made of pine or spruce boughs.

3. An insulated sleeping pad would also work well as would a waterproof poncho.

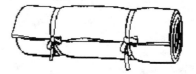

TAKING CARE OF YOUR SLEEPING BAG

1. After each use, open your sleeping bag completely, turn inside out, shake thoroughly and fluff it up.
2. Hang or lay your sleeping bag in the air so the breezes and the sun can dry it out and freshen it.
3. Always carry your sleeping bag in a waterproof carrying bag. This will keep it dry and protect it as well.

THE IDEAL SLEEPING BAG

The "ideal" sleeping bag is the -20° kind used by British commandos. It has a waterproof bottom and a built-in waterproof carrying case. This sleeping bag is sometimes available, both new and used, in The Sportsman's Guide. See Appendix 1.

16

Miscellaneous Survival Tips

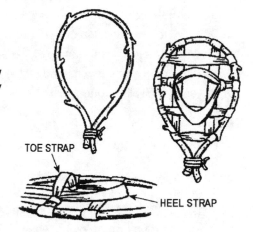

TOE STRAP

HEEL STRAP

IMPROVISED EQUIPMENT

Having snow shoes is a necessity for travel if the snow is deep. A pair can be made with willow or other springy wood and wire, a thong or a cord.

KNIFE SHARPENING

- PUSH BLADE DOWN THE STONE IN A SLICING MOTION
- THEN TURN THE BLADE OVER AND DRAW BLADE TOWARD BODY

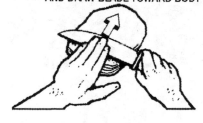

All knives should be kept sharp and ready for use when needed. A well sharpened knife will prevent needless injuries.

WHAT TO DO WHEN LOST

1. Take shelter. If you have a tent then set it up and get in it. If you have no tent, then immediately build an improvised shelter.
2. The kind of shelter you use is unimportant so long as you are able to protect yourself from the weather.
3. Immediately start your fire to keep warm as well as to prepare your food and coffee or other drinks.
4. Get into your sleeping bag to keep warm and to rest. The less energy expended, the less food needed.
5. Check your food supply and supplement it by shooting large animals and trapping or snaring smaller ones.
6. Check to be sure you have waterproof, windproof matches and firestarter (trioxene) in your pocket, your backpack, or inside the butt plate of your rifle.

USE AND CARE OF CLOTHING

Use the word **COLDER** to remind you how to take care of your clothing.

C: Keep your clothing **C**lean.

O: Avoid getting **O**verheated and sweaty.

L: Wear your clothing in **L**ayers and **L**oose.

D: Keep your clothing **D**ry.

E: **E**xamine your clothing for wear and defects.

R: Keep your clothing **R**epaired

TRAVEL TIPS TO HELP YOU SURVIVE

1. Take as many rest stops as you need, maintain a realistic pace, and don't rush or be rushed.
2. Be sure you have enough food and water as well as the means to get more when needed.
3. Take especially good care of your feet.
4. Be sure to load your backpack so it is balanced when on the move.
5. Select the safest route as well as the easiest.
6. Don't try to walk through or over obstacles. Walk around them.
7. Traverse (move crosswise) up slopes.
8. Skirt the edges of canyons and gullies.
9. Never try to walk through a swamp or a mud flat if you can walk around them.

TRAVEL BY RIVER

1. When traveling by river, you save energy and you move faster.
2. When traveling by raft, use a pole in shallow water and an oar in deep water.
3. The best time to travel on a river is during daylight hours or at night when the moon provides light.
4. Try to stay close to the inside edge of river banks where you are less likely to be seen.
5. Be ever alert for danger such as waterfalls, rapids, snags (partially or all under water), and trees and limbs hanging out over the bank.

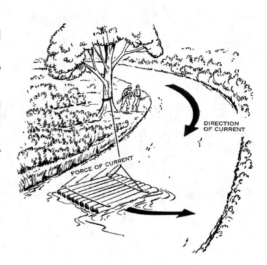

DIRECTION OF CURRENT

FORCE OF CURRENT

BUILD YOUR OWN RAFT

1. A raft can be used to safely cross a deep and fast moving river. Use a "pendulum action" (see previous page) at a bend in the river.
2. Rafting is a safe, fast method of crossing a river as well as getting from one place to another.
3. At right is one of the simplest rafts you can construct. All it takes is a couple of hefty logs and some rope.
4. A sturdy raft can be assembled without any rope or spikes. The only tools needed are a hatchet and a knife.
5. This raft is big enough for three people. Make the size best suiting your needs.
6. Start by cutting four offset inverted notches, one on top and one on the bottom of both ends of each log.

7. These notches are to be broader at the base than at the outer edge of the log.

8. A three-sided wooden crosspiece, about a foot longer than the width of the raft, should be driven through each notch. This will bind your raft together.

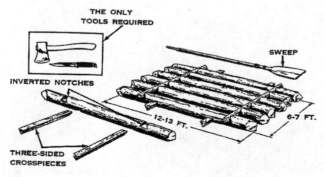

THE ONLY TOOLS REQUIRED

SWEEP

INVERTED NOTCHES

THREE-SIDED CROSSPIECES

12-13 FT.

6-7 FT.

9. Connect all of the notches on one side of the raft before connecting the notches on the other.

10. The overlapping ends of the two crosspieces at each end of the raft should be lashed together to give it additional strength.

11. When your finished raft is pushed into the water, the crosspiece will swell and the logs will be tightly bound together.

12. If the crosspieces are not tight enough, insert thin wedges of dry wood. These will swell when wet and strengthen the crosspieces.

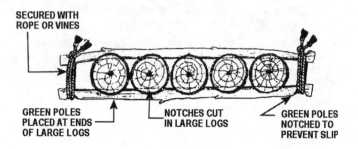

SECURED WITH
ROPE OR VINES

GREEN POLES
PLACED AT ENDS
OF LARGE LOGS

NOTCHES CUT
IN LARGE LOGS

GREEN POLES
NOTCHED TO
PREVENT SLIP

13. Another good way of constructing a raft is with the use of "pressure bars" (see left). Lash these pressure bars at each end to hold your logs securely together.

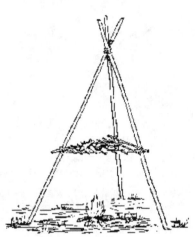

DRYING MEAT BY THE SUN OR A FIRE

1. Meat should be cut in 1/4 inch strips and either dried in the wind, the sun or over a drying fire. The meat strips are placed on a wooden grate and dried until brittle.

2. Fish can also be preserved by smoking. Prepare by cutting off the head and cutting out the backbone. Then spread the fish flat and skewer it in that position.

3. Berries and fruit can also be dried by the sun, the wind, or by using a fire. Simply cut the fruit into thin slices and place directly in the sun or next to a fire.

PROTECTING YOUR FOOD FROM PREDATORS

Some Wilderness Caches

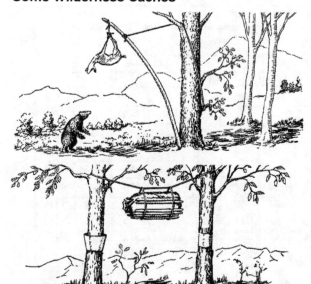

It may be necessary to protect your food from animals. Hang your food at least six feet off the ground. Or you may need to set up a cache to protect it.

PRESERVATION OF FOODS

1. Food can be buried in the snow during the winter months as a temperature of 32 degrees F is maintained.

2. Food wrapped in waterproof material and placed in a stream will remain cold enough to preserve it during the summer months.

PROTECTING YOUR FEET AND LEGS FROM INSECTS

1. Take any available piece of cloth and cut two three-inch wide strips about four feet long.

2. Wrap one strip in spiral fashion (see above) around the top of each shoe or boot. This will keep out most of the insects and sand.

WIND CHILL

1. The wind chill chart given below shows the conditions under which cold weather is highly dangerous and when any exposed skin is likely to freeze.

2. The combined effect of low temperature and wind creates a condition known as "wind chill." You can create your own wind by walking (about 5 mph) or running (10 mph).

3. A wind chill scale must be used to effectively gauge the difference between the temperature and the true impact of the cold.

4. A temperature of -20 degrees F and a wind of 20 mph gives a chill temperature of -75 degrees F.

5. In this situation there is grave danger as your exposed skin can freeze within 30 seconds. Such an injury can disable you as seriously as a broken leg or a bullet.

CHILL FACTOR CHART

WIND SPEED (mph)	LOCAL TEMPERATURE (°F)										
	32	23	14	5	−4	−13	−22	−31	−40	−49	−58
	EQUIVALENT TEMPERATURE (°F)										
CALM	32	23	14	5	−4	−13	−22	−31	−40	−49	−58
5	29	20	10	1	−9	−18	−28	−37	−47	−56	−65
10	18	7	−4	−15	−26	−37	−48	−59	−70	−81	−92
15	13	−1	−13	−25	−37	−49	−61	−73	−85	−97	−109
20	7	−6	−19	−32	−44	−57	−70	−83	−98	−10	−121
25	1	−10	−24	−37	−50	−64	−77	−90	−104	−117	−130
30	−1	−13	−27	−41	−54	−68	−82	−97	−109	−123	−137
35	−1	−15	−29	−43	−57	−71	−85	−99	−113	−127	−142
40	−3	−17	−31	−45	−59	−74	−87	−102	−116	−131	−145
45	−3	−18	−32	−46	−61	−76	−89	−104	−118	−132	−147
50	−4	−18	−33	−47	−62	−78	−91	−105	−120	−134	−148
LITTLE DANGER FOR PROPERLY CLOTHED PERSON			CONSIDERABLE DANGER			VERY GREAT DANGER					
			DANGER FROM FREEZING OR EXPOSED FLESH								

COLD WEATHER CLOTHING

1. Keeping warm is the foremost survival problem in cold weather. The clothing you have with you and how it is worn could determine how long you survive.

2. Clothing worn in cold weather serves only one purpose. It keeps the body heat from escaping by insulating it from the cold outside air.

3. Clothing helps to control your body temperature as inner layers of insulating clothing holds warm air in.

4. Wind resistant outer clothing keeps cold air from penetrating your clothing and carrying away the heat.

YOUR HANDS AND FEET

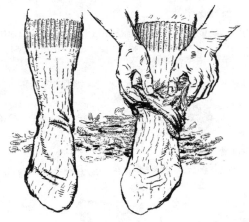

INSULATING SOCKS

1. Your hands and feet require special care since they cool faster than other parts of the body. Keep your hands covered as much as possible.

2. You can always warm your hands by placing them next to the warm flesh under your arm pits, against your ribs, or between your thighs.

3. Feet are more difficult to keep warm because they perspire more readily. Try wearing larger shoes so you can wear at least two pair of socks with them.

4. Make a warm double sock by putting one pair inside another. Fill the open space between your two socks with a layer of feathers, moss or dry grass.

CLOTHING IN HOT WEATHER

1. You must dress to protect yourself from direct sun light, excessive evaporation of your perspiration, and the many annoying insects.

2. Keep your body covered by wearing long sleeved shirts and long pants.

3. Protect the back of your neck from the sun by wearing a piece of cloth to cover it.

LEAVING CLOTHING BEHIND

Sometimes clothing has to be left behind in order to lighten the load. Keep enough for protection against the often chilly nights.

CARRYING YOUR GEAR FOR LONG AND SHORT TERM EXCURSIONS

PART 1 -- BACKPACKS, ETC.

1. There are many varieties of backpacks on the market. Some are good – others not so good. Some have an inner frame – some have an outside frame. Some are extremely light while others are heavy. Some are waterproof while some are not. Some are made with top grade materials while others are more cheaply put together.

2. One of the best is the lightweight A.L.I.C.E. pack. This backpack was developed by the U.S. military:

3. To carry items needed for long treks, and other lengthy wilderness excursions (see Part 2, 3 and 4).

4. They are to be stocked with such things as windproof and waterproof matches, food, poncho, sleeping bag, knife, fire starter, medical supplies, insect repellant, etc.

   a. The military also developed a lighter, more mobile way (web belt and suspenders) to carry essential items. This is designed to carry only those things needed on short-term combat or intelligence excursions.

5. Excellent for civilian use when going hunting, fishing and other short-term hikes or other outdoor ventures (see Part 5).

6. Included are canteen of water, small amount of food, ammunition, survival knife, water purification tablets, etc.

7. Everything carried can be attached to the web belt and suspenders.

   a. Remember – the equipment and provisions to be carried are always determined by the nature of the intended activity.

   b. Carrying a backpack is never an easy task. Carrying one incorrectly makes it even more difficult. Remember:

8. Keep everything in its proper place!

9. Carry important items where they can easily be reached.

10. **Avoid taking unnecessary items such as:**

    a. Excess food.

    b. Unneeded clothing.

    c. Equipment you'll never use.

    d. Various knickknacks and snacks, etc.

11. Don't make a pack mule out of yourself. This is a serious error.

PART 2 — CARRYING A BACKPACK FOR LONG TREKS AND CAMPING

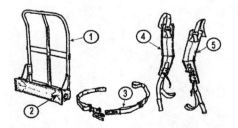

1. Although there are other excellent backpacks on the market, A.L.I.C.E. packs are more fully covered here because of their dependability, popularity, availability, low cost and quality. They are no doubt one of the better backpack buys for the money. Official military issue A.L.I.C.E. packs come in medium and large. Miscellaneous items accompanying these packs consist of:

a. A frame which fits both large and medium packs:
  1. Frame isn't usually required for carrying medium packs.
b. Lower back strap.
c. Waist strap.
d. Shoulder strap without quick release.
e. Shoulder strap with quick release.
f. Two straps for tying down load.
g. Large A.L.I.C.E. pack.
h. Medium A.L.I.C.E. pack.
i. Waterproof cover fits both backpack sizes.

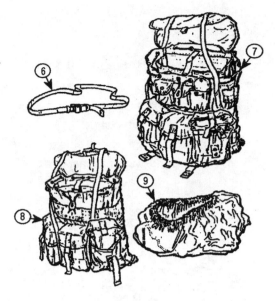

SHOULDER AND WAIST STRAPS

1. Shoulder straps are used with:
  a. Frame used to carry load on attached cargo shelf (when used).
  b. Medium A.L.I.C.E. pack without frame.
  c. Medium or large A.L.I.C.E. pack with frame.

ATTACHING SHOULDER STRAPS

WEBBING THROUGH KEEPER, OVER FRAME, AND THROUGH BUCKLE

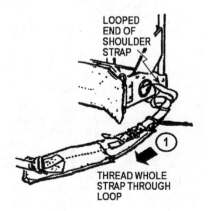

LOOPED END OF SHOULDER STRAP

THREAD WHOLE STRAP THROUGH LOOP

1. One strap with quick release assembly goes on right shoulder.
  a. Other strap goes on left shoulder.
2. To attach shoulder straps to frame:
  a. Insert looped end of strap through **inside** of frame.
    1. Run loop through nylon ring at bottom side of frame.
  b. Thread strap through loop and pull tight.
  c. Insert webbing at top of strap.
  d. Slip through metal keeper on top of frame.
  e. Run through buckle and pull tight.
3. Padded end of strap should be under frame bar.

PROPER WAIST STRAP ATTACHMENT

1. Waist straps have quick release assembly.
2. Pull tab is found on strap attached to frame.
   a. Attach strap with pull tab.
   b. Do on either right or left side of frame.
   c. This depends on which hand is used to pull open.

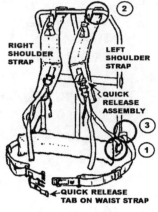

ASSEMBLED FRAME AND STRAPS

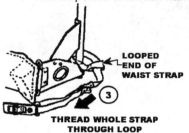

LOOPED END OF WAIST STRAP

THREAD WHOLE STRAP THROUGH LOOP

3. To attach strap:
   a. Insert loop end of waist strap around lower part of frame.
   b. Pull tight.

ATTACHING CARGO SHELF AND CARGO TO FRAME

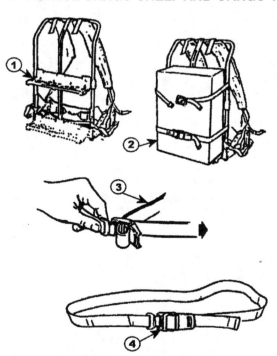

1. Slip cargo shelf on back of frame.
   a. Slide to middle or bottom.
2. Cargo shelves are used for carrying:
   a. Bulky items such as boxes of ammunition.
   b. Cases of food rations (MREs, etc.).
   c. Five gallon can of gas or water, etc.
3. To properly secure load:
   a. Insert hooked.end in buckle as shown.
   b. Pull loose end in direction of arrow to moderately tighten.
   c. Always leave a little slack.
4. Push fastener into closed position to take up slack in strap.
   a. If enough slack isn't left before closing fastener:
   b. Closing puts too much tension on strap.
   c. Fastener won't close properly.
   d. Closed strap may crush contents of container.
   e. Simply pull up strap to release and open fastener.

5. Other Pointers:
   a. Two tie down straps furnished with frame.
   b. Load tied to frame with tie down straps.
   c. Tie down straps hold load to frame.
   d. Proper adjustment is important.
   e. Top strap is wrapped around load and frame .
   f. Do not pull tight.

SHOULDER AND WAIST STRAPS ADJUSTMENT

1. Shoulder and waist straps have adjustment buckles.
2. Adjustment buckles are used:
    a. After straps are attached to loads.
    b. After straps and load are over shoulder.
3. Shoulder straps can be adjusted to lift or lower pack.
    a. Pull down on loose end of webbing to shorten strap:
        1. Pull up on cords to loosen strap.
        2. Easily slip off backpack.
    b. Open buckle and slide toward or away from front.
        1. Slide strap away from front to tighten.
        2. Slide strap toward front to loosen.
        3. Close buckle to hold when finished adjusting.

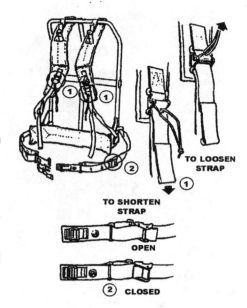

TO LOOSEN STRAP

TO SHORTEN STRAP

OPEN

CLOSED

PART 3 • MEDIUM A.L.I.C.E. PACK - DESCRIPTION

1. The medium A.L.I.C.E. pack:
    a. Is one of the most durable backpacks made.
    b. Outlasts most commercial backpacks by 5 to 1.
    c. Is water repellant.
    d. Can be sprayed to make waterproof.
    e. Can be readily purchased at surplus stores and through most catalogs listed in APPENDIX 1.
        I. Main flap over top of backpack opens by pulling apart two tabs.
        2. Simply close and press flap to seal.
        3. Waterproof backpack cover as well as a multitude of other things can be stored in the main compartment.

MAIN FLAP SHOWN OPEN (1)

SMALL COMPARTMENT IN MAIN COMPARTMENT (2)

MAIN COMPARTMENT CLOSING BUCKLES

EQUIPMENT HANGERS (3)

FRONT

(4) BACK

4. Equipment hangers (webbing with eyelets, webbed loops) for use with hooks or slide keepers are located:
    a. On sides of A.L.I.C.E. pack.
    b. Above pockets on outside of pack.
    c. Pockets are tunneled between main compartment and pockets for carrying a knife scabbard or machete sheath.
    d. By sliding knife or machete down through tunnel, it can be secured to hanger above with slide keepers or hooks.
5. Medium A.L.I.C.E. pack is carried on back by use of shoulder straps. No frame is usually needed.

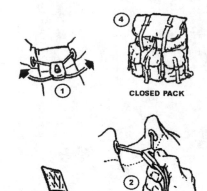

CLOSED PACK

SNAP
BOTTOMS
ON POCKET
CLOSING
STRAPS

CLOSING A LOADED A.L.I.C.E. PACK

1. Once an A.L.I.C.E. pack is loaded, use drawstring buckle to close top flap.

   a. Pull two cord ends to gather top into tight closure.

2. To loosen drawstring, push button up on buckle with thumb. Pull buckle down.

3. Each pocket on A.L.I.C.E. pack is closed by sliding webbing through buckle. Then pull it tight.

   a. Unsnap lower end of tie-down to open pocket.

4. Run tie-down straps through webbed loops on top of main flap.

5. Pull tightly doown over backpack through bottom buckles.

ATTACHING SHOULDER STRAPS TO MEDIUM A.L.I.C.E. PACK

1. Attach shoulder straps the same way they are attached to frame:

   a. Insert looped end of strap through "D" ring at bottom of pack.

      1. Run strap through strap loop & pull tight.

   b. Insert webbing at top of strap.

      1. Run strap through metal loop on top of pack & pull tight.

NOTE: Be sure strap with quick release is over left shoulder.

WARNING:

1. Medium A.L.I.C.E. pack should be carried on frame in extreme cold weather.'

2. Prevents accumulation of perspiration where pack is in contact with person's back.

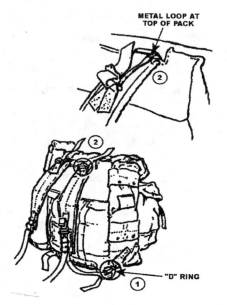

METAL LOOP AT
TOP OF PACK

"D" RING

ATTACHING SLEEPING BAG TO A.L.I.C.E. PACK

1. Sleeping bag can be attached to outside of pack.

   a. Carry on bottom of pack with cargo straps.

2. Pull cargo straps through loops on bottom of pack.

NOTE:

1. Bedroll can be carried the same.

2. Or it can be placed in main compartment of pack.

1. The large A.L.I.C.E. pack is **always** used with a frame.

2. These backpacks are for carrying larger loads.

3. A large pack is bigger than is the medium pack except:

   a. It has three more small pockets at the top.

   b. It has tie-down cords and "D" rings inside main compartment.

   c. This shortens pack if it isn't filled to capacity.

4. The three larger lower pockets are tunneled.

   a. This allows for carrying skis or other special items.

ADJUSTING LARGE A.L.I.C.E.PACK FRAME

1. A special frame is used to carry a large A.L.I.C.E. pack or to carry cargo.

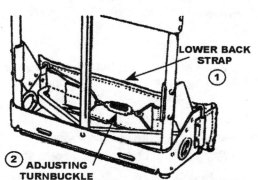

LOWER BACK STRAP ①

② ADJUSTING TURNBUCKLE

2. The frame has adjustable lower back strap.

3. This special strap keeps loaded backpack away from the back.

4. The frame allows air to circulate between pack and the back.

5. This makes it more comfortable to carry large or heavy load.

6. The strap is almost perfectly flat when turnbuckle is screwed tight.

   a. To allow strap to curve in and fit snugly against lower back.

ATTACHING LARGE A.L.I.C.E. PACK TO FRAME

1. Insert bare frame into envelope on back of pack.

2. Attach bottom of pack to frame as shown.

   a. Loop webbing around frame twice before buckling.

   b. Attach shoulder and waist straps as described previously.

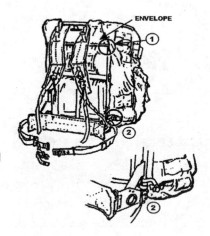

ENVELOPE ①

②

②

QUICK RELEASES FOR SHOULDER AND WAIST STRAPS

1. Sudden removal of a loaded A.L.I.C.E. pack may be required in certain emergency situations.

2. Quick releases are provided on waist and left shoulder straps.

3. Numbers (1), (2), (3), and (4) show how shoulder strap quick- release is assembled.

   a. The metal loop at top of lower end of strap (1) is hooked over the metal loop (2).

   b. Plastic prongs are pushed down (4) so locked assembly is as shown.

   c. (5) through (8) shows how to assemble quick release on waist strap.

   d. For sudden release, first pull tab on waist strap (8).

   e. Follow immediately by pulling up on shoulder strap tab (9)

4. Shift load to right and let pack slide off right shoulder.

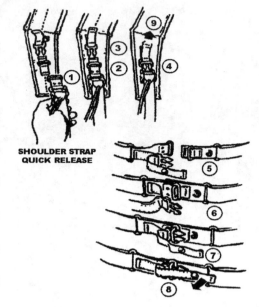

SHOULDER STRAP QUICK RELEASE

PART 5 • CARRYING EQUIPMENT AND SUPPLIES FOR SHORT TERM OUTDOOR ACTIVITIES

HUNTING, FISHING AND HIKING

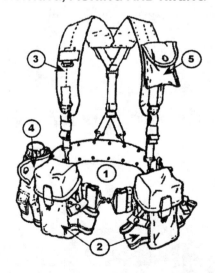

1. An excellent way to carry supplies and equipment for shorter term activities (hunting, fishing, long hikes, shooting practice, etc.) is made up of:

2. Military style web belt.

3. Two small canvas cases or bags to hook on belt.

4. Excellent for carrying ammunition, food or fishing tackle.

5. One pair heavy duty military-style suspenders.

6. Quart size canteen with cover.

7. Small first aid kit.

ASSEMBLING A HUNTING, FISHING OR HIKING SUPPLY CARRIER

1. Try on military-style web belt.

2. Belt should be comfortably snug, not tight!

3. Adjust web belt for a better fit:

   a. Push two metal keepers toward belt buckle.

   b. Spread looped webbing apart and unlock adjusting clamp.

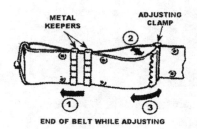

METAL KEEPERS

ADJUSTING CLAMP

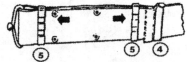

END OF BELT WHILE ADJUSTING

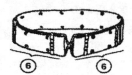

END OF BELT AFTER ADJUSTMENT

4. Slide adjusting clamp toward belt buckle to loosen, away to tighten.

5. Lock adjusting clamp in place by squeezing.

6. One metal keeper should be moved next to adjusting clamp, other next to the buckle.

7. Adjust other end of buckle the same way.

    a. Both clamps should be same distance from buckle.

NOTE: Belt is ready for attaching equipment & supplies.

ATTACHING CANVAS FOOD OR AMMUNITION HOLDERS TO WEB BELT

1. Attach canvas bags close to buckle.

    a. One goes to left of buckle, one to right.

2. Food or ammunition holder is attached to web belt using two slide keepers:

    a. Pull each keeper open and slide one over thickness of webbing.

        1. Keepers should always be vertical.

        2. Make sure keeper bottoms go beyond webbing of belt.

    b. Push keeper slides down and into bottom holes.

    c. Make sure keeper slides are firmly attached in the holes.

        1. A keeper slide in wrong position as shown in sketch can let equipment fall off.

    d. Pockets for a variety of small items are found on either side of canvas ammunition or food holders:

        1. Extra matches.

        2. Fire starter.

        3. Sewing kit.ss

        4. Small magnifying glass.

        5. Wire saw.

        6. Tooth brush

        7. Cayenne pepper.

    e. Be sure to put nylon strap through ring and snap.

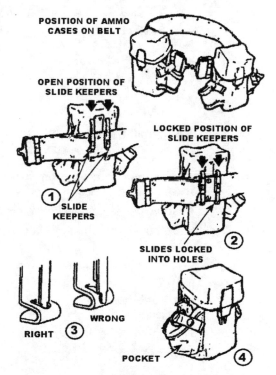

POSITION OF AMMO CASES ON BELT

OPEN POSITION OF SLIDE KEEPERS

LOCKED POSITION OF SLIDE KEEPERS

SLIDE KEEPERS

SLIDES LOCKED INTO HOLES

RIGHT   WRONG

POCKET

ATTACHING SUSPENDERS TO CANVAS BAGS AND WEB BELT

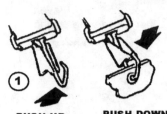

PUSH UP TO OPEN **PUSH DOWN TO CLOSE**

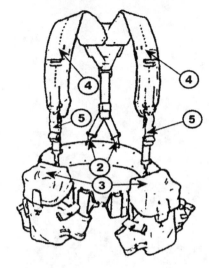

1. Snap hooks are used to attach suspenders to web belt.

   a. Push snap hooks by pushing hooks up and out of side retaining closure to open.

   b. Engage hook in eyelets and snap closed.

2. Attach the rear suspender strap hooks:

   a. To top row of eyelets on belt.

   b. To right and left of eyelet centrally located on back of belt.

3. Attach front suspender strap hooks to strap support eyelet.

   a. This is located on top back of canvas food/ammunition bag.

   b. Front suspender straps are attached to eyelets on either side of belt buckle when canvas food/ammunition bags aren't used.

4. Metal and web loops are found on each shoulder strap. Small items can be attached to these loops:

   a. Flashlight.

   b. Compass.

   c. Pocket knife.

5. After adjustments are made, the elastic loops on each of the adjustment straps are used to secure loose ends of straps.

ATTACHING CANTEEN COVER, FIRST AID DRESSING/COMPASS HOLDER

1. Secure canteen cover on web belt with two slidekeepers on back.

   a. Attach close to canvas food/ammunition bag.

2. Canteen cover has small pocket for water purification tablets.

   a. Velcro fastener is used to secure flap on canteen pocket.

3. The first aid dressing or compass holder is attached with slidekeeper:

   a. To webbing loop on front of suspenders.

   b. Or to belt next to canvas food/ammunition bag.

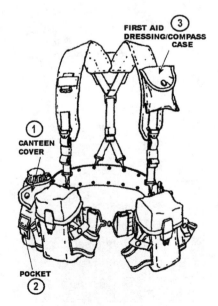

FIRST AID DRESSING/COMPASS CASE ③

① CANTEEN COVER

POCKET ②

ATTACHING HUNTING KNIFE SHEATH

1. Knife sheath goes next to canvas food/ammunition bag on either side of belt.

   a. Attach by its hooks to lower eyelets of web belt.

   b. Small hatchet can also be carried in this manner.

ADJUSTING SUSPENDER STRAPS

1. Try on web belt and suspenders after equipment is attached.

2. Fasten belt buckle.

   a. Adjust length of front and back straps.

   b. Belt should hang evenly at waist.

3. Pull down on loose lower end of strap to tighten.

   a. Lift end of buckle to loosen.

4. After belt is comfortable around waist:

   a. Secure loose ends with the elastic loops.

5. Back strap adjustments are done the same way.

   a. This is best accomplished with the help of another person.

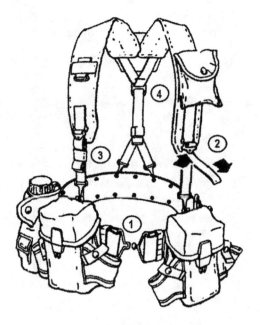

18

Properly Using and Stocking a Backpack

1. Each family member should have a backpack. One of the best is the rugged military A.L.I.C.E. (All-Purpose Lightweight Individual Carrying Equipment) pack.

2. An A.L.I.C.E. pack comes in medium and large. The size depends on the strength of the person who is to carry the pack.

3. A.L.I.C.E. packs are relatively inexpensive when purchased "used" and they are available from military surplus stores and some of the sources given in Appendix 1.

4. A backpack is a necessity to take on a weekend excursion. And they are needed when longer treks are found to be necessary.

5. Each pack should be outfitted with the necessary survival items: at least a three day supply of food, one or more weapons, and a supply of ammunition for each.

Suggestions For Packing Your Backpack

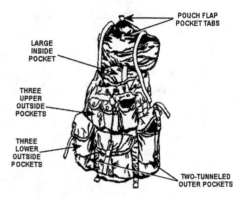

1. The most important thing to remember in packing your backpack is that you want to avoid unpacking it during halts in order to find frequently needed items.

2. Pack your heaviest items at the bottom of your pack and close to the frame. This puts the bulk of the weight on your hips, which is necessary for good balance.

3. Hard or sharp objects should always be placed on the inside. Here they won't rub on the bag or your back.

4. Things often needed should be carried in the outside pockets within easy reach.

5. Maps and other flat items should be placed in the flap pocket.

6. Your sleeping bag can be strapped to the bottom of your backpack or it can be carried under the top flap.

Suggestions For Outfitting Your Backpack

1. Not everyone will feel a need for everything listed. Some won't want to carry nail clippers. One person might want windproof/waterproof matches. Another may prefer a magnesium bar for starting a camp fire. Others may wish to carry Rolaids or Tums while some people won't.

2. Some of the items listed can be put in a backpack, but they'll probably be carried elsewhere. For example a U.S. Air Force survival knife with a sharpener in the sheath is best attached to the belt. So is a canteen.

3. Small fanny (butt) packs are also attached to a belt. These handy items are available in military surplus stores or from the sources listed in Appendix 1.

4. Most often carried in a fanny pack are small items such as a Swiss army pocket knife, a first aid kit, magnifying glass, matches, water purification tablets, etc.

Suggested Backpack Items

Food

MRE Main Dishes [1]

MRE Miscellaneous Food items [2]

Cheese spread	Peanut Butter
Crackers	Bread slice [4]
Apple sauce	Potato sticks

Hot chocolate	Kool-Aid [5]
Coffee packets	Oatmeal bars [6]
Sugar packets	Dehydrated fruit [7]
Creamer packets[3]	Pound cake [8]
Salt packets	Jelly
	3-Day Food Bar (3600 calories)[9]

Medical Items

Garlic oil [10]	Vitamin C	Iodine
Bandaids	Tweezers	Folding scissors
Rolaids/Tums	Chapstick	Tick removal kit
Snake bite kit	Bandages	Tape
Insect repellant	Bug stick	Sunburn lotion

Hygiene

| Toothbrush | Toothpaste |
| Liquid soap tubes [11] | Toilet paper packets |

Fire Starting Items

Trioxene packets [12]	Hexamine tablets [14]
Spark light [13]	Lifeboat matches [15]
Magnifying glass	

Miscellaneous Gear

Hammock [16]	Ivory soap [20]
Shoelaces	Tackle box [21]
Can opener	Fishing pole [22]
Nail clippers	Eating utensils
50' parachute cord [17]	Poncho (military)
Insulated gloves	Mosquito head net
Sewing kit	Flashlight
Compass	Cloth measuring tape
Light sticks [18]	Wire saw [23]
Water purification tablets [19]	Light stick holder [24]
Space blanket [25]	

Footnotes

[1.] Military MRE's (Meals Ready to Eat) are complete individual meals packaged with a seven year shelf life when stored at 70 degrees. Some MRE meals are chicken ala king, pork barbecue, beef stew, etc. Available at gun shows, military surplus stores and sources in Appendix 1.

[2.] All of these food items come in a MRE package as described above in 1.

[3.] Combine one packet of sugar with one creamer in the creamer's foil lined packet. Blend lightly and heat. Makes quick energy cookie. Available in supermarkets.

4. To be sliced down the middle for making a sandwich. See Appendix 1 for sources.

5. Cherry, lemon-lime, grape and orange. See Appendix 1 for sources.

6. Can be chewed dry while on the move or broken up in a bowl of hot water. See Appendix 1 for sources.

7. Peaches, pears and mixed fruit. Eat dry or add water and have a bowl of fruit. Available in MRE packets.

8. Cake choices range from lemon, orange, maple nut and chocolate mint. See Appendix 1 for sources.

9. Easy-to-store, pleasant tasting survival food bar. Five (5) year shelf life. See Appendix 1 for sources.

10. Available in any drugstore or supermarket.

11. Available in any sporting goods section.

12. Trioxane bars are available at gun shows, military surplus stores and from sources in Appendix 1.

13. Spark Light. Air Force survival fire starter. Carried by all pilots. See Appendix 1 for sources.

14. Hexamine tablets are military issue. Available at gun shows and military surplus stores.

15. LifeBoat Matches. Wind and waterproof. See Appendix 1 for sources.

16. Baseball size hammocks available at gun shows, military surplus stores and from sources in Appendix 1.

17. Available at gun shows, military surplus stores and from sources in Appendix 1.

18. 18.Available at gun shows, military surplus stores and from sources in Appendix 1.

19. Available at gun shows, military surplus stores and from sources in Appendix 1.

20. An excellent bait for catching catfish.

21. Small 3"x 5" plastic box with compartments. Available in sporting goods departments.

22. Telescoping fishing pole. Measures 12" when closed. Check local sporting goods departments. Target stores usually stock them.

23. Commonly called a "Commando Saw." Available from sources in Appendix 1.

24. Available at gun shows, military surplus stores and from sources in Appendix 1.

25. Space blankets are windproof, waterproof and warm. Retain 80% or more of your body heat. Available in military surplus stores and from sources in Appendix 1.

Appendix I

Recommended Suppliers of Survival Food & Equipment

Each of the following companies can be depended upon for consistent quality in their offerings. Each has an outstanding variety of good selections. It's a good idea to compare prices for they sometimes vary extensively. Call their number and request a free catalog.

BRIGADE QUARTERMASTERS

1-800-338-4327

Sells quality military surplus and survival-oriented items. Also offers a small selection of long-term storage foods. Excellent prices for high caliber merchandise. One of the Cadillacs in the field. Highly recommended. Call for copy of their classy catalog.

BRUCE HOPKINS
BEST PRICES STORABLE FOODS

1-972-288-0262

Specializes in top quality, long-term survival foods. Their freeze-dried foods are sold at impossible to beat prices. Highly recommended. Call for price list and compare before buying elsewhere. You'll be glad you did!

CABELA'S

1-800-237-4444

This is without a doubt the world's foremost suppliers of outdoors items -- parkas, boots, fishing gear, etc. They're in a class of their own. Highly recommended. Call for top quality catalog.

CHEAPER THAN DIRT!

1-817-625-7557

Has an interesting selection of military surplus and survival items at decent prices. Call for catalog.

ELITE FORCE, INC.

1-800-948-0754

A nice selection of interesting items, some of which aren't usually sold by competitors. Sells excellent survival merchandise -- boots, clothing, bandoleers, tents, etc. Call for catalog.

EMERGENCY ESSENTIALS

1-800-999-1863

Emergency Essentials is exactly what its name implies. Everything they sell is strictly geared to survival. Highly recommended for both the quality and selection of long-term storage foods and other survival merchandise. Call for catalog.

LONG LIFE FOOD DEPOT

1-800-601-2833

Great selections of long-term storage foods at excellent prices. They've been in business 14 years and it's no wonder. Their array of survival foods is most impressive. Highly recommended. Call for their "Current Discount Stock and Price List."

MAJOR ARMY NAVY SURPLUS

1-800-441-8855

Marvelous selection of military surplus items and long-term storage foods. One of the biggest surplus and survival companies in the world. Has huge store in Gardena, California. Always has many special priced items for sale at truly good discounts. Call for catalog.

MASS ARMY NAVY

1-800-343-7749

Specializes in many unique foreign as well as U.S. military surplus items. Excellent quality and fairly priced merchandise. Call for most interesting catalog.

NITRO-PAK PREPAREDNESS CENTER

1-800-866-4876

Specializes in a huge variety of long-term storage survival foods. Also sells a lot of quality survival merchandise. A top company in the field. Call for catalog.

RANGER JOES

1-800-247-4541

Specializes in high quality military and law enforcement gear. Call for copy of their great catalog.

SAFE-TREK OUTFITTERS

1-800-424-7870

Specializes in foods with a long shelf life. Also has an extensive line of top quality survival products. Call for catalog.

THE SPORTSMAN'S GUIDE

1-800-888-3006

Sometimes offers surprisingly low prices on outdoor, foreign and U.S. military surplus, and survival-oriented goods. Included are such things as ammunition, back packs, sleeping bags and boots. Call for colorful, most interesting catalog.

U.S. CALVARY

1-800-333-5102

Sells some of the world's finest military and adventure equipment and supplies. Well worth getting on their mailing list. Call for catalog.

Appendix II

Spiritual Survival – A Helpful Guide

1. CHECKLIST FOR SPIRITUAL SURVIVAL

A. WHEN IN A SURVIVAL SITUATION:
1. Always remember — you are never alone.
2. God is always with you.

B. PRAY FOR GOD'S:
1. Help.
2. Strength.
3. Wisdom.
4. Comfort.

C. RECITE SCRIPTURES:
1. 23rd Psalm (See under "Scriptures").
2. The Lord's Prayer (See under "Scriptures").

D. IF YOU CAN RECALL SCRIPTURAL VERSES, REPEAT THEM:
1. To yourself.
2. To God.

E. PRAY — TALK TO GOD:
1. Ask for God's help.
2. Thank God that He is with you.

F. SING HYMNS TO YOURSELF AND TO GOD (See under "Hymns").

G. REMEMBER:
1. You may not have a Bible.
2. There probably won't be any clergy.

H. NEVERTHELESS, WORSHIP AT SAFE SITE IN OR AROUND SHELTER:
1. Alone, if no one is with you.
2. With others, who happen to be with you.

I. GIVE PRAISE AND THANKS TO GOD:
1. He is bigger than your circumstances.
2. Rejoice that no matter what happens, He will see you through.

J. NEVER DOUBT THE REALITY OF:
1. Heaven.
2. Eternal life.

K. KEEP YOUR TRUST IN:
1. God.
2. Those individuals who are with you.
3. Your own ability to handle stress.

L. NEVER ALLOW YOURSELF:
1. To lose hope.
2. To give up.

M. IF YOU ARE WITH OTHER SURVIVORS:
1. Pray for each other with regularity.
2. Share scriptures and songs (See "Scriptures" and "Hymns").
3. Appoint someone to be chaplain.
4. Try to have regular worship services.
5. Write down scripture and songs that you can remember.
6. Encourage each other and take prayer requests.

N. NEVER FORGET THAT GOD LOVES YOU:
    1. Praise the Lord.
    2. Give Him thanks no matter what happens.

O. KEEP FAITH IN YOUR COUNTRY:
    1. Sing inspirational and patriotic songs (See under hymns).
    2. Recite the Pledge of Allegiance.

2. SCRIPTURE

JOHN 3:16

For God so loved the world, that he gave his only begotten Son, that whosoever believeth in him should not perish, but have everlasting life.

ROMANS 8:28

And we know that all things work together for good to them that love God, to them who are called according to his purpose.

MATTHEW 6:9-13 -- THE LORD'S PRAYER

Our Father who art in heaven, hallowed by thy name.
Thy kingdom come. Thy will be done in earth, as it is in heaven.
Give us this day our daily bread.
And forgive us our debts, as we forgive our debtors.
And lead us not into temptation, but deliver us from evil:
For thine is the kingdom, and the power, and the glory, for ever.
Amen.

PSALM 23

The LORD is my shepherd, I shall not want. He maketh me to lie down in green pastures: He leadeth me beside the still waters. He restoreth my soul: he leadeth me in the paths of righteousness for his name's sake. Yea, though I walk through the valley of the shadow of death, I will fear no evil: for thou art with me; thy rod and thy staff they comfort me. Thou preparest a table before me in the presence of mine enemies: thou anointed my head with oil; my cup runneth over. Surely goodness and mercy shall follow me all the days of my life: and I will dwell in the house of the LORD forever.

3. HYMNS

AMAZING GRACE

A-maz-ing grace! how sweet the sound, That saved a wretch like me!
  I once was lost, but now am found, Was blind but now I see.

'Twas grace that taught my heart to fear, And grace my fears re-lieved;
  How precious did that grace ap-pear The hour I first be-lieved!

Thro' many dangers, toils and snares, I have al-read-y come;
  'Tis grace has bro't me safe thus far, And grace will lead me home.

When we've been there ten thousand years, Bright shin-ing as the sun,
  We've no less days to sing God's praise Than when we first be-gun.

A-men.

BLESSED ASSURANCE

Bless-ed as-sur-ance, Jesus is mine!
  Oh, what a fore-taste of glo-ry di-vine!
Heir of sal-va-tion, purchase of God,
  Born of His spir-it, wash'd in His blood. *(Chorus)*

Per-fect sub-mis-sion, all is at rest,
  I in my Sav-iour am happy and blest:
Watch-ing and wait-ing, look-ing above
  Fill'd with His good-ness, lost in His love. *(Chorus)*

CHORUS

This is my sto-ry, this is my song,
  Prais-ing my Sav-iour all the day long;
This is my sto-ry, this is my song,
  Prais-ing my Sav-iour all the day long.

BATTLE HYMN OF THE REPUBLIC

Mine eyes have seen the glo-ry of the com-ing of the Lord;
  He is tram-pling out the vintage where the grapes of wrath are stored;
He hath loosed the fate-ful light-ning of His ter-ri-ble swift sword;
  His truth is march-ing on. *(Chorus)*

I have seen Him in the watch-fires of a hundred cir-cling camps;
  They have build-ed Him an al-tar in the eve-ning dews and damps;
I can read His righteous sen-tence by the dim and flar-ing lamps;
  His day is marching on. *(Chorus)*

He has sound-ed forth the trumpet that shall nev-er sound re-treat;
  He is sift-ing out the hearts of men be-for His judgment seat;
O be swift, my soul, to an-swer Him! be jub-i-lant my feet!
  Our God is marching on. *(Chorus)*

In the beau-ty of the lil-ies, Christ was born a-cross the sea,
  With a glo-ry in His bos-om that trans-fig-ures you and me;
As He died to make men ho-ly, let us die to make men free,
  While God is march-ing on. *(Chorus)*

CHORUS

  Glo-ry! glo-ry, hal-le-lu-jah! Glo-ry! glo-ry, hal-le--lu-jah!
  Glo-ry! glo-ry, hal-le-lu-jah! Our God is march-ing on.

WHAT A FRIEND WE HAVE IN JESUS

What a Friend we have in Je-sus, All our sins and griefs to bear!
  What a priv-i-lege to car-ry Ev-'ry-thing to God in prayer!
Oh, what peace we of-ten for-feit, Oh what need-less pain we bear,
  All be-cause we do not car-ry Ev-'ry-thing to God in prayer!

Have we tri-als and temp-ta-tions? Is there troub-le an-y-where?
  We should nev-er be dis-cour-aged, Take it to the Lord in prayer:
Can we find a friend so faith-ful Who will all our sor-rows share?
  Jesus knows our ev-ry weak-ness, Take it to the Lord in prayer.

Are we weak and heav-y laden, Cum-bered with a load of care?
  Pre-cious Saviour, still our ref-uge; Take it to the Lord in prayer:
Do thy friends de-spise, for-sake thee? Take it to the Lord in prayer.
  In His arms He'll take and shield thee; Thou wilt find a sol-ace there.

MY COUNTRY 'TIS OF THEE

My country, 'tis of thee. Sweet land of lib-er-ty, Of thee I sing:
  Land where my fa-thers died, Land of the pil-grims' pride,
    From ev-ery moun-tain-side Let free-dom ring!

My na-tive coun-try, thee, Land of the no-ble free, Thy name I love:
  I love the rocks and rills, Thy woods and tem-pled hills;
    My heart with rap-ture thrills Like that a-bove.

Let mu-sic swell the breeze, And ring from all the trees Sweet free-doms song:
  Let mor-tal tongues a-wake; Let all that breathe par-take;
    Let rocks their si-lence break, The sound pro-long.

Our fa-thers' God to thee, Au-thor of lib-er-ty, To Thee we sing:
  Long may our land be bright With free-doms ho-ly light;
    Pro-tect us by Thy might, Great God, our King!

THE STAR SPANGLED BANNER

Oh, say, can you see, by the dawn's ear-ly light,
  What so proud-ly we hailed at the twi-lights last gleam-ing,
Whose broad stripes and bright stars, thro' the per-il-ous fight,
  O'er the ramparts we watched, were so gal-lantly stream-ing?
And the rock-ets' red glare, the bombs burst-ing in air
  Gave proof thro' the night that our flag was still there. *(Chorus)*

Oh, thus be it ev-er when free men shall stand
  Bet-ween their loved homes and the war's des-o-la-tion;
Blest with vic-t'ry and peace, may the heav'n-res-cued land
  Praise the Pow'r that hath made and pre-served us a na-tion!
Then con-quer we must, when our cause it is just;
  And this be our mot-to: "In God is out trust!" *(Chorus)*

CHORUS

Oh, say, does that Star-span-gled Banner yet wave
  O'er the land of the free and the home of the brave?
And the Star-span-gled Banner in tri-umph shall wave
  O'er the land of the free and the home of the brave.

SECTION 2

THE

OFFICIAL

URBAN & WILDERNESS

EMERGENCY MEDICAL

SURVIVAL MANUAL

CONTENTS

Introducing the Official Urban & Wilderness
Emergency Medical Survival Guide

What would you and your family do if one of you fell and fractured a leg while hiking in the mountains? What would you and your family do if while enjoying a weekend camping excursion in the wilderness you were bitten by a rattlesnake? And what would you do on a hunting expedition if one of your party were wounded by an accidental gunshot? What would you do if you are thoroughly soaked by a fall in an icy river and the temperature is dropping to below freezing? What would you do if caught in a snowstorm and someone in your group gets frostbite? What would you do if you slashed your hand with a hunting knife while cutting branches for fuel?

Did you know garlic oil is carried in every Russian soldier's backpack? Why? Because it not only kills germs but it is believed to hasten healing! Did you know that ordinary household laundry bleach can be used as a preventative medicine? A small amount zaps the harmful bacteria in water and makes it safe to drink. And did you know that cayenne pepper sprinkled on a wound will stop bleeding and promote healing?

Emergency medical treatment must sometimes be given someone when no trained professionals are on hand. THE OFFICIAL URBAN & WILDERNESS EMERGENCY MEDICAL SURVIVAL GUIDE was specifically designed for such a situation. Carefully read over all of the various medical procedures. Become familiar with how each should be administered. The time may come when this knowledge must be applied in a real life situation.

Always remember the first and foremost rule of survival medicine. Never panic! It's normal to feel overwhelmed with trepidation and indecision in fear-inducing circumstances. But panic must be avoided at any cost!

Keep a copy of THE OFFICIAL URBAN & WILDERNESS EMERGENCY MEDICAL SURVIVAL GUIDE in your home and another in your vehicle. Every hunter should keep a copy close at hand when out trekking through the woods. The same is true of fishermen, scouts, hikers, backpackers, campers, those in the military and a multitude of others.

PART I

BASIC EMERGENCY SURVIVAL MEDICINE

2

Vital Body Functions

Respiration and Blood Circulation

1. Blood circulation and respiration (breathing – inhaling and exhaling) are both vital body functions.

   a. A person could die if either were interrupted.

   b. But the interruption of these functions isn't necessarily fatal — if appropriate measures are correctly applied in a timely manner.

Respiration Defined

1. Respiration (breathing) is simply how:

   a. Oxygen is taken into the body.

   b. Carbon dioxide is expelled from the body.

2. Respiration is accomplished by means of a canal through which air passes to and from the lungs. This airway consists of:

   a. Nose, mouth and throat.

   b. Voice box, windpipe, bronchial tree.

Airway, Lungs and Chest Cage

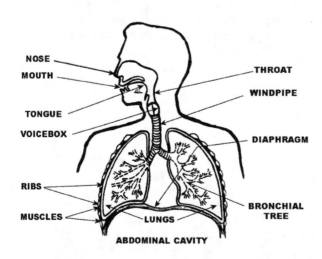

1. The lungs are two elastic organs:

   a. Composed of thousands of tiny air spaces.

   b. Covered by an airtight membrane.

2. Chest cage is formed by muscle connected ribs which join spine in back, breastbone in front:

   a. The chest cage is closed at top by neck structure.

   b. The bottom is separated from abdomen by a large dome-shaped muscle called the diaphragm.

Blood Circulation

1. Blood is circulated through body tissues by the heart pumping and the blood vessels:

   a. Arteries

   b. Veins

   c. Capillaries

2. The heart is divided into separate halves, each acting as a pump.

    a. The left side pumps bright red (oxygenated) blood through arteries and into capillaries.

    b. Nutrients and oxygen pass from blood through capillary walls into cells.

    c. The right side gets deoxygenated blood carried from capillaries through veins.

    d. The blood then goes to lungs. Carbon dioxide is expelled and more oxygen is picked up.

3. Heart Beat: The heart acts as a pump to keep blood circulating continuously through blood vessels and all parts of body:

    a. It contracts and forces blood from its chambers.

    b. It relaxes, allowing its chambers to refill with blood.

    c. This rhythmical cycle of contractions and relaxation is the **heartbeat.**

4. Pulse: The systemic arteries carry spurts of richly oxygenated blood from the left side of heart to all parts of body.

    a. The arteries expand each time blood is forced into them by the heart.

    b. The arteries contract as blood moves further along circulatory system.

    c. This expansion and contraction cycle is the **pulse.** Normal pulse rate averages from 60 to 80 beats per minute.

Conditions Adversely Affecting Vital Body Functions

1. Lack of oxygen quickly leads to death.

2. Human life can't exist without continuous oxygen intake. It's important, therefore, to know how to:

    a. Open airway and restore breathing (*see Chapter 4*).

    b. Restore heartbeat (*see Chapter 5*).

3. Bleeding: an adequate volume of blood to carry oxygen to tissues is necessary to keep person alive. It's important to:

    a. Stop any bleeding to prevent unnecessary blood loss (*see Chapter 6*).

4. Shock:

    a. A condition where blood flow to vital organs and tissues is inadequate.

    b. Blood vessels are overdilated.

5. Can be fatal if not corrected even though injury causing shock wouldn't otherwise result in death.

6. Shock can be caused from:

    a. Blood loss.

    b. Loss of bodily fluids from deep burns.

    c. Blood vessel expansion.

    d. Pain.

    e. Reaction to seeing a wound or blood.

    f. Stress injury.

7. A person must have medical knowledge to prevent shock. Chances are much greater for survival if shock isn't allowed to develop (*see Chapter 8*).

8. Infections: How well a wound is initially protected from contamination determines to a great extent:

    a. The speed of recovery from severe wound.

    b. How quickly wound heals.

9. A person should know how to properly dress wounds in order to avoid infections (*see Chapter 7*).

3

Basic Lifesaving Measures (A B C D)

Emergency Lifesaving Techniques

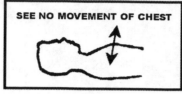

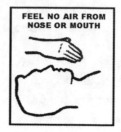

6. Examine

promptly and calmly for:

a. Absence of breathing.

7. What to do:

a. Open person's airway (A) (*See Chapter 4*).

b. Restore person's breathing (See Chapter 4).

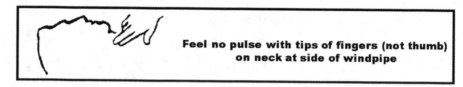

Feel no pulse with tips of fingers (not thumb) on neck at side of windpipe

8. Examine person promptly and calmly for:

a. Absence of heartbeat.

9. What to do:

a. Restore person's heartbeat (A) (*See Chapter 5*).

10. Lack of oxygen intake through breathing and heartbeat will bring death in minutes.

11. Examine promptly and calmly for:

a. Presence of bleeding.

See blood. Spurting blood means bleeding from artery (not vein or capillary).

12. What to do:

a. Stop the bleeding (B) (*See Chapter 6*).

13. The person will soon die unless:

a. There's enough blood to carry oxygen to tissues.

14. Prevent shock (C) (*See Chapter 8*):

a. If shock isn't prevented or corrected, person may die.

b. Death may result even though injury itself wouldn't otherwise be fatal.

15. Dress and bandage wounds to avoid infection (D) (*See Chapter 7*):

a. How well wound heals and recovery of person depends to a great extent upon:

1. How well wound was initially protected from contamination.

Emergency Lifesaving Measures

1. Apply lifesaving measures (A) and (B):
    a. If no sign of breathing, (A) open airway (*See Chapter 4*).
    b. If still no sign of breathing, (A) start artificial respiration (*See Chapter 4*).
        1. Continue until person regains consciousness.
        2. Or continue for at least 45 minutes in the absence of life signs.
    c. If no pulse, weak pulse or erratic pulse, start closed heart massage (*See Chapter 5*).
        1. Continue until person regains consciousness.
        2. Or continue for at least 45 minutes in absence of life signs.
    d. If there's bleeding, apply pressure (*See Chapter 6*).
2. Reexamine person immediately:
    a. From head to toe.
    b. From front to back.
3. Look for:
    a. Fractures, injuries without wounds, etc.
    b. Signs of shock:
        1. Restlessness.
        2. Pale skin.
        3. Rapid heartbeat.
        4. Thirst.
        5. Sweating although skin clammy.
4. Signs of worsening shock:
    a. Gasps or fast breathing.
    b. Staring blankly into space.
    c. Blotchy or bluish skin, especially around mouth.
5. Apply lifesaving measures (C) and (D) promptly.
    a. Apply shock control and prevention measures (C) (*See Chapter 8*).
    b. Dress and bandage wounds to avoid infection (D) (*See Chapter 7*).

DO'S AND DON'T'S TO REMEMBER

Be aware of these basic do and don't rules. They can help quickly make decisions under pressure.

THE DO'S

1. Stay calm. A person who loses composure in a stressful medical situation is of no value to injured person.
2. Stop bleeding! A lack of oxygen intake leads to death in minutes.
3. Reassure person while examining to determine necessity of lifesaving measures. Administer lifesaving measures immediately.
4. Open person's airway. Get breathing started and heart beating!
5. Remember these key letters: A, B, C. They represent **Airway, Breathing & Circulation**. Remember A,B,C, and your mind will never go blank when confronted with a high stress medical situation. Here are examples of what to do:
    a. Keep person's mouth open by placing stick between teeth.
    b. Clear obstructions (such as broken teeth or vomit) out of mouth.
    c. Turn head to side so vomit, etc., will drain rather than obstruct air passage.

6. Re-examine person immediately.

    a. Be methodical and organized.

    b. Try to determine injury type and severity.

    c. Promptly administer necessary lifesaving measures.

7. Do whatever necessary to prevent shock.

    a. First treat injury.

    b. Then cover person to prevent loss of body heat.

    c. Lower extremities should be elevated if:

        1. There's no head injuries.

        2. There's no abdominal injuries.

        3. The injury won't be aggravated.

8. Clean, dress and bandage cuts, abrasions or wounds. Person's recovery depends to great extent on how well injury was initially protected from contamination.

9. Take CPR course. Anybody who participates in any kind of outdoor activities should take such a course.

    a. CPR programs are available from Heart Association and Red Cross.

THE DON'TS

1. Don't move person with fracture until bone has been properly splinted unless absolutely necessary.

    a. Example: you would move injured person from burning building, etc.

2. Don't lay person on back if he has neck or face wound or is unconscious.

3. Don't give water or other fluids by mouth:

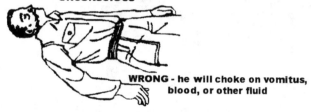

UNCONSCIOUS

WRONG - he will choke on vomitus, blood, or other fluid

    a. To person who is nauseated or vomiting.

    b. To person with abdominal or neck wound.

    c. To unconscious person.

4. Don't permit head of person with head injury be lower than his body.

5. Don't remove person's clothing by tearing or pulling it off.

6. Don't touch or try to clean burn area.

    a. Don't put medication on burn.

7. Don't remove bandages and dressings except to change.

8. Don't give unnecessary medical aid or aid beyond your capabilities.

9. Don't, if medical assistance is close at hand, try to push:

    a. Brain tissue back into head.

    b. Protruding intestines back into body.

10. Don't fail to restock medical bag as items are used. Keep everything in order and ready for emergency.

PART II

LIFESAVING TECHNIQUES

4
Opening the Airway and Restoring Breathing

Open the Airway

AIRWAY BLOCKED BY TONGUE

AIRWAY OPENED BY EXTENDING NECK

   c. It can be accomplished by either:
      1. Thumb jaw-lift method.
   d. Two-hand jaw-lift method.

5. Thumb Jaw-Lift

AIRWAY OPENED
FARTHER BY
ADJUSTING JAW

1. The upper airway must be unobstructed in order for air to flow to and from lungs.

2. Head Tilt Method:
   a. Immediately lay person on back
   b. See that neck is extended.
   c. See that head is in chin-up position.

3. If a rolled blanket, a poncho, or some similar item is available:
   a. Place under shoulders of person to help maintain chin-up position.

4. Jaw Lift Method
   a. If head tilt method unsuccessful, adjust jaw to jutting out position.
   b. This moves tongue further from back of throat.
      1. It enlarges airway passage to lungs.

a. The preferred method for adjusting jaw unless person's injury prevents its use.
b. Put thumb in person's mouth.
c. Firmly grasp lower jaw and lift forward.
d. Don't try to hold or depress person's tongue.

6. Two-Hand Jaw Lift

    a. Use two-hand method if person's

        1. Jaw is tightly clamped closed.

        2. If thumb can't be inserted in mouth.

    b. Using two hands, grasp lower jaw at point below ear lobes.

    c. Forcibly lift jaw forward.

    d. Force mouth open by pushing lower lip toward chin with thumbs.

Artificial Respiration

1. There are two primary methods of administering artificial respiration:

    a. Mouth-to-mouth (preferred way).

    b. Chest pressure arm lift.

2. Mouth-to-mouth can't be used if person:

    a. Has crushed face.

    b. Is in toxic or chemically contaminated environment.

3. The chest pressure arm lift method must be used on person with crushed face.

4. The back pressure arm lift method is best to use in chemical contaminated atmosphere.

Mouth-To-Mouth Resuscitation

1. In this method of artificial respiration:

    a. Inflate person's lungs with air from your lungs.

    b. Blow air into person's mouth.

2. With person lying on back, position yourself at side of head.

    a. Listen for sound of exhaling.

3. Place one hand behind person's neck to keep head in face-up position.

    a. Tilt head back as far as possible with other hand.

    b. Pinch nostrils with thumb and index finger.

    c. Exert pressure on forehead to maintain backward tilt of person's head.

4. Take deep breath and place mouth around person's mouth.

    a. Blow forcefully enough into person's mouth to make chest rise.

    b. If chest rises, sufficient air is getting into person's lungs.

    c. Remove mouth from mouth of injured person.

    d. Listen for sound of exhaling.

    e. Each time injured person exhales:

        1. Pinch his nose again.

        2. Blow four full, quick breaths into mouth.

        3. After this, repeat mouth-to-mouth resuscitation once every five seconds.

    f. As person starts breathing, adjust timing of your efforts to assist him.

        1. A smooth rhythm is desirable.

2. Split-second timing in no longer essential.

Mouth-to-Nose Method

1. This method should be used if mouth-to-mouth breathing can't be undertaken because:

    a. Person has fractured jaw.

    b. Person has severe mouth wound.

    c. Person's jaws are tightly closed by spasms.

2. Mouth-to-nose method is performed in same way as mouth-to-mouth.

    a. Blow into person's nose while pinching his lips closed with one hand.

Chest Pressure Arm-Lift Method

1. This is the treatment to use when person has crushed face.

2. Preliminary steps:

    a. Clear person's airway.

    b. Lay person on back.

    c. Position person's head with face up.

    d. Place rolled blanket under shoulders.

    e. Person's head will drop back in chin-up position.

    f. Stand at person's head, facing feet.

    g. Kneel on one knee.

    h. Place other foot on other side of his head and against his shoulder to steady it.

Chest-Pressure Arm-Lift Procedure

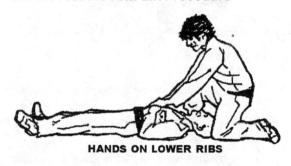

HANDS ON LOWER RIBS

1. Grab person's hands.

    a. Hold them over his lower ribs.

    b. Rock forward.

STEADY PRESSURE DOWNWARD

2. Exert steady uniform pressure almost directly downward.

    a. This pressure forces air out of lungs.

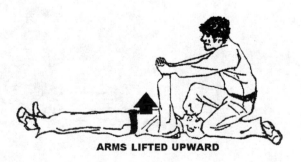

ARMS LIFTED UPWARD

3. Lift person's arms vertically upward.

ARMS BACKWARD AS FAR AS POSSIBLE

4. Stretch arms as far back as possible.

    a. This increases chest size and draws air into lungs.

5. Replace person's hands on chest and repeat entire cycle.

    a. Press.

    b. Lift.

    c. Stretch.

    d. Replace.

6. Keep steady rate of 10 to 12 cycles per minute.

7. As person starts to breathe, time your efforts to assist him.

Back Pressure Arm-Lift Method

1. Place person on stomach with:

    a. Face to one side.

    b. Neck extended.

    c. Hands under head.

2. Kneel at person's head.

    a. Place hands on back.

    b. Heels of hands should be on line just below the armpits.

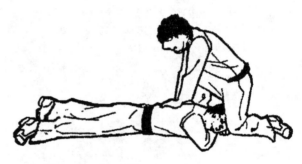

3. Thumbs should be touching.

    a. Fingers should be extended downward and outward.

4. Rock forward while keeping your arms straight.

    a. Exert pressure almost directly downward on person's back.

5. This will force air out of lungs.

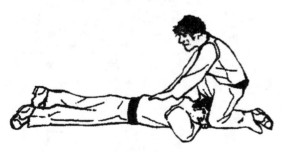

6. Rock back, releasing pressure on person's back.

    a. Grab arms just above elbows.

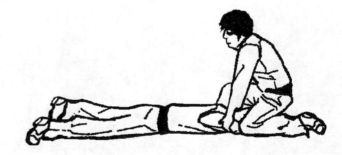

120

7. Continue to rock backward.

    a. Pull person's arms upward and inward towards his head.

    b. Continue until resistance and tension in person's shoulders are noted.

        1. This expands chest, causing air to enter lungs.

    c. Lastly, rock forward and release injured person's arms.

        1. This causes passive expiration (inhaling).

    d. Repeat cycle of PRESS. RELEASE, LIFT and RELEASE 10 to 12 times each minute.

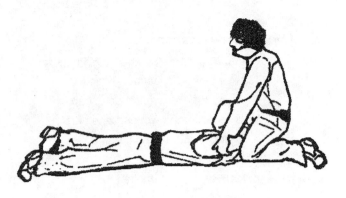

5

Restoring the Heartbeat

If Heart Stops Beating

1. If person's heart stops beating, immediately:
    a. Give closed chest massage.
    b. Give artificial respiration (Chapter 4).
2. When heart stops working, the person:
    a. Is unconscious and limp.
    b. Has no pulse.
    c. Has wide open pupils.

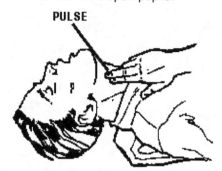

PULSE

Proper Placement of Fingers to Detect Pulse

1. To determine the presence or absence of pulse:
    a. Place finger tips on person's neck.
    b. Put them at the side of windpipe.
        1. If pulse isn't immediately detected, or if pulse is weak and irregular:
    c. Start closed-chest heart massage.
    d. Begin administering artificial respiration.

Closed-Chest Heart Massage

1. A closed-chest heart massage is simply the rhythmical compression of the heart without opening the chest surgically.
    a. It's done to keep blood flowing to brain and other vital organs.
    b. It's continued until heart starts to beat normally again.
2. The heart is located between the breastbone and spine.
    a. Pressure on breastbone pushes heart against spine.
    b. Blood is forced out of heart and into arteries.
    c. The release of pressure allows heart to refill with blood.
3. Closed-chest heart massage should always be combined with artificial respiration (Chapter 4).
    a. It's best to have two people working in unison on injured person.
    b. One person positions himself at person's side and performs heart massage.
    c. The other gets on the opposite side:
        1. Keeping person's head tilted back.
        2. Administering artificial respiration.
4. Injured person must always be in horizontal position when closed-chest heart massage is performed.

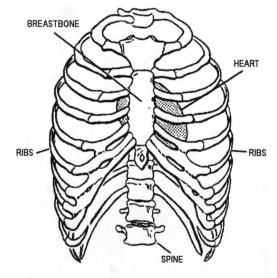

BREASTBONE

HEART

RIBS

RIBS

SPINE

5. The surface on which person is placed must be firm.

    a. A bed or couch is too flexible.

    b. The ground or floor is sufficient.

6. Elevate person's legs about six inches while keeping rest of his body horizontal.

    a. Place yourself close to person's side.

7. Place heel of hand on lower half of person's breastbone.

    a. Be careful not to place hand on soft part of abdomen below breastbone and rib cage.

    b. Place other hand on top of first hand.

8. If you're working on a child

    a. Omit placing second hand over first.

9. If you're working on an infant

    a. Place only fingertips of one hand over breastbone.

Basic Procedures to Follow

Blood forced out of the heart into the arteries by the application of pressure on the breastbone.

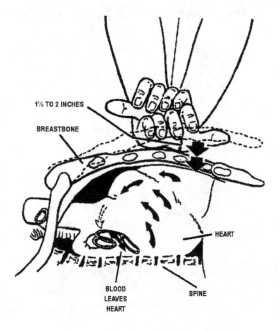

1. With hands in the correct position:

    a. Bring shoulders directly over the person's breastbone.

    b. Keep arms straight.

    c. Press downward.

2. Apply enough pressure to push breastbone down about two inches — but never more than two.

    a. Be careful as too much pressure might fracture breastbone.

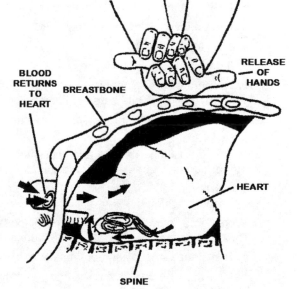

3. Release of pressure to allow the heart to refill with blood.

4. Immediately release pressure.

    a. Pressure on breastbone should be completely released.

    b. It should return to its normal resting position between compressions.

    c. The heel of the hand should not be taken off person's chest during relaxation.

Two People Working in Unison

1. The person administering closed-chest heart massage should compress heart:

    a. 60 times per minute.

    b. Or one compression per second.

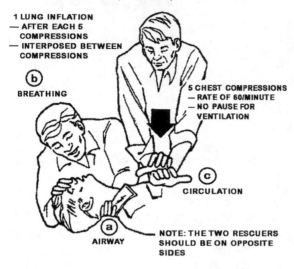

1 LUNG INFLATION
— AFTER EACH 5 COMPRESSIONS
— INTERPOSED BETWEEN COMPRESSIONS

(b) BREATHING

5 CHEST COMPRESSIONS
— RATE OF 60/MINUTE
— NO PAUSE FOR VENTILATION

(c) CIRCULATION

(a) AIRWAY

NOTE: THE TWO RESCUERS SHOULD BE ON OPPOSITE SIDES

Two rescuers administering closed-chest heart massage and mouth-to-mouth artificial respiration

2. Compressions must be:

    a. Regular.

    b. Smooth.

    c. Uninterrupted.

3. The person administering a heart massage doesn't pause for breaths to be blown into airway.

4. Proper timing for 60 compressions per minute can be achieved by:

    a. Counting one 1000, one 2000, one 3000, etc.

    b. The administering person says "one" each time he compresses heart.

    c. He says the thousand number each time he releases pressure.

    d. The same count to 5000 is repeated entire time closed-chest heart massage is administered.

5. The person administering artificial respiration quickly blows air into person's lungs:

    a. After each five compressions by person administering the closed-chest heart massage.

6. A deep breath is blown into person's airway each time other person says "5000."

    a. These breaths are to be interjected without pauses in compression by other party.

7. Avoiding pauses is important as interruption in heart massage results in:

    a. Drop in blood flow.

    b. Blood pressure drop to zero.

8. Two people can administer closed-chest heart massage and artificial respiration in unison when positioned on opposite sides of person.

    a. When either person gets tired, they can switch positions without significant interruption in the 5:1 rhythm.

9. Switching positions can be accomplished by the one performing artificial respiration moving to side of person's chest immediately after inflating the lungs.

    a. He holds his hands in the air next to those of person who continues doing heart compressions.

10. When other person's hands are in place, the one doing the heart compression removes his. This is done after count of 2000 or 3000 in compression series.

    a. The other person then continues with a rhythmic uninterrupted series of compressions.

11. Lastly, the person previously doing compressions then moves to person's head and starts breathing into his mouth at count of 5000.

One Person Working Alone

1. Sometimes there will be only one person to perform both artificial respiration and closed-chest heart massage. Treatment then consists of:

   a. 15 heart compressions followed by two full lung inflations.

   b. This is known as the 15:2 ratio.

2. To make up for time used to inflate lungs, each series of 15 heart compressions must be performed at a faster rate of 80 compressions each minute.

3. The timing for 80 compressions per minute can be achieved by counting aloud as follows:

   a. 1 and 2 and 3 and 4 and 5

   b. 1 and 2 and 3 and 4 and 10

   c. 1 and 2 and 3 and 4 and 15

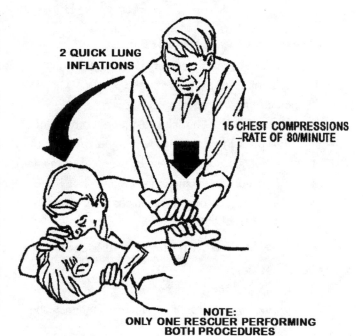

2 QUICK LUNG INFLATIONS

15 CHEST COMPRESSIONS —RATE OF 80/MINUTE

NOTE:
ONLY ONE RESCUER PERFORMING BOTH PROCEDURES

4. After the count of 15, two deep breaths are blown into person's airway without allowing full exhale between breaths.

   a. This is done in rapid succession within a period of 5 to 6 seconds.

   b. Repeat the same count while continuing resuscitation.

When to Stop Trying

1. Closed-chest heart massage and artificial respiration should never be stopped based on discomfort of person or persons applying these procedures.

2. They should be continued:

   a. Until person regains consciousness.

3. Until another person takes over.

4. Until a period of at least 45 minutes after all signs of life have ceased.

6

Stopping Bleeding

Importance of Preventing Loss of Blood

1. An open wound is the most common condition requiring medical assistance.
2. A large loss of blood may lead to shock.
    a. Shock can result in person dying.
3. Immediate action must be taken to prevent great loss of blood.
4. Apply more pressure dressings with digital pressure (fingers, thumbs) to control severe bleeding.
5. Elevate injured limb. Apply pressure with fingers in addition to pressure dressings.
6. Use tourniquet only after pressure dressing has failed to stop bleeding from limb.
7. Never use tourniquet otherwise unless it's only way to stop heavy bleeding from arteries when arm or leg has been crushed or cut off.

Sterile Dressing Applied with Pressure to Bleeding Wound:

1. Helps to clot blood.
2. Acts to compress open blood vessels.
3. Acts to protect wound from further germ contamination.

Searching for Entrance and Exit Wounds:

1. Examine person to determine if there's more than one wound before applying a pressure dressing.
2. A bullet, for example, may have entered one place and come out another.
3. A bullet normally makes smaller wound

when entering than when exiting.

a. Cut clothing and carefully pull away from wound.

b. Tearing clothing off person could result in rough handling of injured part.

4. Don't touch wound. Keep it as clean as possible, but don't try to clean it.

a. This prevents further contamination.

b. Any attempt to clean wound will only lead to more contamination.

c. If wound is found dirty, leave it dirty.

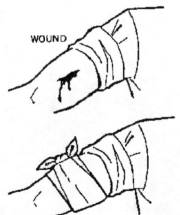

WOUND

5. Cover wound and apply pressure.

a. Use dressing to cover wound. If no dressing available, use any handy cloth.

b. Then apply pressure to wound by using bandage strips attached to dressing.

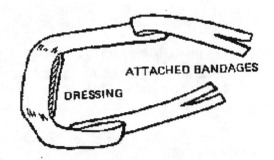

DRESSING

ATTACHED BANDAGES

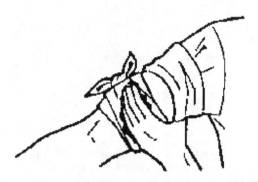

c. If more pressure is required to stop bleeding, put hand over dressing and press hard.

d. Hand pressure may be necessary for five to ten minutes to allow clot to form.

e. When hand is taken away, clot has to be strong enough to hold with only dressing and bandage strips.

f. More pressure can be applied to wound with thick pad or rag on top of dressing.

g. Pad or rag can be firmly secured with strip of material.

6. Don't take bandages off once they've been put on wound.

a. If more bandages are required, simply put new bandages over those already in place.

b. Remember -- partially formed blood clots may be dislodged by removing dressing.

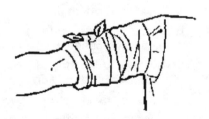

Elevating an Injured Arm or Leg:

1. Raising injured arm or leg above heart slows bleeding.

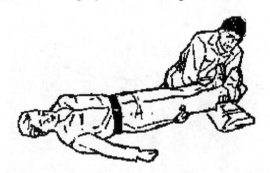

2. Of course, if bone is broken, don't raise it until splint has been properly applied (*see Chapter 12*).

3. Moving unsplinted fracture can cause:

4. Much unnecessary pain.

5. Further damaged muscles, blood vessels and nerves.

6. Increased shock.

Applying Digital Pressure (with Fingers, Thumbs and Hands)

1. Digital pressure should be used to control bleeding if blood spurting from wound (artery).

2. You'll know you've located artery when pulse is felt.

3. Digital pressure shuts off or slows blood flow from heart to wound.

4. Press at point where main artery supplying bleeding area lies near skin surface or over bone (see above illustration).

5. Continue until pressure dressing can be applied.

Tourniquets and Their Application:

1. A tourniquet is a constricting band placed around arm or leg to stop bleeding.

2. Use tourniquet **only** when bleeding can't otherwise be controlled by:

   a. Pressure over wound area.

   b. Pressure over appropriate pressure point (see above illustration).

   c. Elevation of injured part.

3. Use of tourniquet has sometimes caused blood vessel and nerve injury.

4. If left on for too long a period, tourniquet can cause loss of arm or leg.

   a. **Use tourniquets only as last resort!**

5. Bleeding from major artery of thigh, lower leg or upper arm, or bleeding from multiple arteries is often beyond control by usual pressure method.

   a. If dressing becomes blood soaked while under hand pressure and bleeding continues, a tourniquet should unquestionably be applied.

   b. Don't remove tourniquet once it's been put on.

   c. Don't loosen tourniquet after it has stopped the bleeding.

      1. Shock and blood loss may result in death of person.

      2. Get person to the closest medical treatment facility if one is available.

Improvised Tourniquets

1. Tourniquets can be made from:

   a. Gauze or muslin bandages.

   b. Cutup sheets or clothing.

   c. Kerchiefs, etc.

2. To avoid skin damage, an improvised

If blood is spurting from wound (artery), press at the point or site where main artery supplying the wounded area lies near skin surface or over bone as shown. This pressure shuts off or slows down the flow of blood from the heart to the wound until a pressure dressing can be unwrapped and applied. You will mnow you have located the artery when you feel a pulse.

DIGITAL PRESSURE
(PRESSURE WITH FINGERS, THUMBS, OR HANDS)

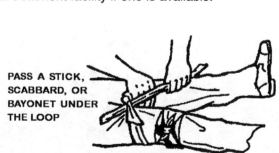

PASS A STICK, SCABBARD, OR BAYONET UNDER THE LOOP

tourniquet should be a minimum of one inch wide after it's tightened.

3. Wrap tourniquet around upper limb and two to four inches above injury.

MAKE A LOOP AROUND THE LIMB WITH ANY FLEXIBLE MATERIAL; TIE WITH SQUARE KNOT

SQUARE KNOT

a. Never put tourniquet directly on wound or fracture.

4. Put tourniquet over pant leg or smoothed shirt sleeve to prevent skin from being pinched or twisted.

    a. Protecting skin reduces pain.

TIGHTEN TOURNIQUET JUST ENOUGH TO STOP ARTERIAL BLEEDING

5. Tighten tourniquet only enough to stop blood from passing under it.

6. Feel for pulse at wrist or foot of injured limb before applying tourniquet.

    a. When pulse is stopped then tourniquet pressure is sufficient.

BIND FREE END OF STICK TO LIMB TO KEEP TOURNIQUET FROM UNWINDING

7. To detect pulse, use two fingers (no thumb) over pressure point in wrist or ankle (see illustration).

8. Don't use thumb because small arteries in thumb may cause false pulse reading.

9. After tourniquet is properly tightened, arterial bleeding (spurting) will stop.

    a. Bleeding from veins in lower part of limb will continue until drained of blood already in them.

    b. Don't continue to tighten tourniquet in effort to stop drainage.

10. Lastly, after tourniquet is securely in place, dress and bandage wound (*see Chapter 7*).

7

Dressing and Bandaging Wounds

General Data on Wound Contamination.

1. Every wound (cut, puncture, etc.) is considered contaminated since infection producing germs are always present:

    a. On skin.

    b. On clothing.

    c. In air.

2. Any instrument (knife, etc.) or missile (bullet, etc.) causing wound carries or pushes germs into wound.

3. Infection is result of germ multiplication and growth after they invade:

    a. A wound.

    b. A break in person's skin.

4. The fact that wound or other injury is contaminated doesn't lessen importance of protecting it from further contamination.

5. When less germs are allowed to get into wound:

    a. There's less possibility of infection.

    b. The chance for recovery is greater.

6. Wounds, therefore, must be dressed and bandaged as soon as possible:

    a. To stop further contamination.

    b. To stop bleeding.

7. Applying Dressing

1. Cut injured person's clothing.

    a. Lift clothing away from wound to avoid further contamination.

2. Remove wrapped dressing from plastic envelope.

    a. Twist to break paper wrapper.

3. Grasp folded bandages with hands.

    a. Don't touch side of dressing to go on wound.

4. Carefully place dressing on wound.

    a. Don't let it touch anything else.

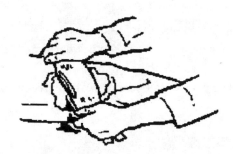

5. Wrap bandages over dressing and around injury.

    a. a. Tie ends securely with square knot.

Basic Kinds of Bandages — General Information.

1. A bandage is useless unless firmly put on wound with ends securely tied or taped in place.

    a. This prevents bandage and dressing from slipping off wound.

    b. Never put bandage on so tight it stops circulation.

    c. If bandage is tied rather than taped, use square knot as it won't slip

2. Tailed Bandages

    a. Tailed bandages can be attached to dressing as in "B — Applying a Dressing."

    b. The two tails are split four to six inches from loose ends.

    c. The tails can be split further as required to bandage particular part of body.

    d. Tailed bandages may be made by splitting strip of gauze (4" x 36") from each end.

        1. Leave center part intact to cover dressing placed over injury.

3. Cravat or Triangular Bandages

    a. Cravat or triangular bandages are made from triangular piece of muslin (37" x 37" x 52).

        1. If this bandage is used without folding, it's known as a triangular bandage.

        2. If folded into a strip, it's known as a cravat bandage.

    b. Cravat and triangular bandages are great for use in emergency:

        1. They are so easily applied.

        2. They can be readily improvised by:

            a.   Cutting up sheet.

            b.   Using piece of shirt.

            c.   Using kerchief.

            d.   Using any other cloth material of suitable size.

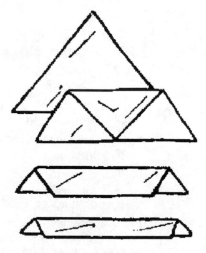

4. To improvise triangular bandage:

    a. Cut square of material somewhat larger than three feet.

    b. Fold diagonally into a triangle.

    c. If two bandages needed, cut material along first fold.

Head Bandages

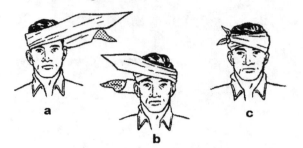

APPLY CRAVAT BANDAGE TO HEAD

1. Bandage person's head as in Chapter 11.

2. Cravat bandage applied to head.

    a. Put middle bandage over dressing (fig. a).

    b. Pass bandage ends in opposite directions completely around head (fig. b).

    c. Tie bandage ends with square knot directly over dressing (fig. c).

3. Triangular bandage applied to head.

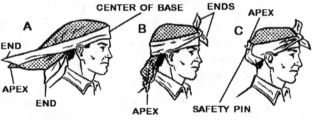

    a. Turn base of bandage up.

        1. Place center of base in middle of forehead (fig. a).

        2. Let apex fall on back of neck, then take ends backward (fig. a).

    b. Cross ends of bandage over apex (fig. b).

        1. Take ends over forehead (fig. b).

        2. Tie with square knot (fig. b).

    c. Tuck apex behind crossed part of bandage (fig. c).

        1. Further secure with safety pin if available (fig. c).

4. Ear Bandages

    a. The cravat bandage should be used to bandage ear as follows:

        1. Put middle bandage over ear (fig. a).

        2. Cross bandage ends.

        3. Take bandage in opposite directions around head.

        4. Tie ends with square knot (fig. b/c).

        5. Place dressing or small padding between ear and side of head if possible.

        6. Such padding will stop ear from being crushed against head when bandage is applied.

APPLY CRAVAT BANDAGE TO EAR

5. Eye Bandages

a. Even though only one eye might be injured, both must be bandaged.

b. Both eyes move in unison. Any movement of uninjured eye causes same movement in injured eye.

   1. Further damage to uninjured eye is the end result.

c. Bandages are applied to eye injuries as follows:

   1. Take upper bandages on each side (fig. a) to back of head and cross them.

   2. Bring bandages around head and tie with square knot (fig. b).

   3. Cross lower bandages on top of head.

   4. Bring longest bandage under chin.

   5. Secure bandages to other bandages and tie in square knot (fig. c).

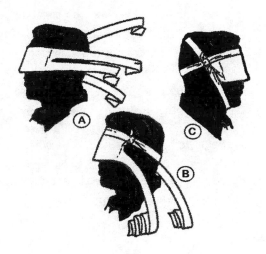

6. Bandaging a Jaw

a. Take removable dentures from person's mouth before bandaging jaw.

   1. Place in person's pocket.

b. Avoid bandaging person's mouth closed.

c. Place an approximately 1/8" thick wad of material between gums.

d. Leave streamers of material attached to the wad and tied to bandage.

   1. This insures that wad doesn't fall into mouth and block airways.

e. Place bandage under chin and carry ends upward.

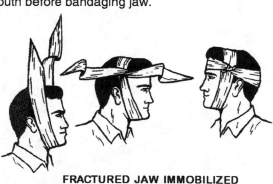

FRACTURED JAW IMMOBILIZED

f. Adjust bandage to make one end longer than other (fig. a).

g. Take longer end over top of head to meet short end at temple. Cross ends (fig. b).

h. Take ends in opposite directions to other side of head.

i. Tie ends with square knot over part of bandage applied first (fig. c).

j. Tie mouth wad streamers to bandage (fig. c).

Bandaging a Shoulder

1. Put one end of bandage across chest.
2. Take other end across back and under arm.
   a. This should be opposite the injured shoulder.
   b. Tie ends of bandage with square knot.
3. Cravat bandage applied to shoulder or armpit injury.
   a. Extend length of cravat bandage by placing one end of triangular bandage along middle of another one.
   b. Fold two bandages into single cravat.
   c. Secure thickest part with one or more safety pins.
4. Place middle of cravat under armpit so front end is longer than back end (fig. a).

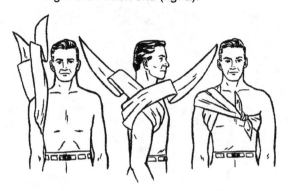

a. Cross ends on top of shoulder (fig. b).
5. Take one end across back and under arm on opposite side.
   a. Take other end across chest.
   b. Tie ends with square knot (fig. c).
6. Place sufficient wadding in arm pit and don't tie bandage too tightly.
   a. This will avoid compressing major blood vessels in the armpit.

Bandaging an Elbow

1. A cravat bandage is put on elbow as follows:
   a. Bend arm and place middle of the cravat at point of elbow (fig. a).
   b. Bring ends up and across and extend ends downward (fig. b).
   c. Take each end around arm and tie with square knot at front of elbow (fig. c).

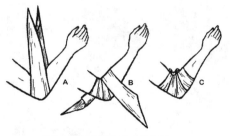

APPLY CRAVAT BANDAGE TO ELBOW

Bandaging a Hand

1. Using triangular bandage.
2. Place hand in middle of triangular bandage with wrist at base of bandage (fig. a).
   a. Check fingers to be sure they are separated with absorbent material.
   b. This prevents skin irritation and chafing.
3. Place apex of bandage over fingers (fig. b).
4. Tuck excess fullness of bandage into pleats on each side of hand (fig. b).
5. Cross ends of bandage on top of hand (Fig. c).
6. Take ends of bandage around wrist (fig. d).
7. Tie ends of bandage in square knot (fig. c).

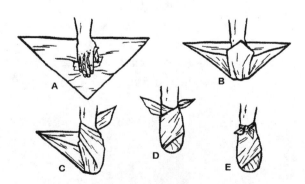

APPLYING TRIANGULAR BANDAGE TO HAND

Cravat Bandage on Palm of Hand

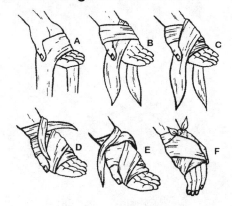

Applying cravat bandage to palm of hand.

1. Lay middle of cravat bandage over palm (fig. a).

   a. Leave ends hanging down each side (fig. a).

2. Take thumb end of bandage across back of hand (fig. b).

   a. Then go over palm (fig. b).

3. b. Lastly, put bandage through hollow between thumb and palm (fig. b).

4. Take other end of bandage across back of hand (fig. c).

5. Extend it upward over the base of thumb (fig. c).

6. Then bring bandage downward across palm (fig. c).

7. Take ends of bandage to back of hand and cross (fig. d).

8. Bring them to front of hand and cross again (fig. e).

9. Tie ends of bandage with square knot at wrist (fig. f).

APPLYING CRAVAT BANDAGE TO KNEE

Bandaging a Knee

1. Apply cravat bandage to knee as shown.

2. Use identical technique as previously described for elbows.

Bandaging a Leg

1. A cravat bandage is applied to leg as follows:

   a. Place center of cravat bandage over dressing (fig. a).

   b. Take one end around and up leg in spiral motion (fig. a).

   c. Take other end around and down leg in spiral motion (fig. a).

   d. Overlap part of each preceding turn of bandage (fig. b).

   e. Bring ends together and tie with square knot (fig. c).

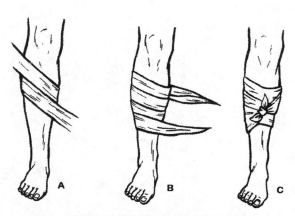

APPLYING CRAVAT BANDAGE TO LEG

Bandaging a Foot

1. A triangular bandage is applied to the foot as follows:

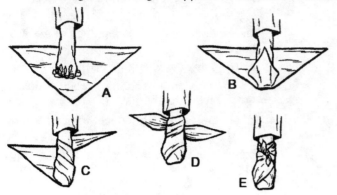

APPLYING TRIANGULAR BANDAGE TO FOOT

a. Place foot in middle of triangular bandage (fig. a).

   1. Heel should be well forward of base.

b. Insure that toes are separated with absorbent material.

   1. This prevents skin irritations.

c. Place apex over top of foot (fig. a).

d. Tuck excess fullness into pleats on each side of foot (fig. c).

e. Cross ends on top of foot and take around ankle (fig. d).

f. Tie bandage ends in square knot at front of ankle (fig. e).

8
Preventing a Person From Going Into Shock

General Shock Prevention Information.

1. Shock may be end result of any injury.

    a. It's more likely to develop in more severe injuries.

2. Some warnings of possible shock include:

    a. Rapid heartbeat.

    b. Pale skin.

    c. Thirstiness.

    d. Restlessness.

3. A person in shock may:

    a. Sweat when skin feels cool and clammy.

    b. Be highly excitable.

    c. Appear to be calm and extremely tired.

4. Signs as shock gets progressively worse:

    a. Person vacantly stares into space.

    b. Skin gets blotchy and turns blue — especially in mouth and lip area.

    c. Person gasps for air and takes small fast breaths.

5. The objective is to administer measures to prevent shock from developing or from getting worse:

    a. Elevate person's feet.

    b. Loosen person's clothing.

    c. Place covers under and over person to prevent chilling.

Maintaining Adequate Respiration and Heartbeat.

1. This may entail clearing person's upper airway (*see Chapter 4*).

2. Position person to insure the drainage of fluids blocking airway.

3. Keep person under close observation to make certain airway remains clear and unobstructed.

4. You may have to give injured person artificial respiration (*see Chapter 4*).

5. May have to give closed-chest heart massage (*see Chapter 5*).

Stopping Bleeding.

1. Bleeding can be controlled by (*see Chapter 6*):

    a. Application of pressure dressing.

    b. Elevating the injured arm or leg.

    c. By the use of pressure points as appropriate (*see Chapter 6*).

    d. Applying tourniquet if needed.

Loosen All Restrictive Clothing.

1. Clothing must be loosened at neck, waist and other areas where it binds the injured person.

2. Loosen boots or shoes, but don't take off.

Try to Reassure Injured Person.

1. Take charge of situation immediately.
2. Show by calm demeanor, self confidence and gentle yet firm actions that you know what you're doing.
3. Let person know you expect him to feel better because you're helping him.
4. If asked about seriousness of injury:
5. Remember — Incorrect or badly timed information can increase person's anxiety.

Splint Any Fractures.

1. If person has a fracture, apply a splint (*see Chapter 12*).

Properly Position Person.

1. The position in which to place injured person always depends on:

    a. Type of wound or other injury.

    b. Whether person is conscious or unconscious.

2. Gently place person on blanket unless he has injury for which special position is prescribed (*see Chapters 11 and 12*).

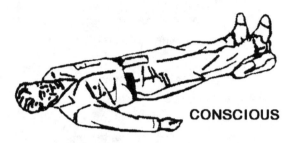

CONSCIOUS

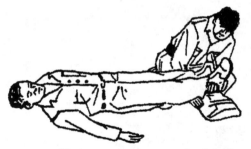

3. If person is conscious:

    a. Place on back on level surface.

    b. Elevate legs six to eight inches to increase blood flow to heart.

    c. This can be accomplished by placing clothing, towels, or other items under his feet.

    d. If person is placed on stretcher, also elevate feet.

    e. Never move person who has fracture until it has been splinted.

4. If person is unconscious:

    a. Place him on side or on abdomen.

    b. Turn head to one side to prevent choking on vomit, blood or other fluids.

More on Positioning.

1. The person with head injury should always lay with head higher than body.
2. The person with face or neck wound should sit and lean forward with head down.

    a. Or get in unconscious position.

3. The person with sucking wound in chest should sit up or lie on injured side.
4. The person with abdominal wound should lie on back with head turned to side.

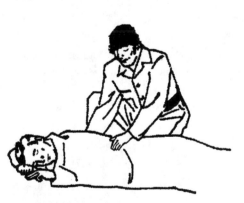

Keep Person Comfortably Warm.

1. Place blanket, poncho liner, etc., under as well as over person.
2. Be careful not to overheat person.
3. Remove all wet clothing.

PART III

SPECIAL MEDICAL MEASURES
9

Moving Someone Who is Seriously Hurt

General Information

1. Moving or transporting injured person is usually the responsibility of medical personnel with special training and equipment.

    a. When emergency arises and no professional assistance is available, you may have to move an injured person.

    b. For this reason, everyone should know how to move injured person without making injury worse.

    c. Transporting person on stretcher is:

        1. Safer than carrying.

        2. More comfortable than carrying both for you and injured person.

2. Easier for you.

    a. Carrying person may sometimes be only feasible method because:

        1. Of rugged terrain.

        2. It may save person's life.

    b. In these situations, the person should be transferred to stretcher:

        1. As soon as one can be improvised.

Handling an Injured Person

1. A person's life can be lost through careless or rough handling while transporting or moving.

2. Prior to moving person:

    a. Evaluate kind and extent of injury.

    b. Check dressings over wounds to see they are adequately reinforced.

3. Make sure fractured bones are properly immobilized.

4. Check fractures to see they are properly supported to prevent them from cutting through skin, blood vessels and muscle.

Carrying an Injured Person

1. The distance person can be transported depends upon a number of factors:

    a. The strength and endurance of those doing carrying.

    b. Weight of person.

    c. The nature of person's injury.

    d. Obstacles encountered while carrying person.

2. Some carrying methods are inappropriate due to nature of injury.

    a. Certain kinds of carries can't be used if person has fractured:

        1. Arm or leg.

        2. Neck or back.

        3. Hip or thigh.

3. Rolling from back onto abdomen.

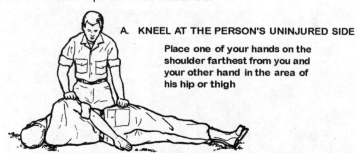

A. KNEEL AT THE PERSON'S UNINJURED SIDE

Place one of your hands on the shoulder farthest from you and your other hand in the area of his hip or thigh

4. Rolling from abdomen onto back.

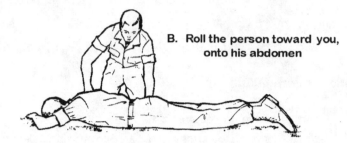

B. Roll the person toward you, onto his abdomen

Fireman's Carry Technique:

1. The fireman's carry is one of the easiest ways for one person to carry another.

2. An unconscious person should first be properly positioned.

    a. The unconscious or injured person is then to be raised from ground in four steps as shown here.

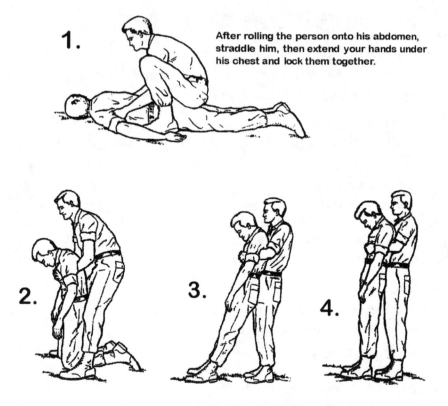

1. After rolling the person onto his abdomen, straddle him, then extend your hands under his chest and lock them together.

2.

3.

4.

3. Step Five of Fireman's Carry

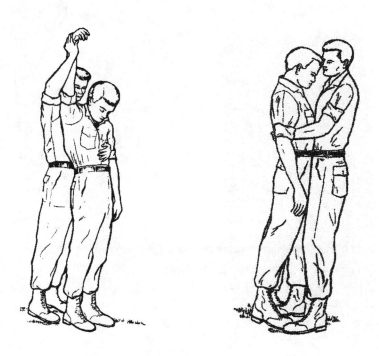

4. Step Six and Seven of Fireman's Carry.

5. Steps Eight and Nine of Fireman's Carry.

6. Alternate Method of Raising Unconscious or Injured Person From Ground.

    a. This is carefully done in three steps as illustrated.

    b. Use this method only when bearer believes it's safer for injured person because of the location of wounds.

    c. When using this method, take care to prevent person's head from snapping back and causing neck injury.

① Kneel on one knee at injured person's head, facing his feet, then extend your hands under his armpits, down his sides and across his back.

② As you rise, lift the person to his knees. Secure a lower hold and raise him to a standing position, with his knees locked.

③ Secure your arms around the person's waist, with his body tilted slightly backward to prevent his knees from buckling. Place your right toe between his feet and spread them 6 to 8 inches apart.

Supporting Carry

1. In this kind of carry, injured person must, while using another person as crutch, be able to:

   a. Walk.

   b. At least hop on one leg.

   c. This carry method is used to transport person as far as he's able to walk or hop.

Raise the person from the ground as in fireman's carry. Using the hand farthest from the injury, grasp the person's nearest wrist and draw his arm around your neck. Place your other arm around his waist.

Arms Carry

1. The arms carry is useful for:

   a. Carrying person a short distance.

   b. Carrying person to stretcher.

Saddleback Carry

1. Only conscious person can be transported by saddleback carry.

2. Person transported saddleback must be able to hold on to neck of person doing carrying.

Pack Strap Carry

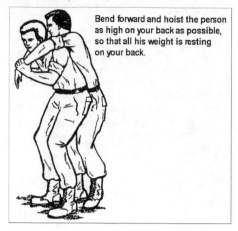

Bend forward and hoist the person as high on your back as possible, so that all his weight is resting on your back.

1. In pack strap carry,   person's weight rests high on carrier's back.

   a. This makes it easier to carry person a reasonable distance.

2. Person doing carrying must hold person's arms in palms down position.

3. This eliminates possibility of injury to person's arms.

Lift person from ground as in fireman's carry. Supporting the person with your arms around him, grasp his wrist closest to you and place his arm over your head and across your shoulder. Move in front of him, while supporting his weight against your back, grasp his other wrist and place this arm over your shoulder.

Belt Carry (Steps One and Two)

4. The pistol-belt carry is the best one-man carry to use over long distances.

   a. Person is secured by belt on shoulders of bearer.

   b. Hands of injured person and bearer are left free to:

   c. Carry gun or knife.

   d. Carry equipment.

   e. Help get up embankments.

   f. Help surmount other obstacles.

5. With hands free and person in place, bearer is then able to:

   a. Go through shrubs.

   b. Go under low hanging branches.

6. Belt Carry (Steps Three and Four).

Link together two belts to form a sling. (If two belts are not available, use other items such as a belt and a rifle sling, two cravat bandages, or any other suitable material that will not cut or bind the person.) Place this sling under the person's thighs and lower back so that a loop extends from each side.

Lie between the person's outstretched legs. Thrust your arms through the loops, grasp the person's hand and trouser leg on his injured side.

Step 3: Roll toward person's uninjured side onto your abdomen, bringing person onto your back.

Step 4:
Rise to kneeling position. The belt will hold person in place.

144

7. Belt Carry (Steps Five and Six).

Step 5: Place hand on knee for support and rise to upright position.

Step 6: Carry person with hands free for climbing banks, carrying rifle, etc.

Belt Drag

Extend two belts or similar objects to their full length and join them together to make a continuous loop.
Roll the injured person onto his back. Pass the loop over his head and position it across his chest and under his armpits. Cross the remaining portion of the loop to form a figure eight. Lie on your side with your back away from the injured person, resting on your right elbow. Slip the loop over your right arm and shoulder, then turn onto your abdomen, thus enabling you to drag the person as you crawl.

1. The belt drag is useful only for short distances.

2. The bearer and injured person can stay close to ground when using this method.

Neck Drag

Extend two belts or similar objects to their full length and join them together to make a continuous loop.
Roll the injured person onto his back. Pass the loop over his head and position it across his chest and under his armpits. Cross the remaining portion of the loop to form a figure eight. Lie on your side with your back away from the injured person, resting on your right elbow. Slip the loop over your right arm and shoulder, then turn onto your abdomen, thus enabling you to drag the person as you crawl.

1. If person is unconscious, his head should be protected from striking ground while being transported.

2. The bearer and injured person can go:

   a. Through culverts.

   b. Under vehicles.

   c. Behind walls or shrubbery.

Cradle Drop Drag

1. This method of transporting injured person is effective for moving them up or down steps.

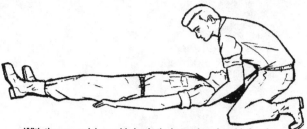

With the person lying on his back, the bearer kneels at his head, then slides his hands, palms up, under the person's shoulders and gets a firm hold under the armpits.

The bearer partially rises, supporting the person's head on one of his forearms. (The bearer may bring his elbows together and let the injured person's head rest on both of his forearms.)

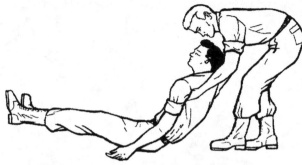

With the injured person in a semi-sitting position, the bearer rtises and drags the person backward.

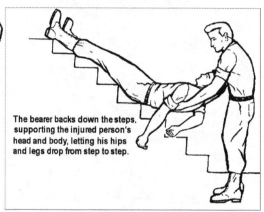

The bearer backs down the steps, supporting the injured person's head and body, letting his hips and legs drop from step to step.

TWO MAN CARRIES

Two bearers help the injured person to his feet and support him with their arms around his waist. They grasp the person's wrists and draw his arms around their necks.

Alternate method for when injured person is taller than bearers.

Two Man Carry

1. Used to transpsort person who is either conscious or unconscious.

2. If person is taller than bearer:

3. Lift legs and let them rest on forearms of bearers.

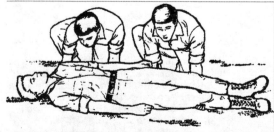

Two bearers kneel at one side of the injured person and place their arms beneath the person's back, waist, hips, and knees.

Two Man Arms Carry

1. Good for transporting person:

    a. For moderate distance.

    b. To place person on stretcher.

2. To avoid becoming tired, bearers should carry person:

    a. High.

    b. As close to chest as possible.

3. This is safest way to carry someone with back injury:

    a. In extreme emergency.

    b. When there's no time to get a board.

4. Two extra bearers should be used when transporting person with back injury:

    a. To keep head and legs in alignment with body.

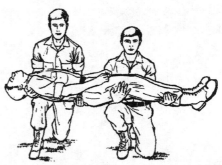

The bearers lift the injured person as they rise to their knees.

As the bearers rise to their feet, they turn the injured person toward their chests.

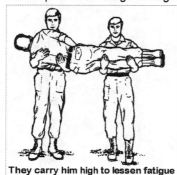

They carry him high to lessen fatigue

Alternate carry

1. Two Man Fore and Aft Carry.

   a. This is excellent carry method for:

      1. Transporting person over long distances.

      2. Placing person on stretcher.

One bearer spreads the injured person's legs, kneels between the legs with his back to the person, and positions his hands behind the knees. The other bearer kneels at the injured person's head, slides his hands under the person's arms and across the chest, locking his hands together.

Alternate position—facing person

The bearers rise together, lifting the person.

Four Hand Seat Carry

Each bearer grasps one of his own wrists and one of the other bearer's wrists, thus forming a pack saddle.

1. This carry method is useful in transporting person with head or foot injury:

    a. For moderate distances.

    b. For placing person on stretcher.

2. Only conscious person can be transported with "Four Hand Seat Carry".

3. Person must be able to support himself by wrapping arms around shoulders of carriers.

The two bearers lower themselves sufficiently for the injured person to sit on the pack saddle; then they have the injured person place his arms around their shoulders for support before they rise to an upright position.

Two Hand Seat Carry.

1. This carry method is useful for two things:

    a. Carrying person for short distance.

    b. Placing person on stretcher.

FRONT VIEW

With the person lying on his back, a bearer kneels on each side of him at his hips. Each bearer passes his arms under the injured person's thigh and back, then grasps the other bearer's wrists.

BACK VIEW

The bearers rise, lifting the injured person.

10

Improvised Litters or Stretchers

General Information

1. A useable litter or stretcher can be improvised from a number of things. Almost any flat surface item of a suitable size can be used as a stretcher in time of need:

    a. Ordinary doors.

    b. Boards.

    c. Ladders.

    d. Cots.

    e. Workshop benches.

    f. Shutters from window.

    g. Simple poles (even tree limbs) tied together.

2. When possible, some sort of padding should be used on stretcher. This could be:

    a. Blankets or quilts.

    b. Sleeping bag.

3. Stretchers or litters can also be improvised by securing poles inside such things as:

    a. Blankets or quilts.

    b. Ponchos and tarpaulins.

    c. Jackets and shirts.

    d. Sacks and bags.

    e. Sheets and mattress covers.

4. Stretcher or litter poles can be improvised from a number of things:

    a. Strong tree branches.

    b. Skis

    c. Rifles and other items.

    d. Drapery rods.

5. If no poles available, simply roll heavy blanket or quilt from both sides toward middle. These rolls can then be used to obtain a firm grip when carrying someone.

Methods of Improvising Stretchers or Litters

Stretcher or litter made with poles and blanket

A. OPEN THE BLANKET AND LAY ONE POLE LENGTHWISE ACROSS THE CENTER

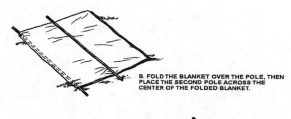

B. FOLD THE BLANKET OVER THE POLE, THEN PLACE THE SECOND POLE ACROSS THE CENTER OF THE FOLDED BLANKET.

C. FOLD THE FREE EDGES OF THE BLANKET OVER THE SECOND POLE

Stretcher or litter made with pole and jackets.

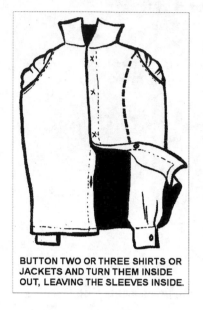

BUTTON TWO OR THREE SHIRTS OR
JACKETS AND TURN THEM INSIDE
OUT, LEAVING THE SLEEVES INSIDE.

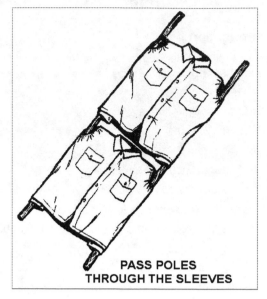

PASS POLES
THROUGH THE SLEEVES

Stretcher or litter made by inserting poles (or tree limbs) through sacks.

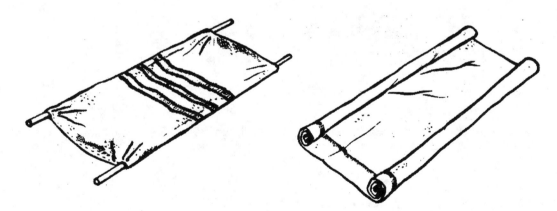

LITTERS MADE BY INSERTING POLES THROUGH
SACKS AND BLANKETS

Stretcher or litter made by rolling a blanket.

11

Treating Severe Injuries and Wounds

Introduction

1. Kinds of wounds and injuries requiring special medical measures:
   a. Face wounds.
   b. Neck wounds.
   c. Head injuries.
   d. Abdominal wounds.
   e. Sucking wounds of chest.
   f. Fractures (*see Chapter 12*).
   g. Burns (*see Chapter 13*).

Head Injuries

General Information

1. Head injuries may consist of any of the following conditions:
   a. Cut or bruise on scalp.
   b. Skull fracture with injury to brain.
   c. Skull fracture with injury to blood vessels of scalp, skull and brain.
2. Serious fractures of skull combined with brain injuries usually occur together.
3. It's possible, though, to have brain injury without having skull fracture.

Signs and Symptoms

1. Head injury with scalp wound is readily detected.
2. Head injury without scalp wound is more difficult to detect.
3. Suspect head injury and take necessary measures if person:
   a. Is unconscious.
   b. Has recently been unconscious.
   c. Has slower than usual pulse (*Chapter 2*).
   d. Has had recent convulsion.
   e. Has headache.
   f. Is breathing slowly.
   g. Is vomiting or feels nauseated.

Lifesaving Measures

1. Leave any protruding brain tissue alone.
2. Apply sterile dressing over brain tissue.
3. Don't remove or disturb any dirt or other foreign matter seen in wound.
4. The person should be lying so head is in a position higher than body.

Securing Dressing to Head with Attached Bandages.

1. Secure dressing in place with bandages (fig. a).

2. Cross the two front bandages under chin. Tie them on top of head (fig. b).

3. Bring right rear bandage under chin. Meet left rear bandage at point above ear. Secure to other bandages (fig. c). Do not cover ear.

4. Take rear bandages in opposite directions to other side of head. Tie above ear to other bandages with square knot (fig. d).

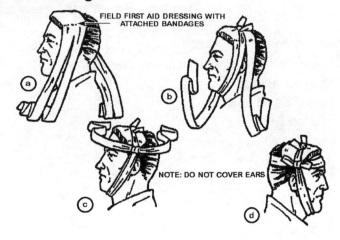

FIELD FIRST AID DRESSING WITH ATTACHED BANDAGES

NOTE: DO NOT COVER EARS

Face and Neck Wounds

General Information

1. Wounds of face and neck bleed profusely.

2. This type of bleeding is hard to control due to the many blood vessels in area.

Lifesaving Measures

1. First stop bleeding which may be causing an obstruction of person's upper airway.

2. Clear the person's airway with your fingers (*see Chapter 4*).

3. Remove from person's mouth any:

   a. Blood or mucous.

   b. Pieces of broken bone or teeth.

   c. Loose bits of flesh.

   d. Dentures.

4. If person is conscious and wants to sit up, have him lean forward with his head down. This permits drainage from mouth.

5. If person is unconscious or doesn't want to sit up, place him on abdomen with head turned to one side.

6. This permits the necessary drainage from mouth.

Chest Wounds

General

1. A chest wound resulting in air being sucked into the chest cavity is especially dangerous.
2. This will cause lung on injured side to collapse.
   a. Person's life will depend on how quickly wound can be made airtight.

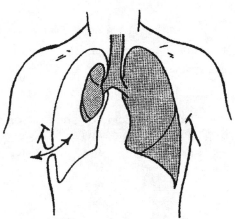

SUCKING SOUND OF THE CHEST

Lifesaving Measures

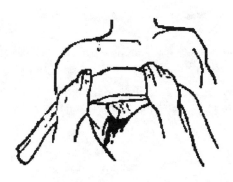

1. Have person exhale (breath out) forcefully, if possible, and hold breath until wound is sealed.
2. Seal wound airtight by applying directly over the wound either:
   a. A piece of plastic wrapper from the dressing.
   b. Or the foil-lined envelope from burn salt solution (in medical bag).
3. Apply dressing over plastic or foil-lined envelope.
4. Have injured person or another person exert pressure on dressing with open hand.
5. Now secure bandages around body.
6. Cover dressing completely with bandaging material torn from clothing, a blanket, etc., or:
   a. Apply folded poncho over dressing and around person's body.

7. Each turn of the bandaging material must overlap the preceding one.
8. This provides firm, evenly distributed pressure over entire dressing.
9. It makes wound airtight.
10. Secure bandages with person's belt.
11. Let injured person sit up if he wishes.
   a. Sitting up relieves abdominal pressure.
   b. Breathing is less difficult.
   c. Diaphragm muscle functions more easily.
12. If person prefers to lie down, have him lie on injured side.
   a. In this way, the lung on uninjured side can receive more air.
   b. The surface on which he is lying serves somewhat as a splint to injured side.
13. Pain is decreased.

153

Abdominal Wounds

General

1. The most serious abdominal wound is one where object penetrates abdominal wall, piercing:

    a. Some internal organs.

    b. Large blood vessels.

Lifesaving Measures

1. Never touch protruding organs such as intestines. Don't push back into wound.

2. Leave protruding organs alone. Simply cover with sterile dressing.

3. Move exposed intestine back into abdomen only when necessary to adequately cover wound.

4. Secure dressing in place with bandages, but not tightly.

    a. Excessive pressure from tight bandages may cause additional injuries.

    b. Internal bleeding can't be controlled by pressure from tight bandages.

5. Don't let person with intestinal injury eat or drink. Both food and water will eventually pass through intestines.

    a. Contamination could in this way be spread throughout abdomen.

6. Person's lips may be moistened to help lessen thirst.

7. Leave injured person on back, but turn head to one side.

    a. Watch closely to prevent choking since he's likely to vomit.

12

How to Handle Bone Fractures

FRACTURES

Basic Information on Fractures

1. Fractures (broken bones) can cause:
    a. Total disability.
    b. A person's death.
2. On the other hand, fractures can usually be treated so person completely recovers.
3. Traction is required on most fractures of long bones to overcome muscle contractions.
4. Much depends on medical treatment injured person receives prior to being moved.
    a. Such treatment includes:
        1. Proper splinting (immobilization) of fractured bone.
        2. Use of lifesaving measures A, B, C and D (*see Chapter 3*) if required.

Two Types of Fractures

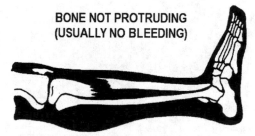

BONE NOT PROTRUDING (USUALLY NO BLEEDING)

CLOSED FRACTURE

Closed fracture:

1. A closed fracture is a break in bone without a corresponding break in skin.
2. Tissue damage under skin is almost always certain.

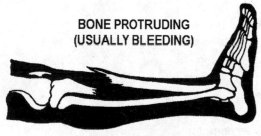

BONE PROTRUDING (USUALLY BLEEDING)

OPEN FRACTURE

Open fracture:

1. Open fracture is break in bone as well as a break in overlying skin.
2. The broken bone can be seen protruding through skin.

Another kind of open fracture:

1. An open fracture is also the damage caused by a bullet, shell fragment or something similar.
2. The object cuts through flesh and shatters the bone.
3. Open fractures are always contaminated and subject to infections.

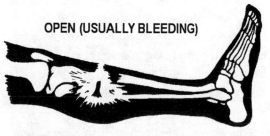

OPEN (USUALLY BLEEDING)

OPEN FRACTURE PRODUCED BY MISSLE

Detecting Fractures

1. A fracture is easily detected:

   a. When bone is seen protruding from skin.

   b. The arm or leg is in unnatural position.

   c. The chest wall is caved in.

   d. The injured person hears bone snap.

   e. When abnormal motion is seen such as arm bending but not at elbow.

      1. Other signs and symptoms of possible fractures:

         a. Tenderness or pain when slight pressure is applied to injured part.

   f. Swelling at sight of injury.

   g. Discoloration of skin around injury site.

   h. Sharp pain felt when attempt is made to move injured body part.

2. Never encourage injured person to move injured body part in order to identify fracture.

   a. Movement of injured body part causes further damage to surrounding tissues.

   b. Such movement also promotes shock (*see Chapter 8*).

   c. If uncertain about bone being fractured, be on safe side and treat as if it is.

Purpose of Using Splints

1. A fracture must <u>always</u> be splinted (immobilized) to prevent sharp edges of bone from moving and cutting:

   a. Tissue and muscle.

   b. Blood vessels and nerves.

2. Splinting a fracture:

   a. Greatly reduces pain.

   b. Helps prevent or control shock.

3. Splinting a closed fracture:

   a. Keeps bone fragments from cutting through skin and causing open wound.

      1. This prevents contamination and possible infection.

Basic Splinting Rules

1. Start by undertaking the following:

   a. First stop bleeding (*see Chapter 6*) if fracture is an open one.

   b. Apply dressing and bandage (*see Chapter 7*) as for any other wound.

2. Always go by the golden rule: <u>"Splint them where they lie."</u> This simply means to splint fracture:

   a. Before movement of person is attempted.

   b. Without any change of position of fractured part.

3. Other rules to follow in splinting:

   a. If joint is bent, don't try to straighten.

   b. If joint isn't bent, don't try to bend it.

   c. If bone is in unnatural position, don't try to change it.

4. If person with leg fracture must be moved before splint can be applied:

   a. Use uninjured leg as splint by tying fractured leg to it.

   b. Then grasp person by armpits and pull in straight line.

   c. Never roll injured person or move sideways.

5. Always apply splint so both joint above and below fracture are immobilized (splinted).

6. Padding should be placed between fracture and splint in order to:

    a. Prevent undue pressure.

    b. Prevent further injury to tissue, blood vessels and nerves.

    c. Padding is especially important to use in armpit and at crotch.

    d. Padding should be put where splint comes into contact with bony parts such as:

        1. Elbow and wrist joint.

        2. Knee and ankle joint.

7. Bind splint with bandages at several places above and below fracture.

    a. Don't bind so tightly that it cuts off flow of blood.

    b. No bandage should be applied directly across fracture.

    c. Always tie bandages against splint with square knot.

8. Always use sling to support:

    a. A splinted arm bent at elbow.

    b. A fractured elbow which is bent.

    c. A sprained arm.

    d. An arm with painful wound.

Splints, Padding, Bandages and Slings

1. Splints can be improvised from:

    a. Poles and boards.

    b. Tree limbs and sticks.

    c. Rifles or knives.

    d. Rolled up magazines or newspapers.

    e. Cut up cardboard boxes.

        1. If nothing can be found to make splint:

            a. The wall of person's chest can be used to immobilize fractured arm.

    f. An uninjured leg can be used to immobilize a fractured leg.

2. Padding can be made from:

    a. A cut up shirt or jacket.

    b. Pieces of sheet or blanket.

    c. Socks.

    d. Leafy vegetation.

3. Bandages can be improvised from:

    a. Handkerchiefs.

    b. Strips of clothing or blankets.

    c. Sheets.

4. Bandages can be secured to fracture by:

    a. Belts.

    b. Rifle slings.

    c. Strips of clothing.

5. Slings can be improvised by using:

    a. Shirts and jackets.

    b. Belts.

    c. Triangular bandages are ideal for this purpose (*see Chapter 7*).

METHODS OF IMMOBILIZING

Fractures of Upper Extremities

1. The following seven illustrations show the proper application of:

    a. Splints.

    b. Slings.

    c. Cravats.

    > 1. The seven are all excellent for immobilizing and supporting fractures of upper extremities.

2. Padding may not be visible in the illustrations. Nevertheless, it's always to be applied along injured part for entire length of splint.

3. Two methods of using triangular bandage to make sling.

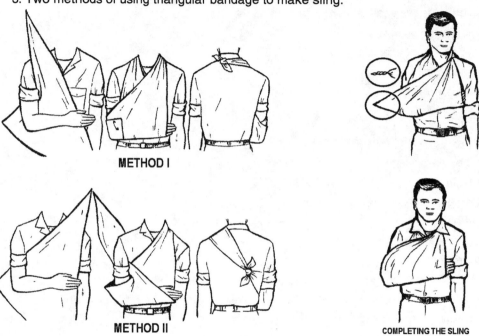

METHOD I

METHOD II

COMPLETING THE SLING

4. Board splints on fractured arm or elbow when elbow isn't bent.

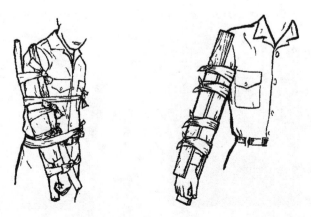

5. Chest wall used as splint for upper arm fracture. Used when regular splints are unavailable.

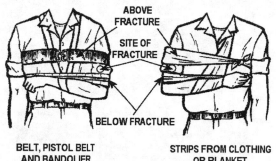

BELT, PISTOL BELT AND BANDOLIER

STRIPS FROM CLOTHING OR BLANKET

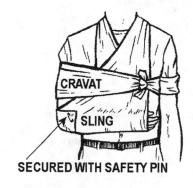

SECURED WITH SAFETY PIN

6. Chest wall, sling and cravat used to immobilize fractured elbow when elbow is bent.

7. Board splint used on fractured forearm.

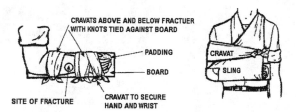

8. Fractured wrist or forearm splinted with sticks. Supported with shirt tail and cloth strips.

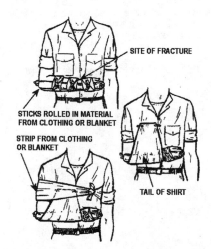

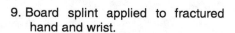

9. Board splint applied to fractured hand and wrist.

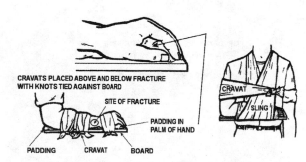

159

Fractures of Lower Extremities

1. The following five illustrations show how splints are applied to immobilize fractures of lower extremities.

2. Padding is always applied along injured part for the length of splint.

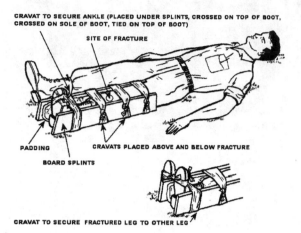

CRAVAT TO SECURE ANKLE (PLACED UNDER SPLINTS, CROSSED ON TOP OF BOOT, CROSSED ON SOLE OF BOOT, TIED ON TOP OF BOOT)

SITE OF FRACTURE

PADDING

BOARD SPLINTS

CRAVATS PLACED ABOVE AND BELOW FRACTURE

CRAVAT TO SECURE FRACTURED LEG TO OTHER LEG

a. Board splints on fractured lower leg or ankle.

b. Board splint on dislocated or fractured knee.

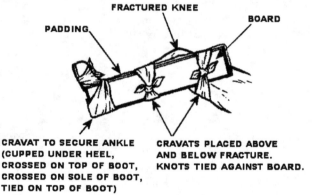

FRACTURED KNEE

PADDING

BOARD

CRAVAT TO SECURE ANKLE (CUPPED UNDER HEEL, CROSSED ON TOP OF BOOT, CROSSED ON SOLE OF BOOT, TIED ON TOP OF BOOT)

CRAVATS PLACED ABOVE AND BELOW FRACTURE. KNOTS TIED AGAINST BOARD.

SITE OF FRACTURE

OR

STRIPS FROM CLOTHING OR BLANKET

BOARDS

CRAVATS PLACED ABOVE AND BELOW FRACTURE

CRAVATS SECURE FRACTURED LEG TO UNINJURED LEG

3. Board splints on fractured thigh or hip.

4. Splints on fractures of lower extremities using poles rolled in blanket.

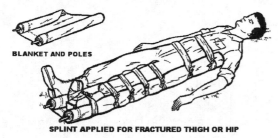

BLANKET AND POLES

SPLINT APPLIED FOR FRACTURED THIGH OR HIP

BLANKET AND POLES

SPLINT APPLIED FOR FRACTURED LOWER LEG, KNEE, OR ANKLE

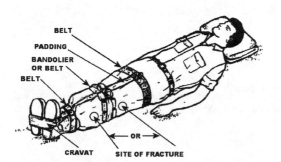

BELT

PADDING

BANDOLIER OR BELT

BELT

CRAVAT   SITE OF FRACTURE

—OR—

5. Uninjured leg used as splint for fractured leg.

Fractures of Jaw, Collarbone and Shoulder

1. Apply cravat to immobilize fractured jaw as shown in Chapter 7.

2. All support bandages go to top of person's head, not to back of neck.

   a. Bandage to back of neck pulls jaw back and interferes with breathing.

   b. Apply two belts, a sling and cravat to immobilize fractured collarbone.

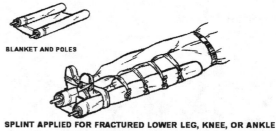

PADDING

BELTS

CRAVAT

SLING

SECURED WITH SAFETY PIN

APPLICATION OF BELTS, SLING AND CRAVAT TO IMMOBILIZE A COLLARBONE

c. Apply sling and cravat to immobilize fractured or dislocated shoulder.

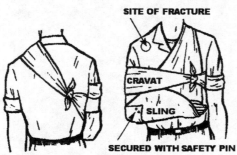

SITE OF FRACTURE

CRAVAT

SLING

SECURED WITH SAFETY PIN

Rib Fractures

1. Rib fractures can be detected by one or more of the following:

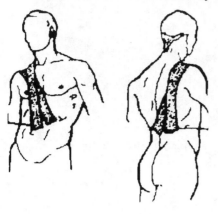

    a. Pain in breathing.

    b. Pain when coughing.

2. Pain in area of fracture.

3. Tenderness around fracture.

4. Pain produced by slight hand pressure on sternum.

5. Fracture can sometimes be felt.

6. If lung punctured, person may cough red frothy blood.

7. Treatment for fractured upper ribs.

8. Clean skin.

9. Paint skin with tincture of benzoin.

10. Have person exhale and hold breath.

11. Apply two three-inch wide adhesive strips across shoulder of injured side.

12. Adhesive strips should extend:

    a. Well down on abdomen in front.

13. To lower back in rear.

Treatment for Fractured Lower Ribs

1. Clean skin.

2. Paint skin with tincture of benzoin.

3. Have injured person exhale and hold breath.

4. Apply three-inch wide adhesive strips and let them extend beyond the midline.

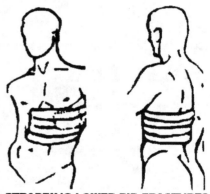

STRAPPING LOWER RIB FRACTURES

Alternate Treatment for Fractured Ribs

1. An alternate method of bandaging fractures of upper and lower ribs is to:

    a. Encircle trunk just below nipples.

2. Use eight inch elastic bandage.

STRAPPING WITH ELASTIC BANDAGE OR MUSLIN DRESSING

Fracture of Spinal Column

1. It's often difficult to be certain person has fractured spinal column.

2. Be suspicious of any back injury, especially if person's back has been:

   a. Bent.

   b. Sharply struck.

3. If person has back injury and lacks feelings in legs, or can't move them:

   a. You can be sure he has severe back injury.

   b. The injury should be treated as a fracture.

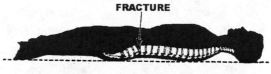

FRACTURE

IN THIS POSITION, BONE FRAGMENTS MAY BRUISE OR CUT THE SPINAL CORD

FRACTURE

BLANKET IS IN PLACE

4. Remember — If spine is fractured, bending spinal column can cause bone fragments to:

   a. Bruise or cut spinal cord.

   b. Cause permanent paralysis.

5. The spinal column must be kept in swayback position to remove pressure from spinal cord.

6. Steps to follow.

   a. Warn person (if conscious) not to move.

   b. Leave person in position he was found.

   c. Don't try to move any body part.

   d. If person is lying face up, slip blanket under arch of back for support.

   e. If person is lying face down, don't put anything under any part of body.

7. If person is lying face up and must be moved:

   a. Tie wrists together with a strip of cloth.

   b. Lay folded blanket across stretcher, litter board or door.

   c. Place folded blanket where arch of person's back is to be placed.

   d. Using four people, carefully place person on stretcher.

      1. Do not bend spinal column.

   e. Three of these people kneel and position themselves on one side of injured person.

   f. The fourth person kneels on opposite side.

BACK: TIE WRISTS TOGETHER LOOSELY, PLACE BLANKET ON LITTER AT ARCH SITE, GENTLY LIFT INJURED PERSON ONTO LITTER WITHOUT BENDING HIS BACK

NECK: STEADY HEAD, AND NECK; GENTLY SLIDE INJURED PERSON ONTO BOARD, SLIP ROLL UNDER NECK, AND IMMOBILIZE HEAD.

163

g. The three people place their hands as in illustration.

h. The fourth person assists in lifting at site of fracture.

i. All four in unison gently lift injured person up about eight inches.

j. The first person slides the stretcher under the injured person.

k. The first person also makes certain the folded blanket is in proper position.

l. All four people in unison carefully lower person onto stretcher.

8. If person is face down.

    a. Person must be moved in same face down position.

    b. Using a four man team, lift person onto a:

        1. Regular litter or stretcher.

        2. Blanket roll litter (*see Chapter 10*).

    c. Keep person's spinal column in swayback position.

    d. If regular litter or stretcher is used, first lay on folded blanket.

    e. The blanket should be at point where person's chest will be placed.

Fracture of Neck

1. A fractured neck is extremely dangerous as bone fragments may bruise or cut spinal cord.

2. If person with neck fracture is conscious, warn not to move.

3. Leave person in exact position in which he was found.

    a. Moving may cause person's death.

    b. If neck in abnormal position, immobilize if in this position.

INJURED PERSON WITH ROLL OF CLOTH (BULK) UNDER THE NECK

4. If person found lying face up:

    a. Keep head still.

    b. Raise shoulders slightly

    c. Slip roll of cloth under neck.

        1. Roll should be thick enough to slightly arch neck (see illustration).

        2. Back of head should be left touching ground.

    d. Don't bend person's head or neck forward.

    e. Don't raise or twist person's head.

IMMOBILIZATION OF A FRACTURED NECK

5. Immobilize person's head as in illustration.

    a. Pad around head with heavy objects such as boots or small packs.

    b. If boots or packs are used, fill with sand, dirt, gravel or stones.

    c. Stuff material in boot tops so contents won't spill out. Tie tightly.

    d. Place boots or packs on each side of head.

6. Moving person with fractured neck.

   a. A minimum of two people required to prepare person with fractured neck to be moved.

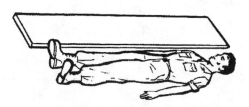

   b. The head and body must always be moved in unison.

      1. Neck must not be allowed to bend.

   c. Lay wide board or door lengthwise beside person.

      1. Board or doors must extend beyond head at least four inches.

7. If person is lying face up.

   a. Number one man steadies person's head and neck between hands.

   b. Number two man places one foot and one knee against board to prevent from slipping.

   c. Number two man then grasps person at the hip and shoulder.

      1. Person is slid onto board.

8. If injured person lying face down:

   a. Number one man steadies person's head and neck between hands.

   b. Number two man gently rolls person over onto board.

   c. Number one man continues to steady person's head and neck.

   d. Number two man:

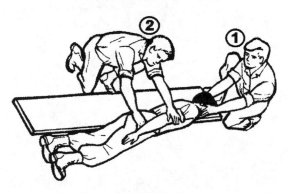

      1. Raises person's shoulders slightly.

      2. Puts padding under neck.

      3. Immobilizes person's head.

9. Person's head and neck can be immobilized with:

   a. Boots.

   b. Stones rolled up in pieces of blanket.

IMMOBILIZATION OF A FRACTURED NECK

10. Secure any improvised supports in position:

    a. With cravat

    b. Or with strip of cloth.

11. This should go under board and over forehead.

   a. Lastly, lift board holding person onto litter or stretcher.

13

Proper Treatment of Burns

CAUSES OF BURNS

Handling Burn Situations

1. Clothes on fire:

    a. Force person down to floor.

2. Smother flames with:

    a. Blanket

    b. Coat

    c. Rug, etc.

3. Scalded person:

    a. Quickly tear off scalded person's clothing.

    b. This is the only way to reduce the time in which the hot fluid damages skin.

KINDS OF BURNS

First Degree Burns

1. Only outer layer of skin burned.
2. Mild swelling, pain, tingling and skin surface turns red.

Second Degree Burns

1. Most but not all skin thickness burned.

    a. Capillary walls damaged.

2. Much more swelling than with first degree burns.
3. More intense pain.
4. Blistering.
5. Skin bright red, blotchy and looks wet.

PROPER BURN TREATMENT

Special Measures for Treating Burns

1. Special treatment measures are called for when burned areas are blistered and charred.
2. Primary treatment goal is to:

    a. Lessen shock.

3. Ward off infections.
4. The A, B, C, D lifesaving measures in Chapters 3 through 6 may have to be initially gone through.
5. Only then can proper treatment for burns be undertaken.
6. Carefully protect burns against further contamination.
7. This lessens possibility of infection.

Dos and Don'ts for Treating Burn Areas

1. Cut away clothing and gently lift material while avoiding touching burn.
2. Don't let person's clothing touch burned area as it's being removed.
3. Don't try to remove pieces of clothing stuck to burn.
4. Don't try to clean burn in any manner.

5. Don't even try to remove dirt or cloth stuck to burn as these are sterile.

6. Don't break blisters.

7. Don't put greasy ointments/medications on burn.

8. Carefully watch for signs of infection.

Sterile Dressings

1. Place sterile dressing over burn area.

   a. Secure in place with bandages.

2. Change dressings as needed to prevent infection.

To Prevent Shock

1. Shock can be prevented by applying measures found in Chapter 8 or those outlined as follows:

2. If burned person is conscious and not vomiting and has no neck or abdominal wound:

   a. Use sodium chloride-sodium bicarbonate packet found in medical kit.

3. Dissolve packet in a quart cold water.

   a. Never use warm water.

   b. Warm salt water often induces vomiting.

4. This solution helps restore body fluids and salts.

5. If sodium chloride-sodium bicarbonate isn't available:

   a. Mix 1/2 teaspoon table salt and 1/4 teaspoon baking soda in one quart cold water.

   b. If only salt is available, use without baking soda.

6. Slowly give solution to burn victim.

   a. Entire quart should be consumed within a half hour.

7. Should victim become nauseated:

8. a. Immediately stop giving solution in order to prevent vomiting (a further fluid loss).

   a. Keep solution available for victim to later drink.

PART IV

TREATMENT FOR COMMON EMERGENCIES

14

Sun or Heat Related Injuries

General Pointers

1. The human body depends on water to cool itself in hot environments.

2. It's possible for person to lose at least one quart of water each hour in severe heat.

3. Water lost must be replaced or person can become a "heat injury."

4. The activity undertaken will determine amount of water necessary to maintain proper body functions.

5. It's a myth that humans can learn to adjust to decreased water intake.

6. When water is in short supply, significant water economy can be accomplished only by limiting physical activity to coolest part of day or night.

WATER REQUIREMENTS FOR ONE PERSON			
ACTIVITY	TYPE ACTIVITY	QUARTS OF DRINKING WATER NEEDED PER DAY	
		BELOW 80°	ABOVE 80°
LIGHT	SITTING AT DESK	4	7
MODERATE	SLOW WALK	6	9
HEAVY	FAST WALK	8	12

Heat Stroke (Sun Stroke)

1. Symptoms:

   a. Skin hot, red and dry.

   b. Involuntary urination and defecation.

   c. Sweating usually stops.

   d. Headache, dizziness and rapid pulse.

   e. Nausea, vomiting and mental confusion.

   f. Shallow, irregular breathing.

   g. Collapse and unconsciousness.

   h. Possible convulsions.

   i. Temperature may rise as high as 106 to 110 degrees F.

2. Treatment:

   a. Closely monitor person's temperature.

   b. Treatment should be promptly administered.

3. The longer body temperature is high, the greater the threat of:

   a. Permanent damage

   b. Death.

4. If person survives until second day, recovery usually occurs.

5. Cool person down as rapidly as possible.

   a. Immerse briefly fully clothed in cold water.

168

    b. Add cold packs to groin/under arms.

    c. Dunk in stream or pond, etc.,

    d. Saturate person's clothing.

6. If cold water treatment isn't possible:

    a. Get person to shaded area.

    b. Remove their clothing.

    c. Keep entire body wet by pouring water over person.

    d. Cool further by fanning person's body.

7. If no water available:

    a. Rub person down with alcohol.

    b. Fan by waving shirt, towel, etc.

    c. Rub arms, legs and trunk briskly to increase circulation to skin.

8. When temperature drops to 102 degrees F.:

    a. Dry person off.

    b. Cover with blanket.

9. Temperature should drop below 97 degrees.

10. When temperature again rises to 97 degrees, start cool down procedures again.

11. Get person to medical treatment facility if one is available.

12. Continue to cool body on the way.

13. When person regains consciousness, give cool salt water to drink.

    a. Have person drink three to five quarts of salt water over 12 hour period.

    b. The proper mix should be one teaspoon ordinary table salt per quart of water.

Sun Blindness (Sunburn of Cornea)

1. The sun's glare or reflection on water, sand, etc., is cause of sun blindness.

    a. Sun blindness can also occur during cloudy weather and when hazy.

2. Symptoms of sun blindness:

    a. A person's ability to see normally is affected.

    b. Ground level variations can no longer be detected while walking.

    c. A burning sensation in eyes.

    d. The eyes feel sandy, gritty and scratchy under lids when closed.

    e. Eyes red and watery.

    f. Eyelids usually swollen, red and difficult to open.

    g. Eyes hurt when exposed to even weak light.

    h. Headaches commonplace.

3. Treatment of sun blindness:

    a. Eyes must be immediately and correctly treated or pain will become unbearable.

    b. The sun blinded person must stay in total darkness.

    c. A sterile bandage or blindfold must be worn for a 18 hour period.

    d. A dark cloth can be placed on damaged eyes to shut out light.

4. Prevention of sun blindness:

    a. Reduce the amount of sun glare striking eyes by blackening nose.

    b. Also blacken cheeks under eyes.

    c. Sun blindness can be prevented by wearing sun glasses when outside.

Improvised Sun Shades

1. Sun glasses not available? Make your own emergency pair out of a:
    a. Piece of scrap rubber.

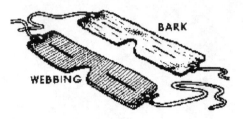

2. Piece of scrap leather.

3. Piece of tree bark.
    a. Make these sun goggles to cover entire width of face.
        1. Trim out narrow slit for each eye.
        2. Attach a string, leather thong or shoe lace to each side.
        3. Tie at the back of head.

Heat Cramps

1. Heat cramps are direct result of excessive loss of salt from body.
2. Symptoms of heat cramps:
    a. Cramps or painful muscle spasms in arms, legs or abdomen.
    b. Cramps may be mild or severe.
    c. Dizziness.
    d. Extreme thirst.
    e. Pale and damp or wet skin.
    f. Person may grimace and thrash about with arms and legs drawn up.
3. Treatment
    a. Move person to shady area.
    b. Loosen clothing.
    c. Get plenty of rest.
    d. Have person drink 1/2 pint cool salt water.
    e. Or swallow packet of salt and wash down with glass of water.
    f. Then have person drink three to five quarts of cool salt water over 12-hour period.
    g. Use 1/4 teaspoon table salt to each quart water.
4. Do not:
    a. Use hot packs on cramping muscles.

Heat Exhaustion

1. Heat exhaustion caused by:
    a. Severe dehydration (or excessive loss of water from body).
    b. Excessive loss of salt from body due to extreme physical activity.
2. Symptoms of heat exhaustion:
    a. Heavy or excessive sweating.
    b. Headache.
    c. Dizziness, vertigo.
    d. Extreme weakness.
    e. Nausea.
    f. Muscle cramps.
    g. Cool, moist, pale, clammy skin.
    h. Mental confusion.
3. Treatment:
    a. Lay person down in cool, shaded area.
    b. Loosen clothing.
    c. Have him drink three to four quarts cool salt water over 12-hour period.
    d. Spray or sprinkle person's body with water to lower body temperature through evaporation.

15

Cold Related Injuries

1. The combination of wind and low temperature creates an adverse condition known as "wind chill."

2. The Wind Chill Table shows:

    a. When cold weather is highly dangerous.

    b. When exposed skin is likely to freeze.

3. For example, a temperature of -20 degrees F. combined with 20 mph wind is extremely dangerous.

    a. It gives a wind chill temperature of -75 degrees F.

4. The danger is grave in above situation.

    a. Exposed skin can freeze within 30 seconds.

    b. Such an injury can disable person as seriously as broken bone or bullet wound.

CHILL FACTOR CHART

WIND SPEED (mph)	LOCAL TEMPERATURE (°F)										
	32	23	14	5	-4	-13	-22	-31	-40	-49	-58
	EQUIVALENT TEMPERATURE (°F)										
CALM	32	23	14	5	-4	-13	-22	-31	-40	-49	-58
5	29	20	10	1	-9	-18	-28	-37	-47	-56	-65
10	18	7	-4	-15	-26	-37	-48	-59	-70	-81	-92
15	13	-1	-13	-25	-37	-49	-61	-73	-85	-97	-109
20	7	-6	-19	-32	-44	-57	-70	-83	-98	-10	-121
25	1	-10	-24	-37	-50	-64	-77	-90	-104	-117	-130
30	-1	-13	-27	-41	-54	-68	-82	-97	-109	-123	-137
35	-1	-15	-29	-43	-57	-71	-85	-99	-113	-127	-142
40	-3	-17	-31	-45	-59	-74	-87	-102	-116	-131	-145
45	-3	-18	-32	-46	-61	-76	-89	-104	-118	-132	-147
50	-4	-18	-33	-47	-62	-78	-91	-105	-120	-134	-148
LITTLE DANGER FOR PROPERLY CLOTHED PERSON	CONSIDERABLE DANGER			VERY GREAT DANGER							
	DANGER FROM FREEZING OR EXPOSED FLESH										

Frostbite (Freezing of Body Tissue)

1. Frostbite is result of tissue freezing from exposure to temperatures below 32 degrees F.

2. The body parts most easily frostbitten are:

    a. Hands, wrists and feet.

    b. Forehead and ears.

    c. Nose, cheeks and chin.

3. Frostbite can be either:

    a. Superficial (only skin tissue affected).

    b. Deep (frozen all the way to bone).

4. The degree of frostbite injury depends on:

    a. The wind chill factor.

    b. How long person was exposed to elements.

    c. How much proper protection was available.

d. If exposure short, frostbite will be superficial.

e. If exposure long, the frostbite will be deep and to bone.

5. Symptoms of superficial frostbite are:

a. Stinging, tingling sensation of affected skin area.

b. Dull aching and stiffness followed by numbness, skin blistering and peeling.

c. Skin turns bright red and then changes to waxy white or pale gray.

d. At this stage frostbitten body part may feel like block of wood.

6. Symptoms of deep frostbite are:

a. Severe stiffness.

b. Numbness rather than painful.

c. Blue skin.

d. Tissue is hard, even brittle with complete lack of sensation and movement.

7. Frostbite treatment:

a. Put frostbite victim in warm sheltered place.

b. Rewarm frostbitten parts with body heat until pain returns:

1. Rewarm face, ears and nose by covering with your hands.

2. Place victim's hands under clothing and against your body.

3. Frostbitten hands may be placed under victim's own armpits.

c. Take off victim's shoes and socks.

1. Place feet under clothing and against abdomen of another person.

2. After victim's feet are warmed, put on dry socks.

d. Give plenty of hot liquids (soup, coffee, tea, etc.).

e. Button or zip clothing closed to prevent further loss of body heat.

f. If "deep" frostbite is suspected, protect body part from further injury.

1. Get person to a medical treatment facility if one is

2. available.

8. Things not to do for a frostbite victim:

a. Don't give alcohol to drink.

1. Alcohol increases body heat loss.

b. Don't allow victim to smoke tobacco.

1. Causes blood vessels to narrow.

c. Don't break or open blisters found on arms and legs.

1. Just cover with loose dry bandage.

d. Don't rub frostbitten areas with snow.

e. Don't soak hands and feet in cold water.

f. Don't rewarm frostbitten hands, feet, etc. by massaging them.

g. Don't expose frostbitten hands, etc. to open fire.

h. Don't neglect frostbite.

1. To do so is to invite gangrene.

i. Don't try to thaw deep frostbite.

1. There's less danger when walking on frozen feet than there is after feet are thawed.

Snow Blindness

1. General pointers:

a. Snow blindness caused by reflection or glare of sun on snow and ice.

b. Snow blindness is more likely to occur when hazy or cloudy than when sun is shining.

c. A mild case of snow blindness may completely disable person for several days.

    1. A mild case will also heal spontaneously in a few days.

2. More severe cases of snow blindness may cause permanent damage to eyes.

a. The pain may be quite severe if snow blindness isn't properly treated.

3. Symptoms of snow blindness:

a. The initial sign of snow blindness is noted when the vision changes.

    1. Variations in level of ground can no longer be detected when walking.

b. Early stages of snow blindness can be recognized by:

    1. Scratchy, gritty or sandy feeling in eyes when closed.

    2. Burning sensation in eyes, especially when closed.

    3. Eyes start watering and are red.

    4. Eyelids are unusually red, swollen and hard to open.

    5. Headaches are common and the eyes hurt when exposed to even weak light.

4. Treatment of snow blindness:

a. Snow blinded person shouldn't rub eyes.

b. The best medicine is to stay in total darkness.

c. Put dry, sterile bandage, a blindfold, or dark cloth over eyes to shut out all light.

    1. Keep this on for 18-hours.

d. Cold compresses on eyes will help relieve pain.

e. An ophthalmic ointment should be applied to eyes hourly to:

    1. help relieve the pain.

5. lessen inflammatory reaction.

6. Prevention of snow blindness:

a. Blacken cheeks, nose and under eyes to reduce glare of reflected sun.

b. Snow blindness can be prevented by simply wearing sun glasses at all times while outdoors.

c. If you have no sunglasses, or lose the ones you had, improvise an emergency pair.

IMPROVISED SUNGLASSES

    1. These can be made of leather, bark, rubber or even a piece of cloth.

    2. Make the width of face.

    3. Cut narrow horizontal strips as shown.

7. Hold improvised glasses in place with string attached to each side.

8. Tie strings at back of head.

Trench Foot

1. General Pointers:

a. Trench foot is the result of prolonged exposure of feet to wetness.

2. This injury usually contracted while temperatures range from freezing to 50 degrees F.

3. The possibility of developing trench foot is greater if feet are inactive.

4. Trench foot can result in loss of toes or parts of feet.

5. Symptoms of Trench Foot:

a. Numbness of feet.

b. Feet may tingle, ache or cramp.

c. If exposure has been prolonged and severe, feet may swell so much the pressure:

    1. Closes blood vessels

6. Cuts off circulation.

7. Treatment of trench foot:

    a. Dry feet thoroughly.

    b. Avoid walking if possible.

    c. Get person to a medical treatment facility if one is available.

Immersion Foot

1. General pointers:

    a. Immersion foot results from feet being submerged in water for prolonged period of time.

    b. Immersion foot is also caused by feet being constantly wet over extended time period.

2. Immersion foot usually comes about after more than 12 hours of water or wet exposure.

3. Immersion foot develops more quickly if water is below 50 degrees F.

4. Immersion foot can also occur when feet are exposed to warm water for longer than 24 hours.

5. Symptoms of immersion foot:

    a. Soles of feet become wrinkled and white.

    b. Standing or walking is extremely painful.

6. Treatment of immersion foot:

    a. Dry feet thoroughly.

    b. Avoid walking if possible.

    c. Get person to a medical treatment facility if one is available.

Hypothermia

HYPOTHERMIA TABLE			
Temperature		Stage	Symptoms
98°F	37°C	Mild	Cold at first, then stops shivering. Rigid muscles
95°F	34°C	Moderate	Poor coordination, impaired speech - slow and slurring. Memory loss. Convulsions.
88°F	32°C	Severe	Unintelligible pulse. Pupils dilated. Shallow breathing. Unconscious. Glassy stare.
82°F	28°C	Extreme	Heart rhythm may change. Heart stops beating.

2. Hypothermia takes place when the temperature of the body's inner core (98.6 degrees F) drops below 98 degrees F (37 degrees C).

1. General pointers:

    a. When body loses heat faster than it can produce heat, the result is called hypothermia.

    b. Hypothermia is simply the lowering of body heat below normal.

        1. Temperature of body (98.6 degrees F.) drops below 98 degrees F. (37 degrees C).

    c. Hypothermia is usually caused by exposure to prolonged or extreme cold temperatures.

    d. Body's temperature control center breaks down and can't produce enough heat.

e. It can't maintain balance when the internal body temperature is about 95 degrees F.

f. Further body temperature decline is quite fast at this point.

g. Such a condition can be dangerous and should immediately be treated.

h. Death usually occurs by time internal body temperature reaches 80 degrees F.

2. Symptoms of hypothermia

a. As body temperature drops, person may become delirious, drowsy or comatose.

b. Pale skin, cold feeling skin are notable.

c. Slower and shallow sounding breathing.

1. Casual observation could lead one to believe breathing has stopped.

d. Blood pressure and pulse difficult to take, and sometimes unobtainable.

e. Person becomes unresponsive to painful stimuli (pin prick, pinch, etc.).

f. Death usually follows due to:

1. Cardiac arrest (heart stops beating).

3. Ventricular fibrillation (extremely rapid non-pumping muscle movement of heart).

4. Preventing hypothermia:

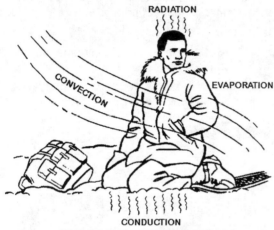

HOW HEAT IS LOST FROM THE BODY

a. Avoid drinking cold water if hypothermia conditions exist.

b. Drink high energy liquids such as hot chocolate, coffee and tea.

c. Eat properly and often.

1. Eat simple sugar food when possible.

d. Avoid overexertion but keep active.

1. Take frequent rest breaks.

e. Be careful to protect yourself from wind, wet and cold.

1. Stay dry and keep warm.

Treatment for hypothermia

1. The primary intent is to raise person's body temperature.

a. Get person warm quickly.

b. Prevent further heat loss from body with more clothing, blankets, etc.

c. Get out of wind and weather and into best available shelter.

d. Undress person and dry off.

e. Put dry clothing on victim.

f. Put blankets, ponchos, boughs, etc., between victim and ground.

g. Get in sleeping bag with victim or bundle in blankets to share body heat.

2. Try to warm victims groin area, neck, sides of chest, etc.

3. Use heated rocks wrapped in cloth, hot water bottle, etc.

4. Feed hot sugar-sweetened drinks, soup, and food loaded with carbohydrates.

5. Never give whiskey, brandy or any other alcohol to drink.

16

Snakes, Spiders and Scorpions

Snakes

Snake Categories

1. Poisonous long fanged snakes are few—rattlesnakes, copperheads and moccasins.

2. The bite from large water moccasin often fatal since venom is extremely poisonous.

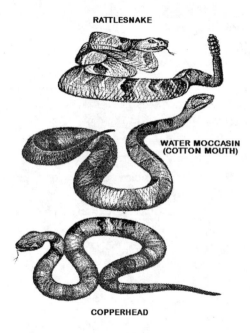

RATTLESNAKE

WATER MOCCASIN
(COTTON MOUTH)

COPPERHEAD

3. The bite from all three of these snakes results in an incredible amount of pain.

   a. Starts with local swelling.

   b. Quickly increases as venom spreads throughout body.

4. An example of a highly dangerous short-fanged poisonous snake is the coral.

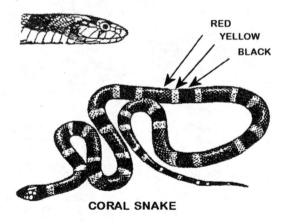

RED
YELLOW
BLACK

CORAL SNAKE

   a. This serpent's venom is the most deadly among poisonous snakes.

Snake Bite Treatment -- Current Method

1. Bite victim should avoid physical exertion.

2. Incision shouldn't be made over bite area.

3. Use vacuum pump suction device if one's available (comes with all snake bite kits).

4. Keep bite area immobilized.

5. Don't immerse bite in cold water and don't put ice pack on bite.

6. Keep bite on leg or elsewhere below heart level.

7. Don't use tourniquet on bitten arm or leg.

8. Observe bitten person for signs of labored breathing. Some venom affects the breathing mechanism.

9. Should victim's breathing stop, initiate artificial respiration (*see Chapter 4*).

10. Get snake bite victim to medical treatment facility if one's available.

Snake Bite Treatment -- Another Way

1. Some believe this to be a better method when no medical facility is close at hand.
2. Don't let bite victim move around. Keep quiet.
3. Remove rings, bracelets and watches from bitten limb.
    a. Taking them off will be painful and quite difficult when swelling starts.
4. Place ice or freeze-pack on bite, if available.
    a. This helps slow spread of venom.
5. Apply lightly constricting band two to four inches above bite, between bite and heart.

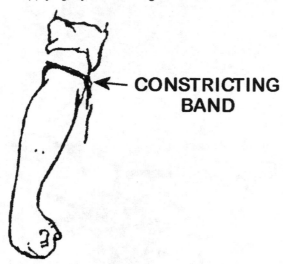

← **CONSTRICTING BAND**

a. Do this between 15 and 30 minutes from time of bite.

b. Use a shoe lace, strip of cloth, string, handkerchief, etc.

c. You should be able to insert finger between band and limb.

d. Loosen every 15 minutes for 90 seconds.

e. It should be tight enough to cut off flow of venom in blood vessels near skin surface.

f. It should not be so tight as to stop pulse or arterial blood flow.

g. When veins under skin pop out, appropriate flow of blood has been halted.

    h. Arterial blood flow is confirmed by feeling pulse in limb below band.

6. Check pulse by placing two fingers, not the thumb, over pressure point on wrist or ankle.

7. If swelling progresses up arm or leg, reapply constricting band ahead of swelling.

8. Use fire from match or piece of firewood to sterilize knife or razor blade.

9. Make single slice 1/4" long and 1/4" deep through each fang mark along line of muscle.

10. This should be done within 30 minutes of bite, the sooner the better.

11. Suction should be applied by vacuum pump suction device — by mouth if none available.

a. These are available in snake bite kits.

12. Never apply suction if open sores are in mouth.

    a. Snake venom can get into your system in this manner.

Nonpoisonous Snake Bites

1. Simply clean and bandage bite area.

Dangerous Spiders

Black Widow

1. The Black Widow (red hourglass on abdomen) has a dangerous bite.

2. Black widow females are ones that bite. Her bite is dangerous and causes:

    a. Rapidly developing intense pain.

    b. Lots of swelling.

    c. Severe abdominal cramps.

    d. Possible death.

BLACK WIDOW SPIDER

Symptoms of black widow Bite:

1. The symptoms of black widow bite are:

    a. Tremors and general weakness.

    b. Sweating and rash.

    c. Nausea and vomiting.

2. The abdominal cramps are so severe they are often mistaken for appendicitis or acute indigestion.

    a. The suffering begins to regress after several hours.

    b. It comes and goes for a couple of days and is then gone.

Treatment for Black Widow Bite

1. Keep bitten person as quiet as possible.

2. Place ice pack around the region of body where bite occurred.

    a. This will help keep venom from spreading.

3. Get bite victim to medical treatment facility without delay, if one's available.

Brown Recluse Spider (has dark brown violin on its back)

1. The bite of Brown Recluse initially causes little or no pain.

    a. Victim is often unaware of spider bite.

Symptoms of Brown Recluse Spider Bite

1. Painful red area with mottled dark bluish center appears a few hours after bite.

2. Rash or small colored spots sometimes occur on victim's skin.

3. After a couple days, discolored area appears that doesn't turn white when firmly pushed with finger.

    a. Area turns darker and shrivels in couple of weeks.

    b. Healthy skin around bite area separates from dead part.

    c. Whole center of wound becomes hard black scab.

    d. Scab soon falls off leaving open ulcer.

    e. Bite victim may still be unaware of what caused ulcer.

    f. Ulcer doesn't heal and grows larger.

4. Serious side effects are experienced so long as ulcer is present. These include:

    a. Fever and chills.

    b. Pain in joints.

    c. Vomiting.

    d. Generalized rash.

    e. Enlargement of spleen.

Treatment for Brown Recluse Spider Bite

1. There is no known antivenom for the bite of a brown recluse.

2. All hard, dead tissue and fibrous tissue beneath surface of skin must be cut away from bite area.

    a. This must be done before healing starts.

    b. If not cut out, the sore will continue to grow until it's several inches in diameter.

Scorpions

1. Scorpion stings are seldom fatal although they are exceedingly painful.

    a. Only stings from larger South American species are highly dangerous.

2. They cause death for stung person if proper treatment isn't promptly administered.

Symptoms of Scorpion Bite

1. Pain, swelling, burning sensation, drooling and discoloration around area of sting.

2. Difficulty breathing, body spasms, prickly sensation around mouth and tongue feels thick.

3. Double vision, blindness, involuntary defecation and urination, hypertension and heart failure.

Treatment of Scorpion Bite

1. If stung by a scorpion, quickly do the following to decrease amount of venom absorbed by body:

    a. Put light constricting band above sting area.

2. Leave for five minutes and then remove.

3. Put any of these directly on sting site as soon as possible to help relieve horrendous pain:

4. A cold compress, piece of ice or mud pack on sting for at least two hours.

5. Coconut meat, if available, can be locally applied to bite area.

6. Mix baking soda and water into thick paste and apply to bite.

7. Do not give bite victim morphine or any morphine derivatives, including Demerol:

8. These actually aid rather than inhibit the effect of scorpion venom.

9. Get bite victim to medical treatment facility without delay if, if one is available:

10. If scorpion sting is on face, neck or genitals.

11. If scorpion is one of the larger and more dangerous South American species.

17

Insect Stings and Bites

General Information

1. Of deaths each year resulting from bites:
    a. 40% are caused by insect bites.
    b. 33% are caused by snake bites.
    c. 18% are caused by spider bites.
    d. 9% are caused by animal bites.

Wasps, Bees, Hornets, Yellow Jackets and Ants.

1. Getting stung by swarm of ants, wasps, bees, hornets or yellow jackets could be fatal.

2. Such stings are seldom fatal unless bitten person is allergic to bite or is stung multiple times.

3. If attacked by these potentially dangerous entities, run into dense foliage.
    a. The twigs and leaves will help brush off the insects.

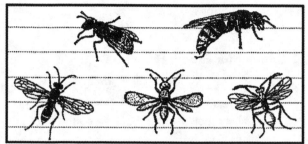

WASPS AND HORNETS

4. Most of this group sting their victims and then leave stinger and venom sac imbedded under skin.

5. Symptoms:
    a. A stinging, burning sensation on skin.
    b. Much swelling and pain.

6. Swelling caused by multiple stings around head, face and neck can be dangerous.
    a. It sometimes inhibits breathing.

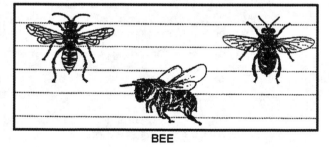

BEE

7. Treatment:
    a. Immediately scrape off barbed stinger with knife. This stops more venom from entering body.
    b. To help relieve sting discomfort:
    c. Rub ammonia over sting area.
    d. Make paste of baking soda and water. Put this on stings.

8. Use cold compresses or ice pack.

Flies, Ticks and Lice

Bot Flies

1. Bot flies are dangerous. The larvae bore into skin and cause painful boil-like swelling.

2. Moist tobacco applied over the bites will kill the larvae.

3. Once dead, larvae can be squeezed out of skin.

Ticks

1. There are two types of ticks:
    a. Wood or hard tick.
    b. Soft tick.

2. These flat, oval-shaped pests are carriers of:
    a. Tick relapsing fever.
    b. Tick typhus.
    c. Rocky Mountain Spotted Fever, an often fatal infection.
3. Symptoms of Rocky Mountain Spotted Fever:
    a. Rash, chills and fever.
    b. Severe pain in arms and legs.
4. Never crush a tick on skin and never pull one off you once it's head is imbedded.
5. A tick sometimes frees itself and is easy to remove if you coat it with such things as:
    a. Alcohol or iodine.
    b. Gasoline or kerosene.
    c. Plain spit or tobacco juice.
    d. The tip of a lighted cigarette held close to a tick's body will cause tick to withdraw head.
6. If tick's head remains imbedded in skin:
    a. Hold tip of knife blade over flame. Then touch tick with hot point.
    b. The tick's imbedded head will withdraw.

Lice

1. Try not to scratch lice bites.
    a. This spreads louse feces into the bites.
    b. Serious infections will result.
2. Clothing can be boiled or hung in direct sunlight for a few hours to rid it of lice.
    a. Be careful to expose seams.
3. Clothing can be placed on ant bed. The ants will quickly eat all the lice.
4. Frequently inspect hairy parts of body for lice. When found:
    a. Spend time in the sun while clothing is being deloused.
    b. Bathe thoroughly with soap and water.
    c. If soap unavailable, wash down with sand or silt from bottom of stream.

Caterpillars, Centipedes and Millipedes

Caterpillars

CATERPILLAR

1. Many caterpillars are covered with hollow venom-filled hairs. If these hairs contact skin, they cause:
    a. Severe burning pain.
    b. Redness and swelling.
2. Painful blistering of skin.

3. Irreversible tissue damage.
4. Treatment:
    a. Scotch tape over sting area is effective in removing broken off caterpillar hairs from skin.

b. Apply paste made of baking soda and water to sting area.

c. Household ammonia patted on sting area will greatly relieve discomfort.

d. Place cold compresses on sting area to also reduce discomfort.

5. Sideline reactions may be experienced such as:

a. Wheezing.

b. Trouble breathing.

c. Loss of consciousness.

Centipedes

1. Centipedes are poisonous with hollow fangs much like snake's fangs.

2. Centipedes rarely bite except when cornered and unable to escape.

3. Person bitten by centipede will experience:

a. Immediate and severe pain.

b. Redness and swelling.

c. Ulcer sometimes appears on skin.

4. Treatment of centipede bite:

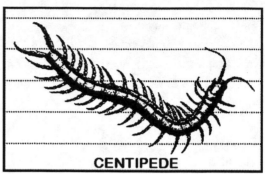

CENTIPEDE

a. Apply baking soda and water paste to sting area.

b. Wipe household ammonia on sting area.

c. Apply cold compresses on bite area to reduce pain and swelling.

Millipedes

1. Millipedes secrete a toxin from their glands.

2. When this fluid touches skin, it produces a:

a. Burning sensation.

b. Terrible itching.

3. Treatment for millipede toxin:

a. Cover affected area of skin with a paste made of baking soda and water.

b. Cold compresses will help relieve burning and itching.

c. Dab household ammonia on affected area.

18

Miscellaneous Medical Conditions

Abdominal Pain

1. Symptoms:

    a. Pain felt in lower right or upper part of abdomen.

    b. May feel nauseated.

    c. May vomit.

    d. Fever not usually initially present.

2. Treatment:

    a. No food or water.

    b. No laxatives!

    c. Don't use heating pad on abdomen.

    d. Ice near appendix area may give some relief from pain.

    e. Get person to medical treatment facility, if available.

3. Symptoms:

    a. Bulge shows part way up groin from crotch.

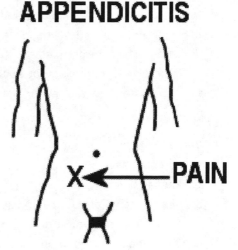

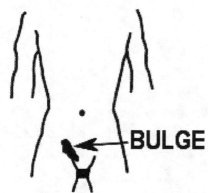

b. Bulge may also appear:

    1. In scar of previous abdominal surgery.

    2. Just below groin at crotch.

    3. At navel.

c. First weeks pain is worst. Discomfort lessens in time.

4. Treatment:

    a. Avoid lifting or other vigorous activities.

    b. Never press on bulge!

    c. If bulge disappears when lying down, still avoid lifting and other vigorous activities.

5. If bulge doesn't disappear when lying down, get person to medical facility if possible.

Beriberi

1. Symptoms:

    a. Progressive muscle weakness (atrophy).

    b. Twitching legs and cramping muscles.

    c. Loss of appetite and paralysis.

2. Treatment:

    a. Beriberi is caused by lack of vitamin B so cure is simple.

    b. Eat more green foods. Eat less rice and white flour products.

    c. Drink lots of tea made by boiling outer layer of bark for five minutes.

Blisters

1. What not to do:

    a. Blisters are sterile until opened. Don't break unnecessarily.

2. Treatment if blister has to be broken:

    a. Wash hands thoroughly with soap and water.

    b. Apply antiseptic on skin and blister.

    c. Sterilize needle by holding in the flame of match or other fire source.

    d. Puncture blister with needle.

    e. Remove fluid from blister by applying pressure with sterile gauze pad.

    f. Cover blister area with sterile dressing.

Boils

1. What not to do:

    a. Never squeeze boil until it pops as it may drive bacteria into blood stream.

    b. Popping boil may also cause internal abscesses or bone infection.

    c. Popping boil is especially unwise if boil is close to nostrils, upper lip or eyes.

        1. Here blood stream leads to brain area.

    d. Never apply warm compresses to boils on face except under medical direction.

2. Treatment:

    a. Relieve discomfort of small boils with warm compresses wet in Epsom Salt solution.

    b. Wet the compresses in mixture of one teaspoon Epsom Salt to one pint warm water.

    c. Change compresses every 15 minutes.

    d. Wipe pus away with sterile alcohol soaked pad if boil breaks.

    e. Always work pad from healthy skin toward boil and its pus.

    f. Lastly, apply sterile dressing over boil.

Carbon Monoxide Poisoning

1. Carbon monoxide poisoning can be severe, long lasting and sometimes fatal.

2. Carbon monoxide poisoning is result of inhaling carbon monoxide — a tasteless, colorless and practically odorless gas.

3. Carbon monoxide stops red blood cells from carrying needed oxygen to the body tissues.

4. Symptoms:

    a. Throbbing in temples, headaches.

    b. Noises in ears, dizziness.

    c. Skin and lips often turn bright red.

    d. Vomiting and convulsions may occur, followed by unconsciousness and death.

5. Treatment:

    a. Move person outside into fresh air.

    b. Give artificial respiration (*see Chapter 4*).

    c. Keep person quiet and transport to medical treatment facility, if there is one.

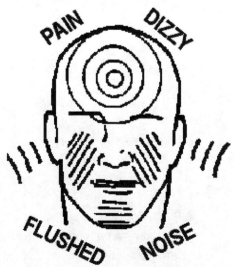

6. A warning:

    a. Any muscular exertion brings on effects of carbon monoxide poisoning more rapidly.

    b. Temperature and humidity extremes also bring on effects more rapidly.

Choking

Trachea Obstructed by Food

  1. Symptoms:

    a. Can't breath.

    b. Can't speak.

    c. Person turns cyanotic (blue).

    d. Person collapses.

  2. Warning:

  3. Person has only four minutes to live if throat obstruction isn't relieved.

  4. Treatment if person is sitting or standing:

    a. Stand behind choking person and wrap arms around his waist.

    b. Make fist with one hand.

    c. Place fist against person's abdomen below rib cage and just above navel.

    d. Grasp wrist of fist with other hand.

    e. Press forcefully into person's abdomen with quick upward thrust.

        1. This causes sharp exhaling of air.

        2. The lodged food may be blown out of airway.

  5. Repeat several times if necessary.

  6. Treatment if person is lying on back:

    a. Face choking person while kneeling astride hips.

    b. Place one hand on top of other.

    c. Place heel of bottom hand on person's abdomen.

    d. This should be placed just below rib cage and slightly above navel.

    e. Press forcefully with heel of hand into person's abdomen with quick upward thrust.

  7. Repeat several times, if necessary.

8. Treatment for choking child:

    a. Hold child over arm or leg with head hanging down.

    b. Give several sharp hand slaps between shoulder blades.

    c. Clear child's throat with index finger.

9. If breathing difficulty continues, start artificial respiration (*see Chapter 4*).

10. Treatment for choking baby:

    a. Hold baby bottom up with head hanging straight down.

    b. Open baby's mouth and pull tongue forward.

    c. If object doesn't come out, reach in back of throat with index finger.

    d. If object still doesn't dislodge, slap baby sharply between shoulder blades.

    e. If object remains lodged in baby's throat, begin artificial respiration (*see Chapter 4*).

Constipation

1. Treatment:

    a. Constipation can be prevented by:

        1. Eating fruit regularly.

2. Drinking lots of water.

3. Frequent exercise.

Diarrhea or Dysentery

1. Treatment:

    a. Diarrhea or dysentery can be avoided or cleared up, whichever the case may be by:

        1. Drinking tea leaves in water.

2. Drinking lots of water regularly.

3. Going on liquid diet for a time.

4. A good regulator to eat can be made from blend of water, chalk and charcoal or charred bone.

Electric Shock

1. Electric shock usually occurs when person accidently touches "live" wire or is struck by lightning.

2. Simply turn off switch if person is still in contact with live wire.

3. Never touch wire or person with bare hands or expect to also be shocked.

   a. Immediately get person away from live wire by using one of more of following:

   b. A dry wooden pole.

   c. Dry clothing.

   d. Dry rope.

   e. Other items that won't conduct electricity.

4. Treatment:

   a. Start artificial respiration (*see Chapter 4*) as shock causes breathing to stop.

   b. Check pulse as shock may cause heart to stop beating.

   c. If no pulse, give closed-chest heart massage (*see Chapter 5*) with artificial respiration.

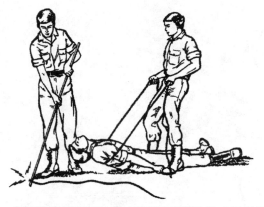

REMOVE PERSON FROM ELECTRICAL SOURCE USING NONCONDUCTIVE MATERIAL

Something Gets in Eye

1. What not to do!

   a. Never rub eye if something gets in it!

2. Treatment:

   a. If particle is under lower lid, gently remove with moistened corner of clean cloth.

   b. If particle beneath upper eyelid, carefully grasp eyelashes of upper lid.

   c. Pull lid up and away from eyeball surface.

3. Hold eyelid in this manner until tears freely flow, flushing out particle.

4. Further treatment:

a. If above technique fails, try to remove particle as shown:

b. Grasp eyelashes between thumb and forefinger.

c. Place wooden match or twig over lid and pull lid over this.

d. Have person look down while you examine inside of lid.

e. Gently remove particle with moist corner of clean cloth.

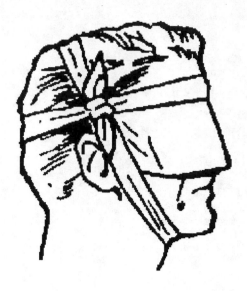

5. If foreign particle is glass or metal, or can't be removed by techniques given above:

   a. Bandage both eyes even if only one is injured.

   b. This prevents movement of uninjured eye which would cause identical movement and further damage to injured eye.

   c. Take to medical treatment center if possible.

6. If battery acid, ammonia or other caustic material gets in eyes:

   a. Flush immediately with lots of water.

   b. To flush right eye, turn head to right side. To flush left eye, turn head to left.

      1. This stops caustic material from washing into other eye.

7. Get person to medical treatment facility, if one is available, to prevent further eye damage.

Foreign Body in Ear or Nose

1. Never try to remove foreign body from ear by probing with stick or other instrument.

2. Water should be used to flush foreign object out of ear.

   a. Never try flushing if object in ear will swell (a seed or wood particle) when wet.

3. An insect flying into ear can be removed by:

   a. Pouring water into ear and drown or immobilize the insect.

   b. Holding flashlight to ear and attracting insect to its light.

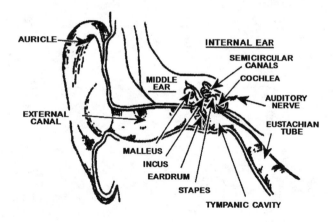

4. Don't probe with stick or other instrument for foreign object in nose.

   a. Probing sometimes jams object tighter up in nose.

   b. Damage to nasal passages may result.

   c. Simply have person gently blow nose to remove object.

Protein Deficiency

1. Associated with diets high in corn and containing little meat, milk, fish or other protein sources.

2. Symptoms:

   a. Appetite loss and muscle waste.

   b. Fluid retention and irritability.

   c. Vomiting and diarrhea.

   d. Loss of weight and strength.

   e. Sore and red swollen tongue.

   f. Severe rash resembling sunburn.

3. Treatment:

a. Eating lots of nuts and grains.

b. Eating lots of meat and eggs or insects.

Scurvy

1. Caused by a serious lack of vitamin C.

2. Symptoms:

    a. Cuts and wounds don't heal as they should.

    b. Gums bleed and teeth loosen and fall out.

    c. Swelling of joints.

    d. Healed wounds reopen.

3. Treatment:

    a. Eat plenty of raw salad greens.

    b. Eat lots of fruit.

    c. Boil evergreen (pine) needles in water for five minutes. Drink the liquid (tea).

Skin Eruptions from Poisonous Plants

1. The sap (juice) from certain plants causes skin eruptions (rashes).

    a. The danger isn't usually a serious hazard.

    b. But under certain conditions, sap from these plants can be extremely dangerous.

    c. Sweating and overheating greatly increases danger of contamination by poisonous plants.

2. There are two general types of these plants:

    a. Plants poisonous to touch.

    b. Plants poisonous to eat.

3. There are three common poisonous plants:

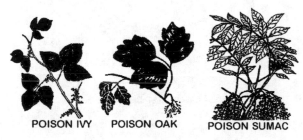

POISON IVY   POISON OAK   POISON SUMAC

    a. Poison ivy, poison oak and poison sumac.

    b. All have small, round, white or greenish-gray fruit.

4. Symptoms of plant poisoning:

    a. Redness on skin and swelling.

    b. Itching and severe burning.

    c. Blistering.

5. Treatment:

    a. After having touched plant, thoroughly wash area with strong soap.

    b. Should skin eruption develop several days after exposure:

        1. Avoid scratching.

        2. Smear paste of wood ashes and water over each infected area of body.

Skin Protection

1. Treatment:

    a. In cold weather, use suntan oil at least once weekly on exposed skin.

2. In hot weather, use suntan oil daily on exposed skin.

19

Pointers on Preventative Medicine

General Pointers

1. There are many ways of reducing the number of illnesses and their causes. Included are:

    a. Proper hygiene.

    b. Insect and rodent control.

    c. Waste disposal.

    d. Care in preparation of food and drink.

Waste Disposal

1. The term "waste" includes all kinds of refuse resulting from living activities of humans.

    a. Human wastes: urine and feces.

    b. Liquid wastes: laundry, bath and kitchen.

    c. Garbage and rubbish.

2. The methods used to dispose of wastes depends upon the situation and location.

    a. Burial and burning are most commonly used while hiking and camping in wilds.

3. Large quantities of liquid and solid waste are generated daily when in wilderness.

    a. Waste must be removed or otherwise handled, promptly and thoroughly.

    b. If neglected, the camp quickly becomes breeding ground for flies, rats and other vermin.

    c. Diseases resulting from filth could become prevalent. They include:

        1. Typhoid

        2. Cholera

4. Paratyphoid

5. Plague

Improvised Waste Facilities

1. Cat hole latrine:

    a. A hole in ground six to twelve inches deep and covered over after being used.

    b. Primarily used when on move.

2. Used for short term stops (one day).

3. Tree branch seat latrine:

    a. The above latrine is simple yet effective.

    b. Set up downwind of camp site if planning to stay awhile.

    c. One is enough for quite a few people.

    d. Can be made windproof with:

        1. Ponchos.

        2. Log walls.

        3. Tree branches.

        4. Snow blocks during winter months.

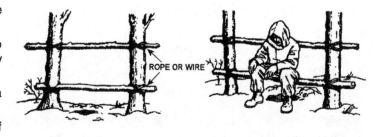

ROPE OR WIRE

4. Burn-out latrine (use in rocky or frozen soil):

    a. Construct using 55 gallon drum.

    b. Bury half the drum or cut drum in half.

    c. Cover with fly proof wooden seat.

    d. Burn out daily or when half full.

    e. Burn until only ash remains and bury ash.

    f. Use two of these latrines in unison -- one to use while other is burning.

BURN OUT LATRINE

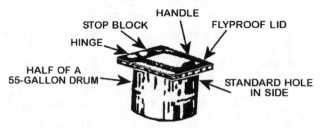

HANDLE

STOP BLOCK · FLYPROOF LID

HINGE

HALF OF A 55-GALLON DRUM →

STANDARD HOLE IN SIDE

URINE SOAKAGE PIT

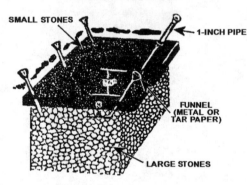

SMALL STONES

← 1-INCH PIPE

FUNNEL (METAL OR TAR PAPER)

LARGE STONES

5. Urine soaking pit:

    a. Dig a hole in ground four feet square by three feet deep.

    b. Place one inch pipes, 36 inches long, at each corner.

        1. Sink eight inches into bottom of pit.

    c. Fill pit with rocks, bricks, etc.

6. Put a funnel on top of each pipe.

7. Trough Urinal

    a. Build urine trough when wood is more readily available than pipe.

    b. Construct v-shaped trough out of tar paper lined wood or sheet metal.

    c. Put splash board down center of trough.

8. Drain urine trough into soakage pit.

TROUGH URINAL

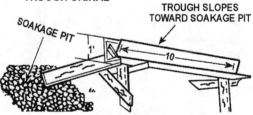

TROUGH SLOPES TOWARD SOAKAGE PIT

SOAKAGE PIT

1'

10

Cleanliness

1. Keeping body clean is essential to good health.

    a. Shower often to avoid skin diseases.

    b. Wash hands:

        1. After using latrine.

        2. Before eating.

2. If unable to shower or bathe:

    a. The face, armpits, hands and crotch should be washed daily.

    b. If soap unavailable, substitute white ashes, sand or loamy soil.

3. If water isn't available, bathe by rubbing down with cornstarch.

    a. Cornstarch removes perspiration and excess oil from skin.

    b. Desitin, A & D Ointment or cornstarch in groin and between buttocks helps control rash.

4. Carefully check body on regular basis for ticks, fleas, lice, etc.

   a. Pick any off body and crush.

   b. Include eggs if possible.

5. Wash and fumigate clothing and equipment with smoke.

   a. This gets rid of ticks and fleas, etc.

6. Bathing with strong soap will get rid of lice infestation, chiggers, etc.

Mouth and Teeth

1. Brush teeth at least once every day.

   a. If toothbrush unavailable, rub teeth with cloth wound around finger.

2. Chew a green twig until end is frayed. It's a decent substitute for a toothbrush.

3. If toothpaste unavailable, brush teeth with:

   a. Soap.

   b. Table salt.

   c. Baking soda.

4. Rubbing teeth with clean finger is another way.

   a. Gums should be stimulated in same manner.

5. Gargle with salt water solution.

   a. Helps clean teeth and toughen gums.

Caring for Feet

1. Keep feet clean.

   a. Wash and dry daily.

   b. Check for blisters and sores.

      1. Tape or Band Aids will help prevent further problems.

2. Massage and powder feet twice daily if possible.

3. Change socks daily or when wet.

4. Keep toe nails short with squared off cut.

5. Avoid infestations of worms and intestinal parasites by never going barefoot.

Hair Care

1. Keep face shaven and hair trimmed.

   a. Helps prevent parasites.

   b. Helps stop bacteria growth.

Clothing Care

1. Launder clothing when it needs cleaning.

   a. Underclothing and socks are especially important to keep clean.

2. If laundering impossible:

   a. Shake out clothing and hang in sun daily.

   b. This destroys mildew and bacteria.

Cold Weather Clothing Wear

1. Wear clothing in layers and loose to permit ventilation.

2. Maximum protection from cold is given by:

   a. Loose layers of clothing with air-space between them.

   b. An outer wind and water resistant garment.

3. Be sure to take off loose inner layers:

    a. During strenuous physical activity.

    b. To prevent overheating.

    c. To prevent accumulation of perspiration.

4. **Note:** wet clothing loses insulating value.

5. Wear shirttail tucked in pants and sleeves down and buttoned to:

    a. Provide protection against insects such as ticks.

    b. Lessen exposure to mosquitoes, sand flies and other disease carriers.

A Summary of Rules to Avoid Illness

1. All water obtained from natural sources should be purified before drinking. Here's three methods:

    a. Boil water for at least 15 minutes.

    b. Add water purification tablets.

    c. Add four drops of ordinary household laundry bleach to each quart of water.

2. Always wash hands before fixing food.

    a. Also, clean under fingernails.

3. Clean and sterilize cooking and eating utensils after each use.

    a. Use boiling water.

4. Clean mouth and teeth at least once daily.

5. Insect bites can be prevented by proper use of:

    a. Insect repellant.

    b. Equipment (head nets, etc.).

    c. Clothing.

6. Kerosene rubbed on neck, wrists and legs at shoe tops prevents infestation from chiggers, mites, etc.

7. Get out of and dry wet clothing as soon as practical.

20

Unusual Medical Treatments That Really Work

General Pointers

1. Medical treatment was often (and still is) undertaken by people who were (are) either:

   a. Far removed from a doctor.

   b. Too poor to consider seeking out one in the first place.

2. Medical expertise grew out of necessity. Their storehouse of medical knowledge came from a need to try things out, to do something!

   a. Things were done because they had to be done.

   b. Risks were taken because there was no other choice.

3. The lack of availability of modern medicines didn't necessarily rule out decent medical treatment.

4. Many people had their own family hand-me-down cures and treatment. Many people still have them today.

   a. Some work nicely while others don't!

   b. Those presented here do work!

Some Primitive Treatments

1. Bleeding:

   a. Ordinary household cayenne (red) pepper sprinkled on a cut or wound can be used to:

      1. Stop the bleeding.

      2. Help cut or wound heal faster.

2. Burns:

   a. Boil dressings/rags for 10 minutes.

   b. Soak dressings/clean rags in tannic acid.

      1. Tannic acid is obtained by boiling bark of hardwood trees for two or more hours.

      2. Also, found in strong tea solution.

   c. Let dressings cool and then lay over burns.

   d. Tannic acid treatment will:

      1. Provide pain relief.

      2. Protect burns against infection.

      3. Speed healing process.

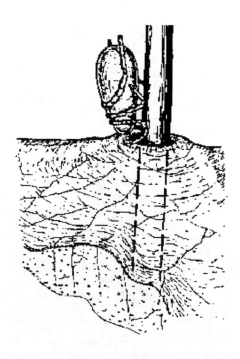

Chiggers:

1. Take pine or other tree sap and smear over red spots on skin.

2. This cuts off chigger's air supply and it dies.

Diarrhea I:

1. Grind or pulverize chalk, charcoal or dried bones until they're powder.
2. Stir handful of powder in cup of water and drink every two hours until diarrhea stops.
3. Add equal portion of apple pulp or pulverized rinds of citrus fruit.
    a. This makes mixture more effective.

Diarrhea II:

1. Tannic acid, found in tea, can help control diarrhea.
2. Prepare pot of strong tea.
3. Drink cup every two hours.
    a. Diarrhea should stop or at least slow.
4. Tannic acid is also obtained by boiling bark of hardwood trees for two or more hours.
    a. Black brew tastes and smells vile, but is effective in stopping diarrhea.

Healing:

1. Garlic oil is good for rubbing on wounds, cuts, scratches, etc.
    a. Promotes faster healing.
    b. Kills bacteria.

Intestinal Parasites and Worms

1. The following home remedies are effective in:
    a. Getting rid of parasites and worms.
    b. Or controlling the degree of infestation.
2. These treatments are not without danger.
    a. They work on principle of changing environment of gastrointestinal tract.
3. Cigarettes:
    a. Eat one to one and one-half cigarettes.
    b. Nicotine kills worms or stuns long enough for them to be passed.
    c. If infestation is severe, treatment can be repeated in 24 to 48 hours, but no sooner.
4. Hot peppers:
    a. Cook hot peppers in soups, meat dishes, rice, etc.
    b. Eat raw hot peppers often.
    c. Hot peppers must be regular part of diet for treatment to be effective.
5. Kerosene:
    a. Swallow two tablespoons kerosene.
        1. Never take more than two tablespoons.
    b. Repeat treatment in 24 to 48 hours.
        1. Never repeat treatment sooner!

Skin Infections

1. The rule of thumb for all skin diseases: "If it's wet, dry it and if it's dry, wet it."
2. Fungal infections:
    a. Keep area clean and dry.
    b. Expose to sunlight as much as possible.
3. Heat rash:
    a. Keep area clean, dry and cool.

b. Put powder on affected area, if available.

Sore Throat

1. Gargle with warm salt water.
2. If tongue is coated, scrape with:
   a. a toothbrush.
   b. a clean stick.
   c. a clean fingernail.
3. Gargle with warm salt water after scraping.

Stings

1. Bee, wasp, hornet, spider, scorpion and centipede bites can be treated in following manner:
   a. Remove stinger.
   b. Cover sting with one of following:
      1. Baking soda and water paste.
      2. Ice or cold compress.
      3. Mud pack.
      4. Coconut meat.
2. Mosquito bites:
   a. Rub antiperspirant on mosquito bites and any other insect bites.
      1. Itching should stop immediately.
   b. Ordinary deodorant won't work. It must be "antiperspirant" kind.

Ticks and Leeches

1. A tick or a leech will usually drop off if:
   a. Covered with wad of moist tobacco.
   b. Coated with grease or oil.
   c. Touched with lit end of cigarette or head of lighted match.
2. Never try to pull off tick or leech.
   a. Head usually remains buried in skin and causes infection.

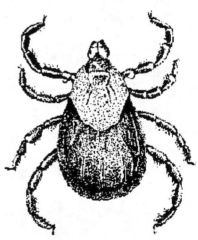

Wound Healing

1. Introducing maggots into wound can be hazardous for a number of reasons:
   a. Wound must be exposed to flies.
   b. Flies are likely to carry even more bacteria to wound.
   c. Maggots also eat healthy tissue once dead tissue has been eaten away.
   d. Maggot invasion of healthy tissue:
      1. Is extremely painful.
      2. May cause hemorrhaging.
      3. May be severe enough to be fatal.

2. Despite dangers, maggot therapy is sometimes a viable alternative:

    a. Where antibiotics are not available.

    b. When wound becomes severely infected and won't heal.

3. Treatment:

    a. Remove bandages and expose wound to flies.

    b. Flies are attracted to fetid odors coming from infected wound.

    c. Flies will land on wound and lay eggs.

    d. Maggots will then hatch from fly eggs.

    e. One exposure to flies is enough to produce more than enough maggots to cleanse wound.

    f. They will eat all infected waste.

    g. Once flies have deposited their eggs, cover wound with bandage.

    h. Remove bandage daily to check for maggots.

    i. If no maggots found within two days after exposure to flies:

        1. Remove bandage.

        2. Again expose wound to flies.

    j. If wound is teeming with maggots:

        1. Remove as many as possible with spoon.

        2. Flush maggots out with sterile water.

        3. Leave only 50 to 100 maggots in wound.

    k. Once maggots are well established:

        1. Again cover wound with bandage.

        2. Monitor maggot activity daily.

    l. Maggots produce froth and are sometimes difficult to see in wound.

        1. Sponge out frothy fluid until all maggots can be seen.

        2. Try to not remove maggots with froth.

    m. All maggots should be removed:

        1. Once they eat all infected tissue.

        2. Before they eat healthy tissue.

    n. Signs that maggots have started eating healthy tissue:

        1. Pain increases in severity at wound.

        2. Bright red blood seen in wound.

    o. Getting rid of maggots:

        1. Flush from wound with sterile water.

        2. Bandage wound when free of maggots.

        3. Check wound every three to four hours for several days to assure maggots are gone.

        4. Pick out remaining few, if any, with sterile tweezers.

4. Once maggots are completely gone:

        1. Bandage wound and treat as any other.

        2. Healing of wound should now be normal.

APPENDIX I

THE ANATOMY — THINGS YOU MUST KNOW

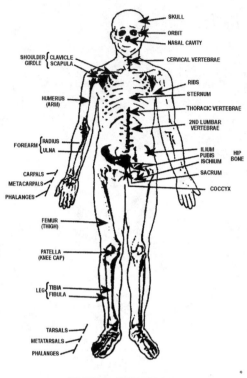

HUMAN SKELETON

SKULL
ORBIT
NASAL CAVITY
SHOULDER GIRDLE — CLAVICLE / SCAPULA
CERVICAL VERTEBRAE
RIBS
STERNUM
HUMERUS (ARM)
THORACIC VERTEBRAE
2ND LUMBAR VERTEBRAE
FOREARM — RADIUS / ULNA
ILIUM / PUBIS / ISCHIUM — HIP BONE
SACRUM
CARPALS
METACARPALS
PHALANGES
COCCYX
FEMUR (THIGH)
PATELLA (KNEE CAP)
LEG — TIBIA / FIBULA
TARSALS
METATARSALS
PHALANGES

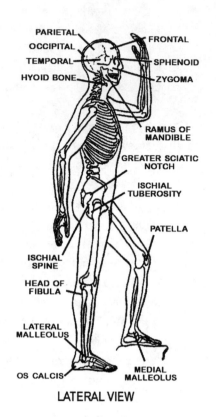

LATERAL VIEW

PARIETAL
OCCIPITAL
TEMPORAL
HYOID BONE
FRONTAL
SPHENOID
ZYGOMA
RAMUS OF MANDIBLE
GREATER SCIATIC NOTCH
ISCHIAL TUBEROSITY
PATELLA
ISCHIAL SPINE
HEAD OF FIBULA
LATERAL MALLEOLUS
OS CALCIS
MEDIAL MALLEOLUS

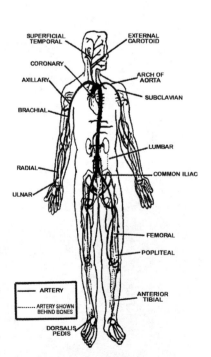

LARGE ARTERIES OF THE SYSTEMIC CIRCULATION

SUPERFICIAL TEMPORAL
EXTERNAL CAROTOID
CORONARY
ARCH OF AORTA
AXILLARY
SUBCLAVIAN
BRACHIAL
LUMBAR
RADIAL
COMMON ILIAC
ULNAR
FEMORAL
POPLITEAL
ANTERIOR TIBIAL
DORSALIS PEDIS

ARTERY
ARTERY SHOWN BEHIND BONES

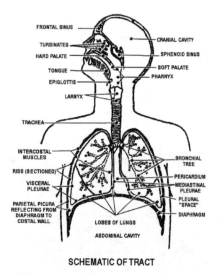

SCHEMATIC OF TRACT

FRONTAL SINUS
TURBINATES
HARD PALATE
TONGUE
EPIGLOTTIS
LARNYX
TRACHEA
CRANIAL CAVITY
SPHENOID SINUS
SOFT PALATE
PHARNYX
INTERCOSTAL MUSCLES
BRONCHIAL TREE
RIBS (SECTIONED)
PERICARDIUM
VISCERAL PLEURAE
MEDIASTINAL PLEURAE
PARIETAL PICURA REFLECTING FROM DIAPHRAGM TO COSTAL WALL
PLEURAL "SPACE"
DIAPHRAGM
LOBES OF LUNGS
ABDOMINAL CAVITY

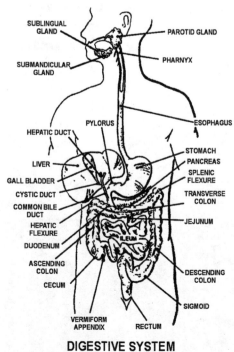

DIGESTIVE SYSTEM

SUBLINGUAL GLAND
PAROTID GLAND
SUBMANDICULAR GLAND
PHARNYX
HEPATIC DUCT
PYLORUS
ESOPHAGUS
LIVER
STOMACH
GALL BLADDER
PANCREAS
CYSTIC DUCT
SPLENIC FLEXURE
COMMON BILE DUCT
TRANSVERSE COLON
HEPATIC FLEXURE
DUODENUM
JEJUNUM
ILEUM
ASCENDING COLON
DESCENDING COLON
CECUM
SIGMOID
VERMIFORM APPENDIX
RECTUM

APPENDIX II

PACKING AN EMERGENCY MEDICAL BAG

CHECK LIST

Bandages and Dressings:

Adhesive pads (for larger wounds).

Adhesive tape roll -- 1/2" or 1" wide (to hold dressings securely).

Band Aid strips and spots -- assorted sizes (for minor wounds, cuts and scrapes).

Butterfly bandages, medium.

Cotton pads, balls and swabs.

Dressing, large or large sanitary napkins.

Elastic bandage ("Ace" type roll) 3", 5", 6".

Eye dressing (prescription antibiotic such as Gentamycin).

Eye pads.

Eye patch (adjustable pressure).

Gauze rolls — 1", 2", 3" (to wrap joints and other difficult to bandage areas).

Gauze squares, sterile (3" x 3" and 4" x 4" size).

Gauze, non-adherent, vaseline or Telfa type. Clings and conforms to shape of wound (for draining wounds, burns, infections and bleeding).

Muslin or triangular bandage (to bind splints, for arm slings, for dressing head wounds and for diapers).

Medicinal Aids

Alcohol preps or wipes (for cleaning cuts, scrapes and minor wounds).

Ammonia inhalant ampules.

Analgesic cream (pain relief for sore joints and muscles).

Antibacterial ointment.

Antibiotic ointment wipes (Betadine).

Antibiotic tablets (prescription item for infections, pneumonia, etc.).

Antacid gas tablets (such as Tums).

Antidiarrheal medication (such as Pepto-Bismol or Immodium or generic equivalent).

Antihistamine (such as Benadryl or generic equivalent).

Aspirin or acetaminophen (Tylenol) (to help reduce fever or pain).

Burn ointment or lotion (for pain from sunburn, insect bites, poison ivy, scratches, etc.).

Cayenne pepper (to stop bleeding).

Dry mustard or ipecac syrup (to induce vomiting).

Epsom salts (to reduce swelling).

Hydrogen peroxide.

Hydrocortisone cream, 1% type (to relieve itching).

Pain medication (prescription item for severe pain, burn, trauma, etc.).

Sodium chloride (to make sterile saline solution for eye wash, eye drops, wound irrigation).

Decongestant (such as Sudafed).

Miscellaneous Supplies

Baking soda (to make paste and spread on insect bites, poison ivy rash, etc.).

Butterfly closures (to hold edges of wound securely together).

Eye drops (such as Visine).

Instant cold pack (to bring down swelling and to reduce pain).

Instant hot pack.

Lip salve (Blistex or Chapstick).

Medical instrument set.

Mild soap (in tube).

Mineral oil (to help remove ticks lodged in ear, cracked skin).

Needles (medium size).

Petroleum jelly in tube (or other lubricant).

Pre-moistened towelettes.

Razor blade (single edge).

Safety pins (assorted sizes).

Salt tablets (for muscle cramps or when perspiring heavily).

Scissors (small pair).

Snake bite kit.

Surgical kit.

Survival blanket.

Thermometer (for temperature taking).

Tissues (such as Kleenex, etc.).

Tongue blades and wooden applicator sticks.

Other Items

Contents list (for medical emergency bag).

Instruction card — artificial respiration, mouth-to-mouth resuscitation.

Instruction card — first aid instruction, choking, drowning, etc.

SECTION 3

THE

OFFICIAL

URBAN & WILDERNESS

EDIBLE PLANT

SURVIVAL MANUAL

CONTENTS

Mallow, Common
Marsh-Marigold (Cowslip)
Mustard, Wild
Nettle, Stinging
Nut Grass
Onion, Wild
Parsnip, Wild
Pennycress, Field
Peppergrass
Pickerelweed
Plantain, Broadleaf
Pokeweed (Poke)
Potato Vine, Wild
Prickly Pear
Purslane
Queen Anne's Lace (Wild Carrot)
Redbud Tree
Rice, Wild
Rose, Wild
Sheep Sorrel (Common Sorrel)
Shepherd's Purse
Solomon's Seal
Spring Beauty
Spiny-Leaved Sowthistle
Sunflower
Sweet Flag (Calamus)
Thistle, Bull
Violet, Sweet
Wappato (Duck Potato)
Water Chestnut
Watercress
Water Lily, Fragrant
Water Plantain
Yellow Dock (Curly Dock)

Bracken Fern
Edible Fern Parts in General
Ostrich Fern

American Beech
Aspen
Birch
Butternut
Chestnut
Cottonwood
Hickory
Maple
Pine
Slippery Elm
Sycamore
White Oak
Willow

PART I

INTRODUCTION TO EDIBLE PLANTS AND SURVIVAL

1

THERE'S NO NEED TO STARVE!

It doesn't take a genius to understand the one most important aspect of surviving. It's having access to drinkable water! Without water a person can't live more than three days.

The second most important thing is food! Men have been known to live more than a month without food. But there's absolutely no need for any person to be deprived of something to eat. Nature is and always has been a good and reliable provider. Everyone should know how to properly use her. Learn to live off the land. It really isn't that difficult.

2

COOKING EDIBLE WILD PLANTS

COOKING TERMS AND WHAT THEY MEAN

BOIL: Put leaves or other plant parts in water and bring to boil. Continue boiling until tender — usually 10 to 15 minutes or less. Boil in as little liquid as possible. Try to not overcook.

DEEP FRY: Drop plant parts in pot or kettle of very hot cooking oil sitting over a fire. This is how French fried potatoes and onion rings are cooked.

FRY: Lay batter coated or uncoated plant parts (sliced root, etc.) in greased frying pan and cook until browned.

SAUTE: Cook in hot, lightly greased frying pan, stirring until browned or tender.

SIMMER: Put leaves or other plant parts in near-boiling water and allow to simmer until tender.

STEAM: Put leaves, stems, or whatever you want to steam, in colander or strainer. Set over boiling water 5 to 10 minutes or until tender. Often the preferred way to cook greens.

STEEP: Put leaves or other plant parts in pot of boiling water. Take off stove and cover. Leave to steep for 10 to 20 minutes. When brewing tea, pour boiling water over crushed leaves in pot, cover, and let steep 5 to 15 minutes.

MAKING CANDY, FRITTERS, PIES AND MORE

CANDY MAKING:

1. Flowers and root pieces of various plants can be "candied" and eaten as a delicious sweet treat. The process is simple:

    a. Whip powdered sugar in egg whites.

    b. Dip flowers or root pieces into this meringue-like blend.

    or

    c. Stir flowers in egg whites and then roll in sugar.

    d. Deep fry in pot of hot cooking oil until slightly browned.

    e. Take out and set on paper to drain and cool.

    f. Sprinkle with sugar-spice mixture.

    or

    g. Take nut meats, roots, etc., and drop into pan of sugar syrup.*

    h. Simmer until plant parts or nut meats are nicely saturated.

    i. Take out of sugar syrup set aside to partially dry.

    j. Roll pieces around in granulated sugar until covered.

* NOTE: Make sugar syrup by blending 2 cups sugar for every cup of water in pot.

COFFEE SUBSTITUTE MAKING:

1. Wash roots or seeds thoroughly or remove nuts from shells.

2. Roast slowly until brittle and dark brown.

3. Pulverize or grind roasted root, seeds or nuts.

4. Put 1 to 2 teaspoons grounds in pot boiling water.

5. Let simmer a short time.

6. Strain before serving.

NOTE: With chicory, for example, use 1-½ teaspoon grounds with each cup water.

FLOUR MAKING:

1. Flour can be made from a variety of plant parts. It's obtainable from any edible plant by doing the following:

    a. Dry plant thoroughly over embers or in sun.

    b. Pulverize or grind plant to powder or meal.

    c. Sift out fiber.

    d. Store flour in air tight container.

2. Flour made from richly flavored cattail pollen is one of best known. Here's how it's done:

    a. Gather starch-filled rootstocks (late fall — early spring).

    b. Wash thoroughly and peel away outer covering.

    c. Crush starchy core in bucket of cold water.

    d. Strain out fibrous material.

    e. Let starch settle to bottom.

    f. Carefully pour off water.

    g. Refill bucket with cold water.

    h. Let starch settle.

    i. Pour off water.

    j. Starch can be used as flour at this point.

3. Flour made with acorns from other than white oak trees:

    a. Boil acorns 2 hours and pour off water.

    b. Soak acorns 12 hours in cold water.

    c. Change soak water twice daily for 4 days.

    d. Now pound acorns to paste.

    e. Spread paste thinly and allow to dry.

    f. Pound into flour.

FRITTER MAKING:

1. Some flowers are suitable for making tasty fritters. Included are dandelions, day lilies, elderberries and wisteria. Here's how:

    a. Make a mildly thick batter by blending flour with water.

    b. Dip fresh flowers in this batter.

    or

    c. Dip fresh flowers in egg whites.

    d. Roll dampened flower in flour or fine meal.

then:

    e. Drop individual flowers in pot of hot cooking oil.

    f. Deep fry until lightly browned and crisp.

    g. Eat while hot or at least while still warm.

JELLY MAKING:

1. Flowers contain no pectin, nor do some fruits, yet they can still be used to make jelly. Making jelly isn't difficult, but it is time consuming. Here's how it's done:

    a. Add quart water for every quart berries in large pot.

    b. Bring to boil and let simmer 15 minutes.

    c. Mash berries and let simmer another 10 minutes.

    d. Strain through cheesecloth several times.

2. Now prepare an equal amount of a high pectin fruit (crabapples, for example) in identical manner.

    a. When finished, combine the two strained juices.

    b. Sweeten to taste.

    c. Heat again until liquid begins to jell.

    d. Pour into sterilized jars and seal.

3. Tasty jam is made same way jelly is made except fruit isn't strained out as with jelly.

PANCAKE MAKING:

1. This recipe uses flour made from ground or pulverized plantain seeds. Other flour made from variety of plant parts can be substituted:

    Plantain flour .... 2 cups

    Sugar .... 3 tablespoons

    Baking powder .... 3 teaspoons

    Salt .... ½ teaspoon

    Milk .... 1 cup

    Cooking oil .... 3 tablespoons

    Eggs 2

2. Blend above ingredients. Add more flour or milk as needed to get proper consistency. Fruit, if available, can be stirred into batter. Cook on hot griddle or in lightly greased frying pan.

PICKLING:

1. To make liquid suitable for pickling, combine and bring to boil:

    Water .... 1 cup

    Vinegar .... 2 cups

    Then stir in:

    Sugar .... 1/2 cup

    Salt .... 1/4 cup

    Ground mustard .... 1 tablespoon

    Celery seed .... 1 tablespoon

2. Boil 10 to 15 minutes while constantly stirring.

PIE MAKING:

Blend following ingredients:

    Berries, grapes, etc. .... 4 cups

    Apples .... 2 cups

Then stir in:

      Maple syrup or brown sugar............4 tablespoons

      Cinnamon........................1 tablespoon

      Raw eggs 2

      Lemon juice....................1 tablespoon

Pour mixture in pie shell. Cover with top crust. If no shell or crust available, just pour mixture in pie pan. Bake at 325 for 20 to 30 minutes.

PUREE MAKING (thick soup):

1. Put 2 cups leaves (or other plant parts) in pot of water.

2. Bring to boil and cook to pulp.

3. Run through sieve.

4. Stir in 2 tablespoons butter or margarine.

5. Simmer long enough to melt butter.

6. Stir in 1/4 cup milk.

7. Season to taste with salt and pepper.

SYRUP MAKING WITH FLOWERS:

1. Put 6 cups flowers in bowl.

    a. Cover flowers with boiling water.

    b. Let seep overnight.

    c. Strain off flowers and throw away.

    d. Stir in 2 cups sugar and bring mixture to boil.

    e. Continue boiling until liquid starts thickening.

    f. Use syrup for pancakes or whatever. Pour what is left in covered jars and store in cool place.

2. Or make syrup with rose hips:

    a. Put quantity of rose hips in pot.

    b. Add boiling water to cover.

    c. Boil until hips become soft and liquid begins thickening.

SYRUP MAKING FROM TREE SAP:

1. Put quantity of sap into large cooking pot.

2. Set over fire and continuously stir.

3. Spoon off scum as it rises to top.

4. Don't allow sap to boil over or burn bottom of pot.

5. Bring to boil.

6. Let simmer until sap thickens and turns to clear amber syrup.

7. Continue simmering until teaspoon of syrup forms soft ball when held under cold water.

8. Remove from fire at this point.

SUGAR MAKING FROM TREE SAP:

1. Continue to simmer when sap begins to thicken and tastes sweet.

2. Cook until syrup starts crystallizing.

3. Syrup will eventually turn to sugar.

PART II

EDIBLE PLANTS — WHAT PARTS TO USE AND HOW

3

FRUITS AND BERRIES FOUND IN THE WILDS

Edible fruit is plentiful in nature and it supplies great food in a survival situation. You're no doubt already aware of many of the wild fruits and berries in the United States. However, to refresh your memory, all the following are readily available, easy to find and are meticulously covered in this chapter.

Blackberry	Grapes	Blueberry	Mulberry
Crabapple	Persimmon	Elderberry	Rhubarb
Serviceberry	Strawberry		

BLACKBERRY (BRAMBLE)

BERRIES:

1. Eat fresh berries alone or in milk for nourishing quick snack.

2. Stir fresh berries into pancake or muffin batter.

3. Cook berries to make jelly or jam (*Chapter 2*).

4. Set aside boiled berry juice to ferment for making vinegar.

5. Crush berries and cook until juice thickens. Use for pancake syrup or fruit sauce.

6. Stew berries to make dessert or pie (*Chapter 2*).

LEAVES:

1. Use uncooked leaves as a salad green.

2. Cook (boil or steam) as vegetable green.

3. Add liberally to soups and stews.

YOUNG SHOOTS:

1. Eat shoots raw as snack or add to green salad.

2. Peel shoots, boil 15 minutes and eat like asparagus.

FLOWERS:

1. Eat raw as a snack or add to green salad.

2. Use for making fritters (*Chapter 2*).

NOTE:

1. Berries dried thoroughly can be stored indefinitely.

2. Blackberries are extremely high in Vitamin C.

3. Huckleberries and red raspberries are prepared in the same manner for eating as blackberries.

BLUEBERRY

BERRIES:

1. Eat fresh berries alone or in milk for quick snack.

2. Stir fresh berries into pancake or muffin batter.

3. Boil berries to make jelly or jam (*Chapter 2*).

4. Set aside boiled berry juice to ferment for making vinegar.

5. Crush berries and cook until juice thickens. Use on pancakes and ice cream.

6. Stew berries to make wholesome dessert.

LEAVES:

1. Use raw leaves as a salad green.

2. Cook (boil or steam) as a vegetable green.

3. Add liberally to soups and stews.

SHOOTS:

1. Eat young shoots raw as snack or add to green salad.

2. Peel and boil young shoots 15 minutes and eat like asparagus.

FLOWERS:

1. Eat raw as a snack or add to salads.

2. Use to make fritters (*Chapter 2*).

NOTE:

1. Berries dried in sun or over embers can be stored indefinitely.

2. Blueberries are extremely high in vitamin C.

3. Best time to find blueberries is early summer through early autumn.

CRABAPPLES, WILD

1. Eat entire crabapple raw.

2. Boil until soft, sweeten to taste and eat as dessert.

3. Steam crabapple and bake in pies (*Chapter 2*).

4. Crush cooked or uncooked crabapple and eat as applesauce.

5. Make into crabapple jelly (*Chapter 2*).

6. Fry slices of crabapple (*Chapter 2*).

7. Saute crabapple slices (*Chapter 2*).

8. Use pieces of crabapple to make fritters (*Chapter 2*).

LEAVES:

1. Boil until tender and eat as vegetable green.

2. Steam until tender and eat as vegetable green.

3. Saute in lightly greased frying pan (*Chapter 2*).

NOTE:

1. Crabapples sliced and dried in sun or over embers can be stored indefinitely.

2. Crabapples are sour and hard unless tree ripened.

3. Save liquid used for cooking plant parts and drink or use for soups and stews.

COOKING POINTERS:

1. Crabapples are almost always tart. Stew with honey, sugar or maple syrup for best results.

2. Crabapples contain abundant pectin for making jelly and jam. Use with other fruits containing no pectin.

ELDERBERRY

1. Eat fresh ripe berries for quick snack. They're edible but smell bad and don't taste very appetizing.
2. Stir fresh ripe berries into pancake or muffin batter.
3. Cook berries to make jelly and jam (*Chapter 2*).
4. Set aside boiled berry juice to ferment for making vinegar.
5. Crush berries and cook until thickened. Use for pancake syrup or fruit sauce.
6. Mash fresh or cooked ripe berries for a nice drink.
7. Stew berries to make dessert or pie filling (*Chapter 2*).

FLOWERS:

1. Make fritters with elderberry flowers (*Chapter 2*).
2. Soak flower clusters in water with lemon slices to make tasty elderberry lemonade rich in vitamin C.

NOTE:

1. Berries dried in sun or over embers can be stored indefinitely. Drying uncooked berries takes away bad odor and taste.
2. Gather flower clusters in late spring for best fritters.
3. Elderberries have no pectin for making jelly and jam. Some other fruit (apples, etc.) must be used with them.
4. Berries high in vitamin C, A, potassium and calcium.
5. Vomiting, diarrhea and nausea are the consequence of eating unripe berries, leaves, roots and stem of this plant.
6. Save cooking liquid and drink or use for soups and stews.

GRAPES, WILD

GRAPES:

1. Eat ripe grapes uncooked as a snack.
2. Add ripe grapes to a salad along with the leaves.
3. Boil down to make jelly and jam (*Chapter 2*).
4. Use to make pie (*Chapter 2*).
5. Make raisins by drying grapes in sun for 3 days or more.
6. Steam or boil 10 to 15 minutes and eat as vegetable.
7. Saute shredded leaves until lightly browned.
8. Stuff steamed larger leaves with grapes and other fruit.

TENDRILS:

1. Eat raw tendrils for quick snack or use in green salad.

NOTE:

1. Best to get leaves when tenderest during early summer.
2. Grapes are loaded with pectin and energy-giving sugar.
3. Drinkable water can be extracted from a grapevine.
4. Save cooking liquid and drink or use for soup or stew.

COOKING POINTERS:

1. Take large leaves and lightly boil or steam to soften.
2. Stuff with meat, vegetables, rice, etc.
3. Roll tightly up and put in colander.
4. Put colander in pot and let sit just above water.
5. Let steam until leaves are limp eat while hot.

MULBERRY, RED

BERRIES:

1. Eat ripe berries alone or in milk for nourishing quick snack.
2. Stir ripe berries into pancake or muffin batter.
3. Stew berries to make jelly or jam (*Chapter 2*).
4. Set aside boiled juice to ferment for making vinegar.
5. Crush berries and cook until juice thickens. Use for pancake syrup or fruit sauce.
6. Mash cooked berries thoroughly and use as fruit drink.
7. Stew berries to make dessert or pie (*Chapter 2*).

LEAVES:

1. Use uncooked leaves as a salad green.
2. Cook (boil or steam) and use as vegetable green.
3. Add leaves liberally to soups and stems.

SHOOTS:

1. Steam young shoots and eat as snack.
2. Peel young shoots, boil, and eat like asparagus.

NOTE:

1. Berries dried in sun or over embers can be stored indefinitely.
2. Don't eat raw shoots or unripe berries as they are mildly hallucinogenic.
3. Pectin is needed to make jelly and jam.
4. Pick shoots in the spring when they are tenderest.
5. Save cooking liquid and drink or use for soups and stews.

PERSIMMON

FRUIT:

1. Eat the sweet, plum-like fresh fruit raw for snack.
2. Strain seeds from pulp to make jelly and jam (*Chapter 2*).
3. Bake strained fruit pulp to make tasty pudding.
4. Make pie with strained pulp (*Chapter 2*).
5. Add fruit pulp to nut bread and muffins, etc.
6. Mash fruit and use as nourishing drink.

NOTE:

1. Fruit ripens in early fall.
2. Fully ripened fruit tastes not unlike a date.
3. Fruit is extremely sour until after first frost.

4. Fruit richly sweet when soft with wrinkled skin.

5. Save cooking liquid and drink or use for soup or stew.

RHUBARB, WILD

STALKS:

1. Eat stalk pieces raw as a rather sour snack.

2. Cut stalks into tiny pieces and add to green salad.

3. Stew or steam chunks until tender, add sugar to suit and eat as delightful fruit dessert.

4. Boil stalks until soft and mushy, add sugar to suit and use for pie filling.

5. Boil down until liquid in pot thickens and is ready for making jelly and jam (*Chapter 2*).

6. Steam stalks until tender, add sugar to suit and eat as dessert or use for pie filling.

LEAVES:

1. Boil leaves 20 to 25 minutes and eat as you would spinach.

2. Steam leaves until tender and eat as spinach.

3. Add liberally to soups and stews.

NOTE:

1. BEWARE: Rhubarb leaves are toxic if eaten raw.

2. Save cooking liquid to drink or use for soup and stew.

COOKING POINTERS:

1. Change water 2 or 3 times when cooking to get rid of some of the sour taste.

2. When short of rhubarb, curly dock leaf stems are excellent substitution for pies, puddings and sauces.

SERVICEBERRY (JUNEBERRY)

BERRIES:

1. Eat fresh berries alone or with milk for nourishing snack.

2. Stir fresh berries into pancake or muffin batter.

3. Cook berries to make jelly or jam (*Chapter 2*).

4. Set aside boiled juice to ferment for making vinegar.

5. Crush, cook until juice thickens and use as pancake syrup.

6. Mash fresh berries thoroughly and use as fruit drink.

7. Stew berries to make dessert or pie (*Chapter 2*).

LEAVES:

1. Use uncooked leaves as salad green.

2. Cook (boil or steam) and eat as a green vegetable.

3. Add liberally to soups and stews.

SHOOTS:

1. Eat young shoots raw as snack or add to green salad.

2. Peel young shoots, boil 15 minutes, and eat like asparagus.

FLOWERS:

    1. Eat raw as tasty snack or add to green salads.

    2. Use for making fritters (*Chapter 2*).

NOTE:

    1. Berries dried in sun or over embers can be stored indefinitely.

    2. Serviceberries are small dark blue, apple-like fruit with an almond-like flavor.

    3. Save cooking liquid and drink or use for soup and stew.

STRAWBERRY, WILD

PLANT:

    1. Steam or boil entire plant and eat.

BERRIES:

    1. Eat fresh berries alone or with milk for nourishing snack.

    2. Stir fresh berries into pancake or muffin batter.

    3. Cook berries to make jelly or jam (*Chapter 2*).

    4. Set aside boiled juice to ferment for making vinegar.

    5. Crush, cook until juice thickens and use for pancake syrup.

    6. Stew berries to make dessert or pie filling (*Chapter 2*).

LEAVES:

    1. Use uncooked leaves in green salads.

    2. Cook (boil or steam) and eat as green vegetable.

    3. Add liberally to soups and stews.

SHOOTS:

    1. Eat young shoots raw as snack or in green salad.

    2. Peel young shoots, boil 15 minutes and eat like asparagus.

FLOWERS:

    1. Eat raw as tasty snack or add to green salads.

    2. Use for making fritters (*Chapter 2*).

NOTE:

    1. Berries dried in sun or over embers can be stored indefinitely.

    2. Berries and leaves are extremely high in Vitamin C.

    3. Save cooking liquid and drink or use for soup and stew.

4

WILD PLANTS USED AS VEGETABLES

Various authorities estimate there are approximately 300,000 plants (those that have been classified) on the surface of the earth. Of these, 120,000 varieties have been determined to be edible.

A person should know what edible plants to look for when in the wilderness. He or she should also be able to properly identify these plants and to properly prepare them for eating.

Such an individual will undoubtedly find enough plant food out there to keep alive over an extended period of time. And he or she may even surprise themselves with a delicious meal.

ALFALFA

LEAVES:

1. Use uncooked leaves as salad greens.
2. Boil until tender and serve as vegetable green.
3. Steam 15 minutes and serve as vegetable green.
4. Add liberally to soups and stews.
5. Pulverize dried leaves and add to breakfast cereal.
6. Crush dried leaves into powder and steep in hot water for nutritious drink.

SPROUTED SEEDS:

1. Eat alone as tasty snack.
2. Use as nourishing addition to green salads.
3. Excellent on sandwiches with meat, cheese, etc.
4. Stir into soups and stews as nutritional supplement.

FLOWERS:

1. Eat raw as snack or add to green salads.
2. Use for making delicious fritters (*Chapter 2*).
3. Saute flowers until lightly browned (*Chapter 2*).
4. Crush dried flowers and add to breakfast cereals.
5. Crush dried flowers and add to soups and stews.

NOTE:

1. Alfalfa seeds are highly nutritious. They contain chlorophyll, amino acids, vitamins A, C, D, K, etc.
2. Drink nourishing liquid used to cook plant parts.

AMARANTH, GREEN

ENTIRE PLANT IS EDIBLE:

1. Cut up and use raw for salad.
2. Boil or steam until tender and eat.
3. Put in soups and stews.

LEAVES:

1. Use uncooked leaves as salad greens.
2. Boil or steam until tender and eat.
3. Add liberally to soups and stews.
4. Grind or pulverize dried leaves to make flour for baking.

SEEDS:

1. Seeds can be eaten raw or dried as cereal.

2. Seeds can be used raw or dried for baking.

3. Seeds can be crushed or ground and eaten as mush for breakfast, lunch or supper.

4. Shiny black seeds, harvested in the fall, can be parched (scorched or burned slightly) and then ground to make flour.

NOTE:

1. This dark flour is good for making waffles, pancakes and for baking homemade bread.

2. This plant has a delicate flavor unlike any other green.

3. When greens taste dry and coarse, add meat drippings.

4. This plant is high in vegetable protein.

5. Drink nourishing liquid used to cook plant parts.

ASPARAGUS, WILD

YOUNG STALKS:

1. Cut into small pieces and eat raw as snack.

2. Cut up raw stems and add to green salads.

3. Boil until tender and eat as vegetable green.

4. Steam about 10 minutes or until tender and eat.

5. Cut up and add liberally to soups and stews.

6. Saute cut up pieces (*Chapter 2*).

7. Deep fry cut up pieces as you would French fries.

8. Use cut pieces to make fritters (*Chapter 2*).

9. Use to make a delicious cream soup.

10. Cut into pieces, stir in eggs and fry as an omelet.

NOTE:

1. Eat only young shoots or stalks as you would cultivated asparagus. Older stalks are toxic.

2. This is one of the finest early spring vegetables.

3. Wild asparagus is best when eaten hot with butter.

4. Wild asparagus is similar to that purchased in stores.

5. Drink nourishing liquid used to cook this plant.

6. Young wild asparagus shoots are difficult to find.

COOKING POINTERS:

1. Use wild asparagus in any recipe calling for asparagus.

2. Wash and peel rough covering near base of stalk before cooking.

3. Tie stalks in small bundles and put into pot of boiling water.

BULRUSH

ROOTS:

1. Boil or steam until tender and eat as potatoes.

2. Boil into a gruel, let dry and use for flour.

3. Bake until tender and eat as baked potato.

4. Roast until done and eat as potato.

5. Roast several hours until dry and pound into flour.

6. Slice up and fry as you would potatoes.

7. Cut into pieces and eat raw in salads with other plant parts.

LEAVES:

1. Tear up raw young leaves and use in green salads.

2. Cook (boil or steam) young leaves and eat as vegetable.

3. Add liberally to soups and stews.

FLOWERS:

1. Dry and grind to make flour.

2. Use for making delicious fritters (*Chapter 2*).

3. Add liberally to soups and stews.

4. Used raw in salads.

BUDS:

1. Add uncooked spring buds to salads.

2. Boil or stew spring buds and eat as vegetable.

3. Add liberally to soups and stews.

POLLEN:

1. Pulverize (pound) to make flour suitable for baking.

WHITE STEM BASE:

2. Cut up and eat raw in salads with other plant parts.

3. Cook (boil or steam) and eat as vegetable.

UNDERGROUND STEMS:

1. Roast and eat like potatoes.

SEEDS:

1. Gather in the fall and grind into meal or flour.

2. Roast seeds and eat as quick snack.

3. Roast seeds and add to salads.

EARLY OR YOUNG SHOOTS:

1. Eat raw by themselves or use in green salad.

2. Boil or steam and eat as a vegetable.

3. Add liberally to soups and stews.

4. Saute shoots until lightly browned (*Chapter 2*).

5. Slice and fry as you would potatoes.

6. Roast shoots and eat as vegetable.

NOTE:

1. Drink nourishing liquid used to cook plant parts.

2. Bulrush roots are loaded with starch and sugar.

3. Flour made from bulrush seeds tastes nutty and sweet.

4. Toast seeds slightly to get rid of insects.

5. Young firm roots are a good source of sugar.

COOKING POINTER:

1. Bruise some young root pieces and cover with boiling water. Boil until water is almost gone to make sweet syrup.

BURDOCK

ENTIRE PLANT IS EDIBLE:

1. Boil or steam until tender and eat as green vegetable.

2. Use in soups and stews.

LEAVES:

1. Eat raw as quick snack.

2. Use uncooked leaves as salad greens.

3. Add liberally to soups and stews.

4. Boil or steam in 2 or 3 changes of water until tender.

5. Saute leaves (*Chapter 2*).

LEAF STEMS:

1. Boil until tender and save broth to make nice tasting soup.

2. Boil 10 to 15 minutes and eat as asparagus.

3. Steam until tender and eat as asparagus.

4. Stir fry in lightly greased pan until lightly browned.

STEM PITH:

1. Eat raw and salted as a nice snack.

2. Add to green salads for taste treat.

3. Boil or steam until tender and eat as vegetable.

4. Bake until tender and eat as baked vegetable.

FLOWER STALKS:

1. Boil core, mash, make patties and fry as you would potatoes.

2. Peel bitter green skin from long plump flower stalk. Slice into pieces and fry, boil, steam, etc., as you would potatoes.

3. Peel rough stalk when flower heads start to form. Eat as you would celery, by itself or in green salads.

BUR:

1. Steam and eat whole bur if young and pliable.

ROOTS:

1. Boil root of young plant after peeling off tough outer skin. Slice into thin strips and serve with butter.

2. Boil or steam strips of roots as you would carrots.

3. Roast roots until nicely browned and serve with butter.

4. Bake root whole until done and eat with butter.

NOTE:

1. Drink nourishing liquid used to cook plant parts.
2. Burdock is high in vegetable protein.
3. Burdock root has a nice sweet taste.

COOKING POINTERS:

1. Boil roots in 2 or 3 changes of water to remove bitter taste.
2. Mash roots after boiling, make into patties and fry.
3. Dandelion, wild onions or leeks and wild carrots (Queen Anne's Lace) should be combined with burdock in beef or chicken broth and simmered until everything is tender.
4. Flower stalks can be made into candy by simmering them in sugar syrup or maple syrup.
5. Barely cover peeled and sliced root with water. Add 1/4 teaspoon baking soda. Simmer 20 minutes. Drain water and cover root with fresh. Simmer until tender for delicious vegetable dish.

CAMAS LILY (WILD HYACINTH)

BULBS:

1. Boil 20 to 30 minutes and eat like potatoes (*Chapter 2*).
2. Slice boiled bulbs, dry in sun and store for future use.
3. Steam until tender throughout and eat like potatoes.
4. Bake in foil at 325 degrees for 45 minutes and eat.
5. Bake wrapped bulbs in pit 2 to 3 days. When finished baking, bulbs will be dark and sugary.
6. Cut up bulbs and add to soups and stews.
7. Slice raw bulbs and fry as potatoes.
8. Deep fry cut up pieces like French fries.
9. Saute bulb slices until lightly browned (*Chapter 2*).
10. Use bulb slices to make fritters (*Chapter 2*).

NOTE:

1. Bulbs edible all year. They taste good and are nutritious.
2. Bulb may be confused with death camas if collected before or after they bloom. Only collect bulbs when plant is in bloom.
3. Bulbs pleasant tasting with gum-like texture.
4. Bulbs extremely high in sugar.

COOKING POINTERS:

1. Indians had special method for cooking bulbs:
2. They were baked in pits until done.
3. Bulbs were then set aside to dry in sun.
4. This method of cooking bulbs got rid of gummy problem.

CATTAIL

1. Boil or steam until tender and eat as vegetable.
2. Use in soups and stews.

ROOTS:

1. Eat delicious and nutritious starchy root raw for quick energy.

2. Dry roots and pound or grind into flour.

3. Peel rootstalk, grate inner portion, boil and eat.

4. Peel raw rootstalk and put tender core in salads.

5. Roast roots in embers of fire and eat (*Chapter 2*).

6. Bake roots until done and eat as baked potatoes.

LEAVES:

1. Boil until tender and eat like asparagus.

2. Steam for 15 minutes and eat as green vegetable.

EARLY SHOOTS:

1. Eat young, tender, cucumber-tasting shoots as raw snack.

2. Eat tender raw shoots in salads.

3. Boil young shoots and eat like asparagus.

4. Saute young shoots (*Chapter 2*).

POLLEN:

1. Mix yellow pollen with water, form loaf and steam as bread.

2. Blend protein rich pollen with regular flour and use for baking bread, biscuits etc.

FLOWER SPIKES OR HEADS:

1. Eat early green spikes raw as highly nourishing snack.

2. Put raw early green spikes in salads with greens.

3. Dry early spikes and grind or pound into flour.

4. Boil pollen spikes 10 minutes in lightly salted water and eat like corn on the cob.

STEM PITH

1. Peel stem, take out pith and eat it raw.

2. Boil or steam pith for 10 minutes and eat while hot.

NOTE:

1. Scrub root when pulled from water and peel before cooking.

2. Roots and lower stem portion of cattail are sweet tasting and loaded with carbohydrates.

3. Crush roots in cold water and remove starch. This starch can later be used to thicken soups, stews and gravies, etc.

4. Cattails are highly nutritious plants. The edible rootstalks contain about 46% starch and 11 % sugar.

5. When yellow pollen is ripe, bend stalk over and shake into bag.

6. Dry pollen over embers or in sun before storing for later use.

7. Cattails top the list of edible wild plants because of its versatility, widespread abundance and ease of availability.

COOKING POINTERS:

1. Add 1 cup honey to 5 cups pollen to make a high energy food.

2. Blend 1 cup pollen to 2 cups pancake mix.

3. Blend 1 cup pollen to 2 cups flour to make bread, etc.

4. If shoots are tough, simply drop into a pot of salted water and let simmer to restore their tenderness.

CHAMOMILE

ENTIRE PLANT IS EDIBLE:

1. Cut up and add to green salad.
2. Boil 10 to 12 minutes or until tender.
3. Steam until tender and eat as vegetable.
4. Put in soups and stews.
5. Saute in lightly greased frying pan (*Chapter 2*).

LEAVES:

1. Eat raw as crispy snack.
2. Use uncooked leaves as salad greens.
3. Boil until tender and eat as vegetable green.
4. Steam 15 minutes and eat as vegetable green.
5. Add liberally to soups and stews.
6. Saute in lightly greased frying pan (*Chapter 2*).

FLOWERS:

1. Eat raw by itself as tasty snack.
2. Add to green salads.
3. Saute flowers until lightly browned (*Chapter 2*).
4. Use for making delicious fritters (*Chapter 2*).

TWIGS:

1. Dry twigs out and use for tasty and crunchy treat.

NOTE:

1. Chamomile is very high in minerals.
2. Always drink leftover liquid when finished cooking.

CHICKWEED

ENTIRE PLANT IS EDIBLE:

1. Boil or steam 2 to 5 minutes and eat like spinach.
2. Put in soups and stews before stems get tough.
3. Young uncooked plants make outstanding salad greens.
4. Stew entire plant in pot with beef, chicken, or other meat.
5. Saute in lightly greased frying pan (*Chapter 2*).
6. Chop up plant, blend with eggs and make delicious omelet.

LEAVES:

1. Boil or steam 2 minutes or until tender and eat like spinach.
2. Use young uncooked leaves as salad greens.
3. Add liberally to soups and stews.
4. Stir into eggs and fry as omelet.

FLOWERS:

1. Put uncooked flowers in salads.

2. Add flowers to soups and stews.

3. Use for making delicious fritters (*Chapter 2*).

4. Stir flowers into eggs and make tasty omelet.

STEMS:

1. Add raw young stems to green salads.

2. Boil or steam tender stems 2 to 5 minutes and eat as vegetable.

3. Saute until slightly browned (*Chapter 2*).

4. Cut up stems and fry in butter.

5. Cut into small pieces, blend with eggs and make omelet.

NOTE:

1. Chickweed has bland taste. Try mixing with stronger-tasting greens such as mustard, dandelion, chicory or watercress.

2. Always save liquid from cooking and use for nutritious drink.

3. Young leaves and stems, when cooked, taste much like spinach.

4. Chickweed is highly nutritious and rich in Vitamin C.

5. Chickweed is tenderer than most other wild greens.

COOKING POINTERS:

1. Seeds obtained in the fall can be used in soups as a thickener.

2. To make unique pancakes:

   a. Drain and then blot 2 cups boiled or steamed leaves.

   b. Add enough leaves to pancake batter to get correct consistency.

   c. Fry and eat with butter and jam or jelly.

CHICORY, WILD

LEAVES:

1. Use tender, young, white parts of leaves for salads.

2. Boil or steam above ground parts of leaves until tender.

3. Saute leaves (*Chapter 2*).

4. Add liberally to soups and stews.

FLOWERS:

1. Eat raw by themselves as snack or add to salads.

2. Saute in lightly greased frying pan (*Chapter 2*).

3. Use to make delicious fritters (*Chapter 2*).

ROOTS:

1. Boil or steam carrot-like roots and eat when tender.

2. Grind or crush roasted roots to make rather bitter coffee substitute (*Chapter 2*).

NOTE:

1. Always drink nourishing liquid used for cooking.

2. When boiling leaves use just enough water to cover.

3. Chicory leaves, highly favorable, can be added to some of the more bland tasting greens such as chickweed and purslane.

4. Pick leaves early as they become bitter with age.

5. Leaves look like dandelion but are thicker and rougher.

COOKING POINTERS:

1. Changes of water one or more times while boiling might be required to get rid of objectionable bitterness of plants.

CLOVERS

ENTIRE PLANT IS EDIBLE:

1. Boil plant and eat as vegetable.

2. Steam plant 10 to 15 minutes and eat.

3. Put plant in soups and stews.

LEAVES:

1. Use uncooked leaves as salad greens.

2. Boil or steam 5 to 10 minutes and eat as green vegetable.

3. Add liberally to soups and stews.

4. Saute leaves (*Chapter 2*).

5. Dry leaves over embers or in sun and pulverize to make flour.

6. Crush dried leaves and add to breakfast cereal.

FLOWERS:

1. Eat blossoms or round flower heads raw as snack.

2. Put blossoms or raw flower heads in salads.

3. Boil blossoms until tender and eat as vegetable.

4. Steam blossoms 15 minutes and eat as vegetable.

5. Add liberally to soups and stews.

6. Saute blossoms (*Chapter 2*).

7. Use for making delicious fritters (*Chapter 2*).

8. Dry blossoms and grind or pound into flour.

STEMS:

1. Raw stems should be eaten raw as snack.

2. Raw stems should be cut up and added to green salads.

3. Cut into pieces, stir into eggs and make omelet.

4. Boil until tender and eat as vegetable.

5. Steam 15 minutes and eat as vegetable.

SEEDS:

1. Crush and boil to make mush.

2. Parch seeds for tasty snack.

3. Eat seeds as a cereal.

4. Dry seeds and grind or pound into nutritious flour.

5. Seeds should be added to soups and stews as flavoring.

NOTE:

1. Clover is very rich in protein.

2. Always drink nourishing liquid used for cooking.

COMFREY

ENTIRE PLANT IS EDIBLE:

1. Boil or steam until tender and eat as vegetable.
2. Put in soups and stews.

LEAVES:

1. Eat early leaves (March and April) uncooked in salads.
2. Boil or steam until tender and eat like spinach.
3. Add liberally to soups and stews.

STALKS:

1. Boil or steam until tender and eat as vegetable.
2. Cut up and add to soups and stews.

FLOWERS:

1. Eat raw as snack or add to green salads.
2. Boil or steam until done and eat as vegetable.
3. Add liberally to soups and stews.
4. Saute flowers until lightly browned (*Chapter 2*).

ROOTS:

1. Boil or steam until tender and eat with butter.
2. Bake or roast until done and ready to eat.
3. Add whole or cut up in soups and stews.

COOKING POINTERS:

1. Older leaves bitter but still edible. Boil in 3 water changes.
2. Hairiness is eliminated somewhat when plant is cooked.
3. Drink the nourishing liquid used for cooking plant parts.

DAISY, OX-EYE

LEAVES:

1. Chew raw leaves by themselves for tasty snack.
2. Put raw leaves in green salads.
3. Boil one minute and eat as green vegetable.
4. Steam lightly and eat as green vegetable.
5. Add liberally to soups and stews.

FLOWERS:

1. Eat raw by themselves for tasty snack.
2. Use as attractive yet nourishing addition to green salads.
3. Add liberally to soups and stews.
4. Blend flowers with batter for pancakes or muffins.
5. Use for making delicious fritters (*Chapter 2*).

STEMS:

1. Eat raw by themselves for quick snack.
2. Stir liberally in soups and stews.
3. Steam lightly and eat as vegetable.
4. Saute lightly (*Chapter 2*).
5. Cut into small pieces, stir into eggs and fry as an omelet.

NOTE:

1. Drink highly nutritious liquid left over from cooking.
2. Also save cooking liquid for making soups and stews.

DANDELION

ENTIRE PLANT IS EDIBLE:

1. Boil 15 to 20 minutes and eat like spinach.
2. Steam until tender and eat as vegetable green.
3. Use entire plant in soups and stews.

LEAVES:

1. Eat raw young leaves by themselves as nourishing snack.
2. Use uncooked young leaves with roots and flowers for salads.
3. Boil for 10 minutes and eat like spinach.
4. Steam until tender and eat like spinach.
5. Add liberally to soups and stews.
6. Add chopped leaves to eggs and fry for delicious omelet.

STALKS/STEMS:

1. Cut up stems and eat as quick snacks.
2. Add to green salads.
3. Boil 10 to 15 minutes and eat as vegetable.
4. Steam until tender and eat as a green vegetable.
5. Add liberally to soups and stews.
6. Cut into small pieces and saute (*Chapter 2*).

FLOWERS:

1. Eat by themselves after twisting green base from sweet petals.
2. Eat raw in salads with leaves and roots.
3. Boil buds for 10 minutes and eat with butter.
4. Saute in lightly greased frying pan (*Chapter 2*).
5. Make terrific fritters (*Chapter 2*).

ROOTS:

1. Eat raw, pleasant tasting roots by themselves as snack.
2. Cut up root and put in salads with leaves and flowers.
3. Slice and boil until tender as you would carrots.
4. Steam sliced root for 15 to 20 minutes or until tender.
5. Roast slowly until dark brown throughout. Pulverize or grind and brew exactly like coffee.

6. Saute slices of root (*Chapter 2*).

7. Fry sliced root as you would potatoes.

8. Thoroughly dry and pulverize root to make flour.

NOTE:

1. Dandelions are more nourishing than any plant found in gardens or the wilds.

2. Dandelions are richer in beta-carotene than carrots and are loaded with vitamin A, calcium and vegetable protein.

3. Dandelion leaves are sweeter in the fall after a few frosts.

4. Gather young leaves before buds and flowers appear. They are more tender and taste better. Leaves get bitter with age.

COOKING POINTERS:

1. Tough, bitter tasting older leaves can be boiled twice to get rid of their bitterness and then added to other cooked greens.

2. Add strong-flavored dandelion greens to bland tasting greens such as chickweed and purslane to make them more edible.

3. Save the liquid used for cooking and use as a soup base.

DAY LILY

UNDERGROUND TUBERS:

1. Cut up crisp raw tubers and eat as healthy snack.

2. Cut up crisp raw tubers and use in salads.

3. Boil 15 to 20 minutes as you would corn on the cob.

4. Steam until tender.

5. Slice and fry until browned as you would potatoes.

6. Cut into strips and deep fry like French fried potatoes.

7. Make tasty fritters (*Chapter 2*).

BUDS:

1. Boil unopened buds 5 minutes and eat like green beans.

2. Steam unopened buds until tender and eat like green beans.

3. Add fresh or dried buds liberally to soups and stews.

4. Deep fry fresh buds as you would French fried potatoes.

5. Use to make fritters (*Chapter 2*).

FLOWERS:

1. Add flowers to green salads.

2. Cook in stews to season.

3. Put in soups and simmer as a thickener.

4. Use fresh flowers to make delectable fritters (*Chapter 2*).

SHOOTS:

1. Boil young or early shoots and eat like asparagus (*Chapter 2*).

2. Steam young shoots until tender and eat like asparagus.

NOTE:

1. Buds and flowers can be dried in the sun and then set aside for later use.

2. Root tubers can be dug up all year round.

3. The day lily is a great source of food, but few people know about this easy-to-find plant.

4. For seasoning in soups and stews, fresh or dried flowers can be used as well as withered flowers still on the plant.

5. Uncooked flowers for use in salads have a strong taste.

6. Tubers make a remarkable tasting vegetable when boiled, steamed or fried.

7. Always save and drink nourishing liquid used for cooking.

8. COOKING POINTERS:

1. Melt cheese or butter over deep fried buds (*Chapter 2*).

EVENING PRIMROSE, COMMON

LEAVES:

1. Peel tender new leaves and use sparingly as flavorful addition to tossed salad.

2. Boil peeled young leaves 20 minutes or until tender.

3. Steam peeled young leaves until tender and eat as vegetable.

FLOWERS:

1. Put liberally into soups and stews.

2. Make delicious fritters (*Chapter 2*).

3. Make into a tasty candy (*Chapter 2*).

SEEDS:

1. Crush seeds, boil and skim oil from surface of water.

ROOTS:

1. Boil roots for at least 10 minutes and eat as vegetable.

2. Steam roots until tender and serve with butter.

3. Slice cooked roots and fry until browned (*Chapter 2*).

4. Cut boiled or steamed roots in strips and deep fry as you would French fries.

5. Slice roots and put liberally in soups and stews.

6. Boil roots until tender and then simmer in sugar syrup for 20 to 30 minutes or until candied.

NOTE:

1. Freshly picked new leaves taste slightly bitter after cooking but are palatable.

2. Fresh roots, usually quite peppery tasting, are milder if dug up in early spring or late fall.

3. Some evening primrose roots are very tasty — somewhat nut-like, and extremely nutritious.

COOKING POINTERS:

1. Leaves should be boiled in 3 changes of water to get rid of some of the objectionable bitterness.

2. Always boil or steam root slices first before frying if the flavor is stronger than desired.

3. Taproots will sometimes have an objectionable strong peppery taste. Here's what to do about this problem:

a. Peel taproots and boil 20 to 30 minutes in 3 water changes.

b. Then let simmer 20 more minutes, serve with butter, vinegar and salt.

FIREWEED (PILEWORT)

YOUNG LEAVES:

1. Eat raw leaves as nutritious snack.
2. Use uncooked leaves for green salads.
3. Boil or steam until leaves can easily be pierced with fork.
4. Add to soups and stews as pot herb.

FLOWER BUDS:

1. Eat clusters of raw buds for a nourishing snack.
2. Add raw bud clusters to any tossed salad.
3. Generously add bud clusters to soups and stews.
4. Boil bud clusters 10 to 15 minutes and eat with butter.
5. Steam bud clusters until tender and serve as vegetable.

YOUNG SHOOTS

1. Eat raw young shoots as a tasty snack.
2. Use as nourishing addition to green salads.
3. Boil in salt water until tender and eat like asparagus.
4. Steam until tender and eat as vegetable.
5. Stir generously into soups and stews.

FLOWER STALKS:

1. Cut up and eat raw as snack.
2. Add to tossed or green salads.
3. Boil 10 to 15 minutes and eat as vegetable.
4. Steam until tender and eat as vegetable.
5. Add generously to soups and stews.

PITH FOUND IN STEMS WHEN PEELED:

1. Eat raw by itself as nourishing snack.
2. Add to tossed or green salads.
3. Put in boiling water until it becomes a thick soup.
4. Steam the pith until it becomes a thick soup.
5. Add as a thickener to soups and stews.

NOTE:

1. Fireweed is an extremely nutritious plant.
2. Far from the most tasty wild plant but it's still edible.
3. Has strong flavor. Must acquire a taste for this plant.
4. Young shoots should be collected in the spring.
5. Bud clusters should be picked before flowers have bloomed.
6. Young stems bear remarkable resemblance to asparagus.

GALINSOGA (QUICKWEED)

LEAVES:

1. Put uncooked leaves in salads.
2. Boil for 10 to 15 minutes and eat as vegetable.
3. Steam until tender and eat with butter, salt and pepper.
4. Add liberally to soups and stews.

FLOWERS:

5. Eat raw as a quick snack.
6. Use uncooked leaves in salads.
7. Saute flowers (*Chapter 2*).
8. Use to make delicious fritters (*Chapter 2*).

STEMS:

9. Boil for 15 minutes and eat as green vegetable.
10. Steam until tender and eat as green vegetable (*Chapter 2*).
11. Cut into pieces and add to soups and stews.
12. Deep fry cut up pieces like French fries.
13. Cut into small pieces, mix with eggs and make omelet.

NOTE:

1. Excellent dishes when served with butter or vinegar.
2. Serve with more flavorful greens to make up for bland flavor.
3. Save cooking liquid to drink or use for soups and stews.

GOOSEGRASS (CLEAVERS)

LEAVES:

1. Use Uncooked spring or early leaves in tossed salads.
2. Boil 10 to 15 minutes and eat as vegetable green.
3. Steam until tender and eat as vegetable green.
4. Add liberally to soups and stews.

SHOOTS:

1. Cut up and put in green salads.
2. Cook shoots (boil or steam), let cool and use in salads.
3. Boil young shoots 10 to 15 minutes or until tender.
4. Steam young shoots until tender.
5. Add shoots generously to soups and salads.

SEEDS:

1. Roast seeds and then chew for nutritious, tasty snack.
2. Add roasted seeds generously to salads.
3. Stir roasted seeds into soups and stews.
4. Grind or crush roasted seeds and use as coffee substitute.

NOTE:

1. Cooled cooked shoots (boiled or steamed) are delicious when eaten with cooked, wild asparagus.

2. Ground and roasted goosegrass seeds far surpass chicory as a coffee substitute (*Chapter 2*).

3. Older, more mature leaves are tough and must be boiled or steamed to make them edible.

GOURD, WILD (LUFFA SPONGE)

GOURD:

1. Boil whole gourd 15 to 25 minutes or until tender.

2. Steam entire gourd until tender.

3. Roast whole gourd until tender.

4. Bake whole gourd until tender to prick of fork.

LEAVES:

1. Boil young leaves 10 minutes or until tender.

2. Steam young leaves until tender and eat as vegetable.

3. Saute young leaves in lightly greased frying pan (*Chapter 2*).

FLOWERS:

1. Eat flowers raw for quick snack.

2. Add to green salads.

3. Toss into soups and stews.

SEEDS:

1. Roast seeds over embers or in oven and eat like peanuts.

SHOOTS:

1. Boil shoots 10 minutes or until ready to eat.

2. Steam shoots until tender and ready to eat.

3. Saute in lightly greased frying pan (*Chapter 2*).

NOTE:

1. Wild gourds grow like cucumbers, watermelons and cantaloupes.

2. Seeds are highly nutritious and improved with salting.

3. Cook gourds (boil or steam) when half ripe for best taste.

GREENBRIER

ENTIRE PLANT IS EDIBLE:

1. Boil or steam until tender and eat as vegetable.

2. Use in soups and stews.

LEAVES:

1. Eat young leaves raw as quick snack.

2. Use uncooked leaves as salad greens.

3. Boil or steam older leaves until tender and eat as spinach.

4. Use liberally in soups and stews.

SHOOTS:

1. Eat young shoots raw as crispy snack.
2. Use crispy young shoots in salads.
3. Boil or steam until tender and eat like asparagus.
4. Cool older shoots after cooking and put in salads.

TENDRILS:

1. Eat tendrils raw as a crispy snack.
2. Use tendrils to liven up salads.
3. Boil or steam with leaves and shoots until tender.

TUBEROUS ROOTS:

1. Wash and crush to obtain jelly-like material. Use in soups and stews as thickener.

NOTE:

1. Drink nourishing liquid used for cooking plant parts.

GROUNDNUT

TUBERS:

1. Peel and eat raw like potatoes for snack.
2. Peel and boil 20 to 25 minutes in heavily salted water as with turnips and potatoes.
3. Wash, peel and roast like potatoes.
4. Wash, peel, slice and fry in bacon grease (if available).
5. Wash and then bake 45 to 60 minutes until tender.
6. Deep fry strips or thin slices of raw tuber like French fries.

SEEDS:

1. Remove from pod and eat bean-like seeds as snack.
2. Remove seeds from pods and add to salads.
3. Remove bean-like seeds from pod and fry until crisp.
4. Boil bean-like seeds until tender and eat.
5. Steam seeds until tender.

PODS:

1. Roast at 375 degrees for 20 to 25 minutes, remove seeds, and fry pod in oil until browned.

NOTE:

1. Tubers are pleasant tasting — sweet and turnip-like.
2. One of the best wild foods available. Widely used by early settlers and Indians.
3. Tubers are made up of 15% protein.
4. Drink nourishing liquid used for cooking plant parts.

HOG PEANUT

FLOWERS:

1. Eat raw as quick snack.
2. Put raw flowers in green salads.
3. Add liberally to soups and stews.
4. Saute in lightly greased frying pan (*Chapter 2*).

STEMS:

1. Cut up into small pieces and add to salads.
2. Boil 10 minutes or more and eat as green vegetable.
3. Steam until tender and eat as green vegetable.
4. Put into soups and stews.
5. Saute in lightly greased frying pan (*Chapter 2*).
6. Cut into small pieces, blend with eggs and fry as omelet.

SEEDS (BEANS):

1. Boil seeds 15 to 20 minutes or until tender.
2. Steam seeds until tender and eat as vegetable.
3. Add liberally to soups and stews.
4. Dry seeds thoroughly and grind or pound into flour.

NOTE:

1. One seed can be found in each pod just below ground level.
2. Drink nourishing liquid used for cooking plant parts.
3. The light brown seeds (beans) from the underground pods are dryer than most.
4. Cooked seeds are good eaten with butter, salt and pepper.

HORSERADISH TREE

LEAVES:

1. Eat fern-like leaves raw in salads.
2. Boil fern-like leaves 10 to 12 minutes.
3. Steam fern-like leaves until tender.
4. Saute in lightly greased frying pan (*Chapter 2*).
5. Add liberally to soups and stews.

FLOWERS:

1. Add uncooked flowers to salads.
2. Add liberally to soups and stews.
3. Saute flowers (*Chapter 2*).
4. Boil 3 to 5 minutes or until tender.
5. Steam until tender and eat as vegetable.

SEED PODS:

    1. Cut in short strips, drop into pot of water, boil 10 minutes and eat like string beans.

    2. Cut in strips, steam until tender and eat like string beans.

    3. Fry until slightly browned.

    4. Saute in lightly greased frying pan (*Chapter 2*).

    5. Add liberally to soups and stews.

NOTE:

    1. Seed pods can be chewed uncooked when fresh.,

    2. Drink nourishing liquid used for cooking plant parts.

    3. Dry roots thoroughly and grind to make pungent seasoning.

HYSSOP HEDGE-NETTLE

TUBERS:

    1. Cut up raw tubers and eat as tasty crispy snack.

    2. Cut up raw tubers and add to green tossed salad.

    3. Boil tubers until tender and eat as vegetable.

    4. Steam tubers 10 to 15 minutes or until tender.

    5. Add liberally to soups and stews for flavoring.

    6. Saute in lightly greased frying pan (*Chapter 2*).

    7. Use for making delicious fritters (*Chapter 2*).

    8. Cut up and blend with raw eggs to make omelet.

NOTE:

    1. Look for crisp white tubers in soil around shriveled stems.

    2. Tubers have pronounced bitter flavor. Also tastes minty.

INDIAN CUCUMBER ROOT

TUBERS:

    1. Eat crispy tubers uncooked as snack.

    2. Cut up crispy tubers and add to salads.

    3. Boil tubers as you would potatoes.

    4. Steam tubers until done and serve like potatoes.

    5. Add tubers to soups and stews.

    6. Slice up tubers and fry like potatoes.

    7. Saute in lightly greased frying pan (*Chapter 2*).

NOTE:

    1. Tubers have delicate cucumber taste.

    2. Tubers are to be gathered during the summer.

    3. Berries on this plant are not to be eaten.

LADY'S THUMB (REDLEG)

LEAVES:

1. Put raw chopped leaves in green salads.
2. Boil 10 minutes and eat like spinach.
3. Steam until tender and eat like spinach.
4. Add liberally to soups and stews.

SMALL BULBS (ON LOWER PART OF FLOWER STALKS):

1. Eat raw as tasty snack,
2. Put uncooked bulbs in green salads.
3. Add liberally to soups and stews.

YOUNG ROOTS:

1. Bake until tender and eat like baked potatoes.
2. Roast until thoroughly done and eat like potatoes.
3. Slice raw and fry like potatoes.
4. Cut in strips and fry like French fries.
5. Boil roots 10 to 15 minutes and eat like potatoes.
6. Steam roots until done and eat like potatoes.
7. Saute in lightly greased frying pan (*Chapter 2*).

NOTE:

1. Serve boiled leaves with vinegar, if available.
2. Roots are tastiest when roasted.
3. Young leaves make decent wild spinach-like dish.
4. Save liquid used for cooking to drink or for making soup.

LAMB'S QUARTER (WILD SPINACH)

ENTIRE PLANT IS EDIBLE:

1. Boil or steam 10 to 15 minutes and eat as vegetable.
2. Crush plant and extract juices to make nutritious juice.
3. Add entire plant to soups and stews.

LEAVES:

1. Eat raw when tender in salads.
2. Boil or steam 10 minutes and eat as you would spinach.
3. Add liberally to soups and stews.
4. Saute in lightly greased frying pan (*Chapter 2*).
5. Mix with eggs, scramble and fry as delicious omelet.
6. Dry and crush or grind to make flour.

FLOWERS:

1. Add liberally to soups and stews.
2. Use to make delicious fritters (*Chapter 2*).
3. Blend with fresh eggs and cook in frying pan as omelet.
4. Dry thoroughly and grind or pulverize to make flour.

SHOOTS:

1. Add raw shoots to green salads.

SEEDS:

1. Dry seeds thoroughly and grind or pound to make meal or flour.

2. Boil 10 minutes or until soft to make nourishing cereal.

3. Pulverize seeds to make mush for breakfast.

ROOTS:

1. Dry thoroughly and grind or crush to make flour.

2. Boil or steam until tender and eat as vegetable.

NOTE:

1. Need lots of this plant when cooking as bulk is diminished.

2. Very high in vitamin A and C, calcium, minerals and protein.

3. Save the liquid used for cooking to drink and making soup.

4. One of the best wild plants to eat. Has no harsh taste.

5. Add dark meal made from this plant to regular flour (half and half) when making muffins, bread, biscuits or even pancakes.

LEEK, WILD (RAMP)

LEAVES:

1. Eat raw by themselves as snack.

2. Chop into pieces and add to green salads.

3. Boil leaves in salted water and eat as greens.

4. Steam leaves until tender and eat as greens.

5. Add cut up leaves to soups and stews.

6. Cut into small pieces and blend with eggs to make omelet.

7. Saute in lightly greased frying pan (*Chapter 2*).

BULBS:

1. Eat raw by themselves as nourishing snack.

2. Chop up and add to tossed salads for taste treat.

3. Boil or steam and eat as vegetable.

4. Saute whole or chopped (*Chapter 2*).

5. Add whole or chopped to soups and stews.

6. Use to make delicious fritters (*Chapter 2*).

7. Chop up and blend with eggs to make omelet.

NOTE:

1. Save cooking liquid for drinking or making soups, etc.

2. Wild leeks have a very strong onion smell.

3. Try putting finely chopped bulbs in mashed potatoes.

COOKING POINTERS:

1. Wild leeks are wonderful for stuffing game before cooking.

2. Cook wild leeks with meats and other vegetables.

LETTUCE, WILD

LEAVES:

1. Eat raw by themselves as snack.
2. Use uncooked in salads with other greens.
3. Boil 10 to 15 minutes with 2 water changes.
4. Steam until wilted and serve with butter, salt and pepper.
5. Pour boiling water on leaves. Drain off water after 5 minutes. Sprinkle leaves with fried bacon crumbles for mouth-watering salad.
6. Saute leaves in lightly greased frying pan (*Chapter 2*).

FLOWERS:

1. Eat raw as snack.
2. Use uncooked in salads.
3. Boil 10 minutes with 2 water changes and eat.
4. Steam until tender and eat as vegetable.
5. Saute in lightly greased frying pan (*Chapter 2*).
6. Add liberally to soups and stews.
7. Use for making delicious fritters (*Chapter 2*)

ROOTS:

1. Slice and chew raw as nutritional gum.
2. Slice and boil until tender.
3. Slice and steam until tender.
4. Slice and fry until browned.
5. Slice and saute in lightly greased frying pan (*Chapter 2*).
6. Cut into small pieces and use to make fritters (*Chapter 2*).

NOTES:

1. Leaves and flowers when eaten raw are often bitter.
2. Dry roots in sun or over embers and set aside for later use.
3. Eating lots of raw wild lettuce can cause indigestion.
4. Wild lettuce leaves, always slightly bitter, are much better when mixed and eaten with other greens.
5. Save cooking liquid and drink or use for making soup or stew.

COOKING POINTERS:

1. Boiling leaves and flowers in two water changes greatly reduces any bitter taste.
2. To also alleviate bitterness, leaves and flowers can instead be parboiled.
3. Flowerheads added to a casserole give the dish a unique flavor.

LOTUS LILY

LEAVES:

1. Boil unrolling young leaves 10 to 15 minutes.
2. Steam unrolling young leaves until tender.

3. Add unrolling young leaves to soups and salads.

STEMS:

1. Remove rough outer layer and boil until tender.
2. Steam until tender after removing outer layer.
3. Remove rough outer layer and add to soups and stews.

TUBERS:

1. Bake until done as sweet potatoes are baked.
2. Boil until tender as you would sweet potatoes.
3. Steam until tender as you would sweet potatoes.
4. Slice and fry until browned as you would potatoes,

SHOOTS:

1. Boil shoots until tender and eat like spinach.
2. Steam shoots until tender and eat like spinach.
3. Add shoots liberally to soups and stews.

SEEDS:

1. Eat young seeds raw as quick snack.
2. Boil seeds, salt and eat like nuts.
3. Roast seeds until browned and eat like nuts.
4. Roast or bake seeds until dry and pound or grind into flour.
5. Fry as you would popcorn and eat as snack.

ROOTSTOCK:

1. Boil 20 to 25 minutes and eat like sweet potatoes.
2. Mash boiled rootstock and fry like potato patties.
3. Slice raw rootstocks and fry like potatoes.

NOTE:

1. Ripe fruit are nut-like and must be cracked to get seeds.
2. Seeds are edible when ripe after removing bitter embryo.
3. Rootstock may grow to 50" long with many tuberous enlargements.

COOKING POINTERS:

1. Substitute flour or meal made from seeds for part of flour called for in bread, roll and muffin recipes.
2. Peel and mash boiled tubers and add milk, butter and salt. Blend well and serve when smooth like mashed potatoes.
3. Can also fry the above as delicious patties until brown.

MALLOW, COMMON

ENTIRE PLANT IS EDIBLE:

1. Boil 20 minutes or more and eat as vegetable.
2. Steam until tender and eat as vegetable.
3. Use in soups and stews.

LEAVES:

1. Eat raw when young and tender in salads.
2. Boil or steam until tender and eat with butter as a vegetable.
3. Saute in lightly greased frying pan (*Chapter 2*).
4. Add liberally to soups and stews as thickener like okra.

SHOOTS:

1. Use shoots in salads with raw leaves and fruit.
2. Boil, steam or stew shoots and eat as vegetable.
3. Add liberally to soups and stews.
4. Saute in lightly greased frying pan (*Chapter 2*).

FRUIT:

1. Eat raw as good snack, or use with leaves and shoots in salads.
2. Boil, steam or stew until tender and ready to eat.

ROOTS:

1. Boil roots to make thick stock for soups.
2. Slice boiled root and fry with chopped onion in lightly greased frying pan until nicely browned.
3. Saute until lightly browned (*Chapter 2*).
4. Cut into small pieces and use to make fritters (*Chapter 2*).
5. Cut up root and add liberally to soups and stews.

NOTE:

1. Drink highly nutritional liquid used for cooking plant parts.

COOKING POINTERS:

1. Boil leaves, shoots and roots in water until liquid thickens. Can be used as substitute for egg white in meringue.
2. For delicious candy, cover peeled and sliced roots with water.
    a. Boil 20 to 30 minutes until tender.
3. Drain off water and add sugar to taste.
4. Again boil and stir until extremely thick.
5. Beat mixture and drop spoonfuls on non-stick surface.
6. When cool, roll each piece in powdered sugar.

MARSH-MARIGOLD (COWSLIP)

YOUNG LEAVES:

1. Boil 10 to 15 minutes or until tender in 3 changes of water.
2. Steam 15 minutes or until tender and eat as vegetable.
3. Use liberally in soups and stews.

FLOWERS:

1. Boil buds 10 to 15 minutes in 3 changes of water.
2. Steam 15 minutes or until tender and eat as vegetable.
3. Stir into soups and stews as nourishing supplement.
4. Use to make tasty fritters (*Chapter 2*).

STEMS:

1. Boil 10 to 15 minutes in 3 changes of water and eat as greens.
2. Steam 20 minutes or longer and eat as greens.
3. Cook until done and use for making pickles (*Chapter 2*).
4. Add liberally to soups and stews.
5. Saute in lightly greased frying pan (*Chapter 2*).

NOTE:

1. Don't eat any of this plant raw. It contains an acrid poison — poisonous glucoside — eliminated only by cooking.
2. Do not drink liquid used for cooking these plant parts!
3. Best leaves are obtained before flowers bloom in spring.
4. Few plants will cook up so delightfully in the spring.
5. Eat cooked leaves and stems with butter, salt and pepper.

COOKING POINTERS:

1. This plant, when cooked, requires several changes of water because of its extreme bitterness.

MUSTARD, WILD

LEAVES:

1. Eat chopped pieces of leaves for slightly bitter snack.
2. Eat finely chopped leaves raw in salads.
3. Boil older leaves 30 minutes in 2 water changes.
4. Steam until tender and eat as vegetable green.
5. Use liberally in soups and stews.

FLOWERS:

1. Eat uncooked as snack.
2. Add to salads, soups and stews.
3. Simmer clusters of closed flower buds 3 to 5 minutes. Do not overcook or they will be ruined!
4. Steam clusters of closed flower buds until done and eat as broccoli-like vegetable.
5. Saute in lightly greased frying pan (*Chapter 2*).

STEMS:

1. Eat raw as crispy snack.
2. Eat uncooked in salads.
3. Boil until tender and eat as vegetable.
4. Steam for a few minutes and eat as vegetable.
5. Add to soups and stews as pungent spice.
6. Saute in lightly greased frying pan (*Chapter 2*).

SEEDS:

1. Dry thoroughly and use to add zest in soups and stews.
2. Finely grind between 2 stones to make yellow mustard.
3. Add tender fresh seed pods to salads as a great garnish.

NOTE:

1. A wild mustard plant is loaded with vitamins A, Bs and C. It also contains iron, phosphorous, potassium, etc.

2. Flower buds are extremely rich in protein.

3. Do not pick any upper stem leaves when picking flower buds. They are terribly bitter and not edible.

4. Leaves become objectionably bitter as they mature.

5. Green and tender seed pods should be picked only while flowers are in bloom.

6. Save cooking liquid and drink or use to make soup or stew.

COOKING POINTERS:

1. Mustard leaves lose their bulk and shrink when cooked as does spinach and some other greens.

2. Wild mustard requires more cooking time than most other greens.

3. Always cook mustard leaves (greens) at least 30 minutes to cut their pungent biting taste.

4. Flavorful mustard greens should be blended with any of the more tasteless greens such as chickweed or purslane. This will make the bland greens as well as the mustard greens more edible.

5. Boiled flower bud clusters are a great broccoli substitute.

NETTLE, STINGING

ENTIRE PLANT IS EDIBLE:

1. Boil entire plant until tender and eat as vegetable.

2. Steam whole plant until tender and eat as vegetable.

3. Boil entire older plant 30 to 40 minutes, strain out plant, and use cooking liquid for soup stock.

4. Put in soups and stews.

LEAVES:

1. Boil young leaves until tender and eat as green vegetable.

2. Steam young leaves 10 minutes and eat as green vegetable.

3. Add liberally to soups and stews.

4. Make a delicious puree (*Chapter 2*).

5. Dry and store for later use. Easily reconstituted by steaming or allowing to simmer.

YOUNG SHOOTS:

1. Eat raw for tasty quick snack.

2. Add to green salads.

3. Boil 10 to 15 minutes or less until tender.

4. Steam until tender and eat as green vegetable.

5. Put into pot of boiling water and simmer until tender. Serve with butter, salt and pepper.

6. Add liberally to soups and stews.

7. Saute in lightly greased frying pan (*Chapter 2*).

8. Stir into eggs and fry as omelet.

NOTE:

1. Best to gather stinging nettle when less then 12" tall.
2. Pick tender green shoots in the spring.
3. Stinging nettle has higher protein content than any other known wild or domestic plant. Also high in vitamin A and C and minerals.
4. One of the tastiest and nutritious plants in the U.S.
5. Always save liquid used for cooking. Add lemon juice and sugar enough to sweeten for refreshing, vitamin loaded drink.

COOKING POINTERS:

1. Liquid used for cooking can also be used as soup by adding salt, pepper and a little vinegar.
2. Cook stinging nettle leaves exactly as spinach is cooked.
3. Stinging qualities of this plant disappear upon cooking.
4. Cook with dandelion greens, wild leeks or wild onions, and wild carrots (Queen Anne's Lace). Eat as vegetable stew.

NUT GRASS

TUBERS:

1. Eat raw as tasty snack.
2. Cut up and put in salads.
3. Boil until tender, peel and eat like small potatoes.
4. Steam in a little water until tender, peel and eat.
5. Bake until soft and eat like baked potato.
6. Bake until done, mash and fry as patties until browned.
7. Slice and fry like potatoes.
8. Cut into strips and deep fry like French fried potatoes.
9. To make delightful coffee substitute:
    a. Roast to dark brown color.
    b. Grind or pound the sweetish, almond flavored tubers to powder.
10. Dry in oven or sun and pound or grind to a powder. Mix half and half with flour for making tasty breads, cakes and cookies.
11. Saute in lightly greased frying pan (*Chapter 2*).
12. Add liberally to soups and stews.
13. Cut up and use to make delicious fritters.
14. Cut in small pieces, stir into eggs and fry as omelet.

NOTE:

1. To make highly nutritious cold drink:
    a. Cut up tubers and soak in water 3 days.
2. Take tubers out and crush in fresh water.
3. Strain and sweeten to taste.
4. Tubers have nice sweet, nutty taste.
5. Save cooking liquid and drink or use to make soup or stew.

ONION, WILD

ENTIRE PLANT IS EDIBLE:

1. Boil 10 minutes and eat with other vegetables.
2. Steam until tender and eat with other vegetables.
3. Put in soups and stews for flavoring and to eat.
4. Cut into small pieces and eat raw as snack.
5. Cut up and add to salads.
6. Saute in lightly greased frying pan (*Chapter 2*).

LEAVES:

1. Eat raw as a tasty snack food.
2. Tear up and put in salad with other greens.
3. Boil 3 to 5 minutes and eat as green vegetable.
4. Steam until tender and eat as green vegetable.
5. Add liberally to soups and stews.

FLOWERS:

1. Eat raw by themselves as snack.
2. Put in green salads.
3. Boil or steam and eat as vegetable.
4. Use to make delicious fritters (*Chapter 2*).

BULB:

1. Eat uncooked for nice tasting snack.
2. Cut up raw and use in green salads.
3. Boil or steam and crush when soft to make onion soup.
4. Boil or steam until tender and eat as vegetable.
5. Boil until tender. Then cream as vegetable dish (*see below*).

NOTE:

1. Eating lots of wild onions reduces cholesterol.
2. Wild onions taste, look and smell like regular onions.
3. Wild onions are the most common edible wild bulb.

COOKING POINTERS:

1. To make creamed wild onions:
    a. Put 1 tablespoon butter in warm frying pan and melt.
    b. Add dash of pepper, 2 tablespoons flour and 3/4 teaspoon salt.
    c. Put in 1 cup cut up onion leaves and bulbs.
    d. Stir in milk and let simmer until mixture thickens.
2. To make onion rings:
    a. Slice bulb and dip slices in egg.
    b. Roll in flour or meal.
    c. Drop in pan of hot oil.
    d. Deep fry until nicely browned.

PARSNIP, WILD

LEAVES:

1. Boil until tender in 2 or 3 water changes.
2. Steam until tender and eat as any other green vegetable.
3. Add liberally to soups and stews.

SHOOTS:

1. Boil until tender in 2 or 3 water changes.
2. Steam until tender and eat as any other green vegetable.
3. Add liberally to soups and stews.

ROOTS:

1. Eat cut up pieces of raw root as quick snack, or add to salads.
2. Boil 30 to 45 minutes or until tender.
3. Steam until tender as you would cultivated parsnips.
4. Slice up and fry like potatoes.
5. Bake until soft and eat with butter, salt and pepper.
6. Saute pieces in lightly greased frying pan (*Chapter 2*).
7. Add liberally to soups and stews.
8. Use for making delicious fritters (*Chapter 2*)

NOTE:

1. Wild parsnips smell similar to cultivated parsnips.
2. Flavor improves immensely if wild parsnips are not gathered before the first frost.
3. Young leaves and shoots are best if picked in the spring.

COOKING POINTERS:

1. Cook exactly as you would cultivated parsnips.

PENNYCRESS, FIELD

YOUNG LEAVES:

1. Tear up uncooked leaves and use for salads with other greens.
2. Boil in two water changes for 15 to 30 minutes.
3. Steam in two water changes for 15 to 30 minutes.
4. Add liberally to soups and stews.

SHOOTS:

1. Break into small pieces and add to salads with other greens.
2. Boil in 2 water changes for 15 to 30 minutes.
3. Steam in 2 water changes for 15 to 30 minutes.
4. Add shoots liberally to soups and stews.
5. Cut into small pieces, stir into eggs and fry as an omelet.
6. Saute in lightly greased frying pan (*Chapter 2*).

SEEDPODS:

1. Dry seedpods and use as seasoning somewhat like pepper.
2. Dry seedpods and grind or pound to make flour.

NOTE:

1. Field pennycress is naturally quite bitter, it is classified in the mustard family.
2. Field pennycress is best when eaten like spinach.
3. Serve cooked greens and shoots with butter, salt and pepper.
4. Boiled shoots make an excellent pot herb.
5. Young shoots and leaves should be gathered in the spring.
6. Save cooking liquid and drink or use to make soup or stew.

PEPPERGRASS

LEAVES:

1. Pungent young leaves can be used sparingly in salads with other greens.
2. Boil 10 to 20 minutes with 2 to 3 water changes.
3. Steam until tender with 2 to 3 water changes.
4. Add leaves liberally to soups and stews after first boiling and changing water 2 times.

SHOOTS:

1. Raw young shoots can be sparingly used in salads.
2. Boil shoots 10 minutes. Dump water out. Refill pot. Bring to boil again. Simmer 15 minutes more. Serve as green vegetable.
3. Boil in at least 2 water changes before using as potherb.
4. Add liberally to soups and stews.

SEEDPODS:

1. Use peppery green seedpods sparingly in salads.
2. Add to soups and stews as seasoning.

NOTE:

1. Seeds have a distinctive peppery taste.
2. Peppergrass is another bitter member of the mustard family.
3. Peppergrass leaves are loaded with vitamin A and C as well as iron and protein.
4. Leaves and shoots must be boiled with changes of water in order to tone down the high degree of objectional bitterness.

PICKERELWEED

YOUNG LEAVES:

1. Chop up fine and add raw pieces to green salads.
2. Boil or steam until tender and eat as green vegetable.
3. Stir fry in lightly greased pan (*Chapter 2*).
4. Add liberally to soups and stews.

SEEDS:

1. Eat starch-filled seeds by themselves as snack.
2. Add fresh seeds to green salads.
3. Dry seeds over embers or in sun and add to various cereals.
4. Roast seeds until crisp and brown and pulverize into flour.
5. Add raw or roasted seeds to soups and stews.

PLANTAIN, BROADLEAF

ENTIRE PLANT IS EDIBLE:

1. Soak in salt water 15 minutes and then boil in covered pot 10 to 15 minutes or until tender.
2. Soak in salt water 15 minutes and then steam until tender.
3. Use in soups and stews after soaking 15 minutes in salt water.

LEAVES:

1. Chop up tender young leaves and add to salads.
2. Boil young leaves in covered pot 10 to 15 minutes or until tender and eat like spinach.
3. Soak older leaves in salt water 15 minutes and then boil or steam in covered pot until tender.
4. Saute in lightly greased frying pan (*Chapter 2*).
5. Use as delicious filler for soups and stews.

SEEDS:

1. Eat raw by themselves as snack or use in salads.
2. Dry seeds over fire or in sun and eat for snack.
3. Roast until browned and eat for nutritious snack treat.
4. Pulverize or grind dry seeds into flour (*Chapter 2*).

NOTE:

1. Gather young leaves if possible. Older leaves too stringy.
2. Young plantain leaves taste somewhat like mushrooms.
3. Plant is loaded with vitamin A and C as well as many minerals.
4. Save cooking liquid and drink it or use for soups and stews.

COOKING POINTERS:

1. Try not to overcook leaves as they get tough and stringy.

POKEWEED (POKE)

YOUNG LEAVES:

1. Boil 10 to 15 minutes or until tender in 3 water changes.
2. Steam until tender and eat like any other vegetable green.
3. Add liberally to soups and stews.

YOUNG SHOOTS:

1. Boil 15 to 25 minutes in at least 2 changes of water. Lastly add fresh boiling water, cover pot and simmer until tender.
2. Add young shoots liberally to soups and stews.
3. Saute in lightly greased frying pan (*Chapter 2*).

STEMS:

1. Remove skin and leaves from young stem. Throw skin away and boil or steam leaves as greens.

2. Simmer stem until tender and eat as you would asparagus.

NOTE:

1. Large root isn't edible. It is poisonous.

2. Mature leaves aren't edible. They are poisonous.

3. Berries aren't edible. They are poisonous.

4. Seeds aren't edible. They are poisonous.

5. Young shoots with red tinge aren't edible. They are poisonous.

6. Young shoots are loaded with vitamin A.

7. Collect young shoots in early spring before they are 6" to 8" tall or leaves have unfolded.

POTATO VINE, WILD

ROOTS:

1. Peel and boil 20 to 25 minutes until tender.

2. Peel and steam until tender.

3. Bake until soft as you would sweet potatoes.

4. Roast until done as you would sweet potatoes.

5. Peel, slice thin and fry like regular potatoes.

6. Peel, slice in strips and deep fry like French fries.

7. Saute peeled pieces (*Chapter 2*).

8. Cut up and liberally add to soups and stews.

9. Dry root pieces over embers or in sun. Set aside to use later.

NOTE:

1. Never eat raw roots as they are a purgative (laxative).

2. Some of these roots are extremely bitter.

3. Bitterness can be reduced by changing the water several times while boiling.

4. The large vertical roots are much like yams.

5. Serve potato vine roots with butter, salt and pepper as you would yams.

6. Older roots are much too fibrous or woody to eat.

7. All potato vine roots have a tough skin on the outside.

8. Slice open baked or boiled root and sprinkle with brown sugar while still hot, then add butter on top of this.

9. Save cooking liquid and drink or use to make soup or stew.

PRICKLY PEAR

FRUIT:

1. Slice top off egg-shaped fruit, peel back skin and eat pulp.

2. Stew fruit pulp and eat as dessert.

3. Boil fruit pulp until it thickens suitably to be syrup.

4. Remove pulp and make jelly or jam (*Chapter 2*).

STEM:

1. Slice stem lengthwise, cut into strips and eat raw as a snack.

2. Boil or steam strips until soft and eat like string beans.

3. Roast strips over fire, peel and eat.

4. Deep fry strips like onion rings (*Chapter 2*)

SEEDS:

1. Dry seeds over embers or in sun. Grind or pound to make flour.

2. Use pulverized dry seeds as soup thickener.

NEW PADS:

1. Slice into string bean-like strips and eat raw as snack.

2. Boil or steam until tender and eat as nourishing vegetable.

3. Stir fry or saute in lightly greased frying pan (*Chapter 2*).

4. Use to make fritters (*Chapter 2*).

5. Roast over open fire and eat as vegetable dish.

6. Bake until lightly tanned and eat.

NOTE:

1. Remove bristles by rubbing with damp cloth or handful of grass.

2. Emergency water (actually a slightly bitter and sticky juice) can be squeezed or sucked from the stems.

3. Save cooking liquid and drink or use for soup or stew.

PURSLANE

ENTIRE PLANT IS EDIBLE:

1. Cut up and use uncooked in salads.

2. Makes a delicious pot herb for various meat dishes.

3. Boil 10 minutes or until tender and eat as vegetable dish.

4. Steam until tender and eat as delicious vegetable.

5. Add to soups and stews as vegetable and thickener.

LEAVES:

1. Put raw, sharp-tasting leaves sparingly in salads.

2. Boil until tender as a green vegetable.

3. Steam until tender as a green vegetable.

4. Saute until lightly browned (*Chapter 2*).

SHOOTS:

1. Pick, wash and eat raw as quick snack.

2. Pick, wash and put raw in salads.

3. Boil until tender and eat as green vegetable.

4. Steam until tender and eat as green vegetable.

5. Add liberally to soups and stews.

STEMS:

1. Boil until tender and eat like asparagus.

2. Steam until tender and eat like asparagus.

3. Saute in lightly greased frying pan (*Chapter 2*).

SEEDS:

1. Dry tiny black seeds in shady place for 7 days. Grind or pulverize for use as flour.

NOTE:

1. Purslane is rather bland. Add chicory, mustard greens or dandelions to make it more edible.
2. The shoots of this plant are one of the best wild shoots when cooked or when used raw in salads.
3. The leaves and stems are a rich source of vitamin A and C as well as iron, calcium and phosphorus.
4. To obtain more seeds, simply dry plant and shake in a bag.
5. Save cooking liquid and drink or use to make soup or stew.

COOKING POINTERS:

1. Boil thicker stems 10 minutes and drain. Dry in sun and store in brown paper bag for later use as thickener for soups and stews.
2. Blend ground seeds with equal amount of regular wheat flour for bread-making. The result is a hearty loaf with a unique texture.

QUEEN ANNE'S LACE (WILD CARROTS)

ENTIRE PLANT IS EDIBLE:

1. Excellent in soups and stews as it imparts carrot-like flavor.
2. Chop up and use for making green salad.
3. Boil until tender and eat as green vegetable.
4. Steam until tender and eat as green vegetable.

LEAVES:

1. Eat uncooked in salad.
2. Boil until tender and eat as greens.
3. Steam until tender and eat as greens.
4. Saute leaves until lightly browned (*Chapter 2*).

FLOWERS:

1. Eat flowers for quick a snack.
2. Eat flowers raw in salads.
3. Saute flowers (*Chapter 2*)
4. Use for making delicious fritters (*Chapter 2*).

SEEDS:

1. Dry over embers or in sun and set aside for crunchy snack.

ROOTS:

1. Boil or steam until tender as you would garden carrots.
2. Use as a vegetable in soups and stews.

NOTE:

1. If root doesn't smell like a carrot, don't eat it! It could be a dangerous hemlock look-alike.

2. This plant is high in fiber content and vitamin A.

3. Save cooking liquid and drink or use for soup or stew.

REDBUD TREE

FLOWERS:

1. Eat raw as quick snack.

2. Add fresh flowers to green salads.

3. Saute in butter until lightly browned (*Chapter 2*).

4. Add liberally to soups and stews.

FLOWER BUDS:

1. Eat raw as quick snack.

2. Add to green salads.

3. Boil 5 minutes and eat as vegetable.

4. Steam until tender and eat as vegetable.

5. Saute for 10 minutes in butter (*Chapter 2*).

6. Make pickled buds (*Chapter 2*).

7. Add liberally to soups and stews.

8. Use to make delicious fritters (*Chapter 2*).

YOUNG PODS:

1. Saute flat bean-like pod (*Chapter 2*).

NOTE:

1. Butter, if available, is better when wanting to saute or stir-fry flowers, flower buds and young pods.

2. Save cooking liquid and drink or use to make soup or stew.

COOKING POINTERS:

1. Flowers, flower buds and young pods can be blended together and sauteed or stir-fried for delicious vegetable dish.

RICE, WILD

RICE GRAINS:

1. Boiling is simplest method of cooking rice:

     a. The proportions: 2 cups water for each cup rice.

2. Add 1 teaspoon salt and 1 tablespoon butter.

3. Bring to boil and stir.

4. Lower heat, cover pan and simmer 30 to 45 minutes.

5. Makes about 4 cups (4 to 6 servings) of cooked rice.

6. Eat boiled rice with milk and sugar for snack or cereal.

7. Roast rice grains and grind or pound into fine flour.

8. Blend rice flour with cooking oil to make small cakes:

     a. Add nuts, fruit, meat or anything else to make tasty dish.

     b. Bake or fry cakes until nicely browned.

     c. Wrap in large green leaves and carry for future use as needed.

9. Pop rice by putting in sieve and deep frying until kernels pop.

NOTE:

1. If purplish or pinkish highly poisonous fungi is found among the seeds of rice plants, go elsewhere to gather rice.

COOKING POINTERS:

1. Cooked wild rice is excellent for use as stuffing for animals.
2. Wild rice vastly improves any recipe calling for domestic rice.
3. For extra tender rice add ½ cup more water and cook 5 minutes longer.
4. 1 pound uncooked rice (2-⅔ cups) makes 10 cups cooked rice.
5. For baking, substitute ground rice flour for ⅓ of the wheat flour called for in recipe.

ROSE, WILD

LEAVES:

1. Eat raw in salads.
2. Boil until tender and eat as greens.
3. Steam for a few minutes and eat as greens.
4. Saute in lightly greased frying pan (*Chapter 2*).
5. Add to soups and stews.

FLOWERS:

1. Eat uncooked flowers for snack after snipping off green or white bases.
2. Add fresh, sweet tasting rose petals to salads.
3. Boil flowers and eat as vegetable.
4. Steam flowers and eat as vegetable.
5. Use rose petals to make candy (*Chapter 2*).
6. Use rose petals to make jelly (*Chapter 2*).
7. Saute flowers and/or petals (*Chapter 2*).
8. Add flowers to soups and stews.

ROSE HIPS (FRUITS):

1. Eat raw by themselves for quick and nourishing snack.
2. Put uncooked hips in salads.
3. Jelly or jam can be made with hips and sour apples (*Chapter 2*).
4. Use hips to make delicious fritters (*Chapter 2*).
5. Saute hips in lightly greased frying pan (*Chapter 2*).
6. Add to soups and stews.
7. Dry hips in sun and set aside for later use.
8. Grind or grate dried hips to get powder rich in essential minerals and vitamin C.
9. Sprinkle hip powder on all kinds of cereals.
10. For excellent vitamin supplement: Cover hips with boiling water. Stir in sugar to taste. Set aside for 2 hours. Strain.

SHOOTS

1. Boil tender young shoot and eat as vegetable.
2. Steam tender young shoot and eat as vegetable.

3. Saute in lightly greased frying pan (*Chapter 2*).

4. Cut into pieces, stir into eggs and fry as omelet.

5. Add to soups and stews.

STEMS:

1. Eat uncooked stems as quick snack.

6. Put raw cut up stems in green salads.

7. Boil until tender and eat as vegetable.

8. Steam until tender and eat as vegetable.

9. Add to soups and stews.

NOTE:

1. Peel stems before eating raw or cooking as vegetable.

2. Rose hips have a delicate apple-like flavor.

3. Save cooking liquid and drink or use for soups and stews.

4. Dark roses taste stronger. Lighter flowers more subdued.

5. Rose hips are outstanding survival food. Excellent for preventing scurvy.

6. Rose hips are rich in vitamin C, much higher than oranges.

7. To use rose hips, just cut away the mature orange-red fruit. Remove seeds and eat raw.

8. Seeds are rich in vitamin E.

SHEEP SORREL (COMMON SORREL)

LEAVES:

1. Eat raw for thirst quenching snack.

2. Add raw leaves to salads and liven them up.

3. Boil or steam 3 to 5 minutes in 2 water changes until tender.

4. Add liberally to soups and stews.

5. Saute with eggs and butter (*Chapter 2*).

6. Steep 30 minutes or longer in boiling water. Strain. Sweeten to taste. Chill and drink as nutritional juice.

FLOWERS:

1. Eat raw for quick snack.

2. Put uncooked flowers in salads.

STEMS:

1. Nibble raw stems for quick snack.

2. Add to salads to liven them up.

3. Boil until tender and eat as vegetable.

4. Steam a few minutes and eat as vegetable.

5. Saute in lightly greased frying pan (*Chapter 2*).

SEEDS:

1. Add sour tasting seeds to salads.

NOTE:

1. Sheep sorrel has a mildly sour vinegary taste.

2. Eat in small quantities only. Contains oxalic acid.

3. The absorption of calcium may be blocked if large amounts of sorrel is eaten over a long period of time.

4. This plant is loaded with vitamin C.

SHEPHERD'S PURSE

ENTIRE PLANT IS EDIBLE:

1. Cut up plant and eat raw for snack or use in salads.

2. Boil until tender in 2 water changes and eat as vegetable.

3. Steam plant until tender and eat as vegetable.

4. Use entire plant in soups and stews.

LEAVES:

1. Eat raw leaves by themselves for snack or use to make salad.

2. Boil dandelion-like leaves until tender in 2 changes of water and eat as highly nutritious green vegetable.

3. Steam leaves until tender and eat as vegetable greens.

4. Add leaves liberally to soups and stews.

5. Dry leaves and pulverize to use as flour.

FLOWERS:

1. Eat raw as quick snack.

2. Add uncooked flowers to salads.

3. Saute until lightly browned (*Chapter 2*).

SEED PODS:

1. Boil heart-shaped seed pods until tender.

2. Steam heart-shaped seed pods until tender.

3. Dry seed pods and pulverize to use as flour.

4. Dry seed pods and grind to make meal for cereal.

5. Dry seed pods and use for pepper-like seasoning in soups.

NOTE:

1. Some leaves may be too peppery tasting for salads.

2. Save cooking liquid and drink or use for soups and stews.

SOLOMON'S SEAL

YOUNG SHOOTS:

1. Chop uncooked shoots into small pieces and blend with other leafy plants in salads

2. Boil 15 to 20 minutes and eat as you would asparagus.

3. Steam until tender and eat as you would asparagus.

4. Saute in lightly greased frying pan (*Chapter 2*).

5. Cut in small pieces, stir into eggs and fry as omelet.

ROOTS:

1. Boil 20 to 30 minutes and eat like potato.

2. Steam until tender and eat like potato.

3. Bake 30 minutes or until tender and eat like potato.

4. Roast on embers or in oven until tender and eat like potato.

5. Cut up and add to soups and stews.

6. Slice and fry until lightly browned like potatoes.

7. Cut up and saute in lightly greased frying pan (*Chapter 2*).

NOTE:

1. Best to gather roots in fall.

2. Baked, boiled or roasted, these roots taste much like parsnips.

3. Rootstocks are loaded with starch.

4. Save cooking liquid and drink or use to make soup or stew.

COOKING POINTERS:

1. Remove leafy heads from young shoots before cooking. The leafy heads become bitter when cooked.

SPRING BEAUTY

LEAVES:

1. Eat crispy leaves as vitamin-rich snack.

2. Chop leaves and add to green salads.

3. Boil 3 to 5 minutes in salted water and eat as greens.

4. Steam until tender and eat as any other green vegetable.

CORMS (ENLARGED STEM BASE):

1. Boil until fork punctures tough outer skin. Peel off skin. Eat with butter, salt and pepper like potatoes.

2. Boil 15 minutes. Peel tough outer skin. Mash like potatoes.

3. Bake until soft, peel off tough outer skin and eat with butter.

4. Peel off tough outer skin, slice and fry until browned.

5. Saute slices of corm in lightly greased frying pan (*Chapter 2*).

TUBERS:

1. Eat tubers raw as starchy snack or cut up and add to salads.

2. Boil 10 to 20 minutes in salted water with jacket intact. Then peel off jacket and eat whole or mash as you would a potato.

3. Boil in salted water until tender, slice thin and fry.

4. Bake tubers until soft, peel and eat like baked potato.

NOTE:

1. Leaves are edible but not outstanding when cooked.

2. Tubers and corms taste like new potatoes and cooked chestnuts.

3. A corm is nothing more than an enlarged bulb-like stem base. It's solid rather than in layers like an onion.

SPINY-LEAVED SOWTHISTLE

YOUNG LEAVES:

1. Cut off spines. Tear up leaves if not extremely bitter. Mix with other more bland greens in salad.

2. Cut off spines. Boil leaves 5 minutes in very little water. Change water and boil 15 minutes or until tender.

3. Cut off spines. Steam until tender. Eat as spinach.

4. Add liberally to soups and stews.

5. Cut off spines, chop up leaves, blend with eggs and fry for outstanding omelet.

FLOWERS:

1. Eat by themselves after twisting green base from petals.

2. Eat raw in tossed salads with leaves, shoots and roots.

3. Saute flowers and buds (*Chapter 2*).

4. Use to make terrific fritters (*Chapter 2*).

5. Crush dried flowers and add to breakfast cereals.

NOTE:

1. Flower petals have pleasantly sweet taste.

2. Cooked leaves are good served with butter or vinegar.

3. Many people enjoy bitter taste of this plant when cooked.

4. Save cooking liquid and drink or use to make soup or stew.

COOKING POINTERS:

1. If leaves are too bitter, simply mix with lamb's quarter or other more bland greens such as purslane or chickweed.

SUNFLOWER

SEEDS:

1. Eat raw, unshelled seeds as a highly nutritious quick snack.

2. Dry seeds over embers or in sun, or eat inside of roasted nut for tasty snack.

3. Eat boiled seeds for nutritious snack.

4. Boil crushed seeds to extract the pale yellow vegetable oil. Simply skim oil from surface of water.

5. Roast and pulverize or grind into flour.

6. Roast and pulverize or grind shells to make coffee (*Chapter 2*).

7. Extract nut meats from seed shells. Grind to paste. Add maple syrup or honey to make great peanut butter substitute.

8. Eat raw, roasted or dried seeds as cereal or with cereal.

FLOWERS:

1. Use to make delicious fritters (*Chapter 2*).

NOTE:

1. Sunflower seeds are high in the B vitamins.

2. Parched seeds can be pound or ground between 2 stones as Indians used to do in making bread.

COOKING POINTERS:

1. Roasted nuts from sunflower seeds can be used in any recipe calling for nuts of any kind.

2. Substitute part of flour called for in bread, roll and muffin recipes with flour made from sunflower nutmeats.

SWEET FLAG (CALAMUS)

YOUNG SHOOTS:

1. Eat raw as pleasant smelling and tasty snack.
2. Use uncooked as excellent addition to salads.
3. Boil 3 to 5 minutes and eat as green vegetable.
4. Steam until tender and eat as green vegetable.
5. Add liberally to soups and stews.

ROOTSTOCK:

1. Use to make delicious candy.
   a. Peel rootstock and cut into 1" lengths.
2. Boil 1 to 2 hours until tender. Change water 4 or 5 times.
3. Now make the candy (*Chapter 2*).
4. Cut into pieces and add to soups and stews.

NOTE:

1. Leaves and rootstock have pleasant spicy smell.
2. Inner part of young stalks are packed with sweet half-formed leaves. That's why they're so great as snacks and in salads.
3. Dried roots can be pounded to powder and used as a natural insecticide.
4. Save cooking liquid and drink or use to make soup or stew.
5. Be extremely careful not to confuse sweet flag with blue flag.
6. Blue flag is poisonous. Sweet flag is not.
7. Blue flag leaves are a dull, bluish-green and have no smell.
8. Sweet flag leaves are glossy, yellow-green with a spicy smell.

THISTLE, BULL

ENTIRE PLANT IS EDIBLE:

1. Boil until tender and eat as vegetable.
2. Steam until tender and eat as vegetable.
3. Add to soups and stews.

LEAVES:

1. Eat raw as snack after cutting away barbed spines.
2. Add to green salads after stripping away barbed spines.
3. Boil until tender after cutting away barbed spines,
4. Steam until tender after cutting away barbed spines.

YOUNG STEMS:

1. Peel and eat raw as snack. They are crispy, crunchy and exceptionally tasty.
2. Peel and add to salads.
3. Boil peeled stems until tender and eat as vegetable.
4. Steam peeled stems until tender and eat as vegetable.
5. Stir into soups and stews.
6. Fry until nicely browned as you would potatoes.

7. Saute stems (*Chapter 2*).

EARLY SHOOTS:

    1. Eat as tasty snack.

    2. Add to green salads.

    3. Boil until tender and serve as green vegetable.

    4. Steam until tender and serve as green vegetable.

    5. Fry until browned as you would potatoes.

    6. Saute in lightly greased frying pan (*Chapter 2*).

ROOT:

    1. Peel, boil until soft and eat as potato.

    2. Peel, steam until soft and eat as potato.

    3. Peel, slice and fry until nicely browned.

    4. Roast until brown, let cool and pound or grind into flour.

    5. Boil until roots turn to mush. Set aside to dry. Then grind or pound into flour.

    6. Stir root pieces into soups and stews.

    7. Saute pieces in lightly greased frying pan (*Chapter 2*).

NOTE:

    1. Save cooking liquid parts and use for soups and stews.

    2. Most nourishing part of the thistle is flower stem when peeled.

    3. Leaves , stems and shoots are somewhat like celery.

    4. The roots of new thistles are one of best wild survival foods.

VIOLET, SWEET

ENTIRE PLANT IS EDIBLE:

    1. Boil 2 to 3 minutes and eat as green vegetable.

    2. Steam 2 to 3 minutes and eat as green vegetable.

    3. Use liberally in soups and stews.

LEAVES:

    1. Eat raw as snack.

    2. Eat raw with flowers and buds in green salads.

    3. Boil in covered pot 3 to 5 minutes and eat as vegetable.

    4. Steam in covered pot until tender and eat as green vegetable.

    5. Add to soups and stews as an okra-like thickener.

    6. Saute in lightly greased frying pan (*Chapter 2*).

    7. Use to make jelly and jam (*Chapter 2*).

FLOWERS:

    1. Pick blossoms from plant and eat raw as tasty snack.

    2. Eat raw in salads with raw young leaves.

    3. Boil 2 to 3 minutes and eat with leaves or other greens.

    4. Steam until soft and eat with leaves or other greens.

    5. Use liberally in soups and stews.

6. Saute in lightly greased frying pan (*Chapter 2*).

7. Use to make candy (*Chapter 2*).

ROOTSTOCK:

1. Boil until tender and eat as vegetable.

2. Steam until soft and eat as vegetable.

3. Roast until done and eat as vegetable.

4. Bake until soft and eat as vegetable.

5. Slice and fry until browned.

6. Slice and saute in lightly greased frying pan (*Chapter 2*).

7. Cut in pieces and use to make fritters (*Chapter 2*).

8. Use liberally in soups and stews.

NOTE:

1. Violets are extremely high in vitamin A and C.

2. Save nourishing liquid used for cooking plant parts and drink or use for soups and stews.

3. Violet leaves are quite bland and are best when combined with mustard, dandelion or other stronger tasting greens.

4. Violet greens are eaten much like spinach.

COOKING POINTERS:

1. Add flowers to flour when baking for unusual flavor.

2. Barely cover plant parts when boiling.

WAPPATO (DUCK POTATOES)

TUBERS:

1. Eat raw as snack although mildly bitter.

2. Boil or steam until tender and eat like potatoes.

3. Boil until tender, peel, mash and eat as mashed potatoes.

4. Boil until tender, peel and mash. Make patties and fry like hash browns.

5. Roast until lightly browned and eat as potato.

6. Saute in lightly greased frying pan (*Chapter 2*).

7. Bake like potatoes, peel when done and eat.

8. Slice and fry like potatoes.

9. Dry tuber and grind or pound into flour.

10. Cut up and make a creamed or scalloped potato-like dish.

11. Cut into strips and deep fry like French fried potatoes.

NOTE:

1. Tubers form in the fall.

2. Tubers have mildly unpleasant taste when eaten raw.

3. To get tubers, dig down in mud around plant and cup hands around tubers. Pull slowly up and out.

4. Tubers are starchy and highly nutritious.

5. To dry tubers for storage:

a. Boil 30 minutes, drain well and cut into 1/4" to 1/2" slices.

b. Lay in warm oven or the sun to thoroughly dry.

6. To use at later date:

    a. Soak in water 15 minutes.

    b. Bring to boil and let simmer 20 to 30 minutes.

WATER CHESTNUT

FLOATING LEAVES:

1. Tear into small pieces and use in salads.

2. Boil or steam 10 minutes and eat as green vegetable.

3. Add liberally to soups and stews.

SEEDS (NUTS AND MEAT):

1. Boil 10 to 15 minutes, eat as quick snack or add to salads.

2. Steam until tender and eat as snack or add to salads.

3. Roast until crunchy and eat as quick snack.

4. Roast and grind or pound into flour substitute.

NOTE:

1. Water chestnut is extremely high in nutrients.

2. Save cooking liquid and drink or use in soups and stews.

WATERCRESS

LEAVES:

1. Eat raw as a quick snack or add to salads.

2. Boil or steam until tender and eat like spinach.

3. Add liberally to soups and stews.

4. Saute in lightly greased frying pan (*Chapter 2*).

5. Tear into pieces, add to eggs and fry as omelet.

STEMS:

1. Cut up and eat raw as snack or add to salads as garnish.

2. Boil or steam until tender and eat as vegetable.

3. Add liberally to soups and stews for flavor.

BLOSSOMS:

1. Eat raw as delicious snack or add to salads as garnish.

2. Make complete salad with raw blossoms, leaves and stems.

3. Add to soups and stews.

4. Put peppery flavored blossoms on sandwiches for taste treat.

NOTE:

1. Only plant parts growing above water are edible.

2. Watercress, rich in vitamin C and minerals, is one of most desirable wild greens known.

3. Most people prefer to eat watercress raw than cooked.

261

COOKING POINTERS:

1. Stuff fish with watercress (after cleaning), then smoke, fry or bake. Add garlic if available.
2. Makes good soup by itself, especially when potatoes are added.
3. Adds flavor when mixed with bland tasting greens.

WATER LILY, FRAGRANT

LEAVES:

1. Eat young unfurled raw leaves as snack.
2. Add young unfurled raw leaves to green salads.
3. Boil young unfurled leaves 5 to 10 minutes and eat.
4. Steam young unfurled leaves until tender and eat as vegetable.
5. Add liberally to soups and stews.

FLOWERS:

1. Eat white flower petals by themselves as excellent snack.
2. Eat unopened raw buds for delicious snack or use in salads.
3. Use white petals in salads with all other flower parts.
4. Boil unopened buds 5 to 10 minutes and eat with butter.
5. Steam unopened buds until tender and eat with butter.
6. Saute buds in lightly greased frying pan (*Chapter 2*).
7. Use for making delicious fritters (*Chapter 2*).

TUBERS:

1. Peel, slice and then boil tubers until tender.
2. Mash boiled tubers and eat like mashed potatoes.
3. Dry mashed tubers and use as flour for baking.
4. Bake like small potato until soft.
5. Cut sweetish tuber in pieces and add to soups and stews.
6. Slice thin and fry like potatoes.
7. Roast until nicely browned.

SEEDS:

1. Fry and pop large seeds as you would popcorn and eat.
2. Seeds make excellent breakfast with milk and sugar.
3. Parch (dry by heat) seed kernels and grind or pound. Use as delicious substitute for cornmeal creamed-corn.
4. Crack seed shells and cook nutmeats as you would rice.

NOTE:

1. Tubers are about the size of a hen's egg.
2. Seeds are high in protein, oil and starch.
3. Tubers contain the most starch from autumn to Early spring.
4. Roots peel easily after boiling, steaming or roasting.

5. Tubers must be loosened from the mud with your feet. Gather them up as they float to the surface.

6. Save cooking liquid and drink or use to make soup or stew.

COOKING POINTERS:

1. If tuber flavor is objectionable, boil in 3 changes of water.

WATER PLANTAIN

LEAVES:

2. Eat by themselves as snack or add to salads.

3. Soak in salt water 15 minutes and boil in covered pot.

4. Soak in salt water 15 minutes and steam until tender.

5. Add liberally to soups and stews.

ROOTSTOCKS:

1. Boil until tender and eat like potatoes.

2. Steam until tender and eat like potatoes.

3. Bake until soft and eat like baked potato.

4. Cut up and add liberally to soups and stews.

5. Slice and fry like potatoes.

6. Mash boiled rootstock, make patties and fry until crisp.

SEEDS:

1. Dry in sun and grind or pound into flour.

2. Make delicious pancakes with plantain flour (*Chapter 2*).

3. Roast seeds and eat as nutritional treat.

NOTE:

1. Save cooking liquid and drink or use for soups and stews.

2. Plantain is rich in minerals as well as vitamin A and C.

3. Run finger up spike to easily collect seeds.

4. Gather young leaves if possible as older leaves are stringy.

COOKING POINTERS:

1. Try to not overcook leaves as they get tougher.

2. Older leaves have woodsy taste. Best to eat with cream sauce after being pureed and strained to get rid of obnoxious fibers.

YELLOW DOCK (CURLY DOCK)

LEAVES:

1. Eat young leaves as snack before they become bitter.

2. Use uncooked leaves as a green in salads.

3. Boil young leaves 10 minutes and eat like spinach.

4. Steam young leaves until tender and eat like spinach.

5. Add liberally to soups and stews.

6. Stir fry in 2 or 3 tablespoons butter and water.

SEEDS:

1. Roast lightly and eat as nutritional snack.

2. Hull and grind or pound into palatable flour. Do this when seeds appear tan rather than green.

ROOTS:

1. Boil until tender as you would carrots.

2. Steam until tender as you would carrots.

3. Roast until done as you would carrots.

4. Cut up and add liberally to soups and stews.

NOTE:

1. Boiled or steamed young leaves taste like beet greens.

2. Extremely high in vitamin A, minerals and vegetable protein.

3. Save cooking liquid and drink or use for soups and stews.

4. Curly dock loses little of its bulk when boiled or steamed.

COOKING POINTERS:

1. If leaves are bitter when boiled, use 2 to 3 water changes to eliminate bitterness and make them tender.

2. Use seed flour by blending half and half with regular flour.

5

FERNS AS A GOOD FOOD SOURCE

Ferns are abundant in moist areas of all climates. They are especially easy to find. Look in gullies, on stream banks, in forested areas, along the sides of hiking trails and on the edge of woods.

Ferns, by and large, are a safe plant to cook and eat. Some are distastefully bitter and certainly not palatable. Yet, no fern is known to be poisonous.

BRACKEN FERN

STALK OR STEM:

1. Eat fresh stems raw in small quantities as snack.
2. Cut up raw stems and add small amounts to a green salad.
3. Boil stems and eat as you would asparagus.
4. Steam and eat as vegetable green.
5. Add to soups and stews.
6. Cut in pieces, stir into eggs and fry as an omelet.
7. Saute stalks-stems (*Chapter 2*).
8. Deep fry cut up pieces (*Chapter 2*).

FIDDLEHEADS (YOUNG SHOOTS – LEAVES BEFORE THEY UNFURL):

1. Eat fresh shoots raw in small quantities as snack.
2. Eat small amounts in green salad.
3. Boil and eat as vegetable.
4. Steam and eat as vegetable.
5. Add to soups and stews.
6. Stir into eggs and fry as delicious omelet.
7. Saute fiddleheads (*Chapter 2*).
8. Deep fry fiddleheads (*Chapter 2*).
9. Use for making fritters (*Chapter 2*).

NOTE:

1. It's best to not eat this fern raw in large quantities on a regular basis. Cooked fern is always fine to eat.
2. Gather top 4" to 6" of young fern leaf. Break off and toss aside the curled wooly covered tips.
3. Drink nourishing liquid used to cook plant parts.

EDIBLE FERN PARTS IN GENERAL

STALK OR STEM:

1. Eat raw as snack or cut up and add to green salad.
2. Boil in salted water and eat like asparagus.
3. Steam and eat as green vegetable.
4. Add to soups and stews.
5. Cut in pieces, stir into eggs and fry as an omelet.
6. Saute stalks-stems (*Chapter 2*).

7. Deep fry cut up pieces (*Chapter 2*).

FIDDLEHEADS (YOUNG SHOOTS-LEAVES BEFORE THEY UNFURL):

1. Eat fresh shoots-leaves raw as snack.

2. Eat fiddleheads in a green salad.

3. Boil or steam and eat as vegetable.

4. Add to soups and stews.

5. Stir into eggs and fry as delicious omelet.

6. Saute fiddleheads (*Chapter 2*).

7. Deep fry fiddleheads (*Chapter 2*).

8. Use for making fritters (*Chapter 2*).

NOTE:

1. Break off stalk and draw closed hand over it. This removes the inedible fuzz or wool-like material.

2. Select young stalks during April and May for best eating. Break off as long as they remain tender.

3. Drink nourishing liquid used to cook plant parts.

OSTRICH FERN

STALK OR STEM:

1. Eat fresh stems raw as snack.

2. Cut up raw stems and add to a green salad.

3. Boil or steam until tender and eat like asparagus.

4. Add to soups and stews.

5. Cut in pieces, stir into eggs and fry as an omelet.

6. Saute stalks (*Chapter 2*).

7. Deep fry cut up pieces (*Chapter 2*).

FIDDLEHEADS (YOUNG SHOOTS – LEAVES BEFORE THEY UNFURL):

1. Eat fresh shoots raw as snack.

2. Eat raw in a green salad.

3. Boil and eat as vegetable.

4. Steam and eat as a vegetable.

5. Add to soups and stews.

6. Stir into eggs and fry as delicious omelet.

7. Saute fiddleheads (*Chapter 2*).

8. Deep fry fiddleheads (*Chapter 2*).

   a. Use for making fritters (*Chapter 2*).

NOTE:

1. Collect fiddleheads (coiled shoot) during April and May while under 6" high.

2. Wash or scrape off inedible brown scales off in cold water.

3. Drink nourishing liquid used to cook plant parts.

6

TREES — AN EXCELLENT FOOD SUPPLY

The inner bark of some trees — the layer next to the wood -- can be eaten raw or cooked. Avoid the outer bark. It contains large amounts of tannin and is extremely bitter.

Flour can be made by pulverizing the inner bark of a number of trees – aspen, birch, cottonwood, pine, slippery elm and willow.

One outstanding example of a tree food source is the pine.

The inner bark is high in vitamin C. The nuts, needles, twigs and sap are all edible. The nuts (eaten raw or roasted) grow in woody cones hanging near the tips of the branches. When mature, they fall out of the ripe cone.

SYRUP TREES

Certain kinds of trees are known to provide nourishing sap. They are invaluable to a person stranded in the wilderness who is desperately trying to stay alive. The sugar maple is the best known and most widely used source of sap for making sweet syrup and sugar. Others used for this are the birch, butternut, hickory and sycamore. Each has to be tapped and the sap collected in buckets or cans. The sap is then boiled down to syrup and further down if sugar (*Chapter 2*) is the desired end result. The sap from the above mentioned trees produce a syrup comparable to maple syrup.

These trees are also excellent sources of drinking water in a survival situation. This is especially important to know if you happen to be in an area where the water is polluted or in short supply.

BIRCH TREE

INNER BARK:

1. Eat raw strips as good emergency rations.
2. Dry and pound or grind into flour for making bread.
3. Cut in strips and boil as noodles in soup or stew.

TWIGS:

1. Dry thoroughly and eat as crispy treat.

BUDS:

1. Eat uncooked as quick snack or add raw to green salads.
2. Cook (boil or steam) and eat as vegetable.
3. Add liberally to soups and stews.
4. Saute buds (*Chapter 2*).
5. Use for making fritters (*Chapter 2*).

LEAVES:

1. Cook (boil or steam) until tender.
2. Saute leaves (*Chapter 2*).

SAP:

1. Drink nutritious sap in spring as it comes from tree.
2. Boil sap down to make delicious syrup and sugar (*Chapter 2*).

NOTE:

1. Drink nourishing liquid used to cook tree parts.
2. The sap from every kind of birch tree is edible.
3. Sap flow is copious but only half as sweet as maple.
4. Dry inner bark in sun and put in sealed jars.

BUTTERNUT TREE (WALNUT FAMILY)

NUTS:

1. Eat raw as snack or add to salad made of various greens.
2. Use nuts in baking — cakes, pies, breads, etc.
3. Roast nuts on embers of fire or in oven.
4. Add nut meats to soups and stews.
5. Mash boiled nuts like potatoes when ready to eat.
6. Boil crushed nuts, skim oil from surface of water and use as cooking oil.
7. Cool the oil and use as excellent butter substitute.
8. Grind or crush nuts to make peanut butter substitute.
9. Use nut meats to make candy (*Chapter 2*).
10. Pulverize or grind roasted nuts to make excellent coffee substitute (*Chapter 2*).
11. Grind or pulverize dried nuts and eat as grits or gruel.

SAP:

1. Drink uncooked sap in spring as it comes from tree.
2. Boil sap down to make delicious syrup and sugar (*Chapter 2*).

NOTE:

1. Nuts and sap are both top notch survival foods.
2. Gather nuts in autumn as they fall from trees.
3. Store nuts in their own shells whenever possible.
4. Nut meats get rancid and inedible if exposed to moisture, heat and light.

HICKORY

NUTS:

1. Eat hickory nuts raw as nutritious snack.
2. Add nuts to salad made up of various greens.
3. Use nuts in baking — cakes, pies, breads, etc.
4. Roast nuts on embers of fire or in oven.
5. Boil the kernel that lies within the shell.
6. Mash boiled nuts like potatoes when ready to eat.
7. Boil crushed nuts and skim oil from surface of water.
    a. Hickory nut oil makes a great cooking oil.
    b. Cool the oil and use as a butter substitute.
8. Grind or crush nuts to make peanut butter substitute.
9. Use nut meats to make candy (*Chapter 2*)

SAP:

1. Drink uncooked sap as it comes from tree.
2. Boil sap down to make delicious syrup (*Chapter 2*).
3. Boil down further to make sugar (*Chapter 2*).

NOTE:

1. Nuts and sap are both top notch survival foods.
2. Hickory trees are to be tapped for sap in late winter.
3. Gather nuts in autumn as they fall from trees.
4. Separate some nuts from their shells by boiling.
5. Boiled hickory nuts can be ground and mixed with cornmeal, potatoes or sweet potatoes for a tasty dish.

MAPLE TREE

INNER BARK:

1. Eat raw strips as good emergency food.
2. Dry and pound into flour for making bread.
3. Cut in strips and boil as noodles in soup or stew.
4. Cook (boil or steam) until tender and eat as vegetable.

TWIGS:

1. Dry in sun and eat as crispy treat.

SEEDS:

1. Eat raw, roasted, boiled or dry seeds as quick snack.

YOUNG LEAVES:

1. Eat raw by themselves for snack or use as salad green.
2. Boil (or steam) until tender and eat as vegetable.
3. Saute leaves (*Chapter 2*).

SAP:

1. Drink nutritious sap in spring as it comes from tree.
2. Boil sap down to make delicious syrup (*Chapter 2*).
3. Boil down further to make maple sugar (*Chapter 2*).

NOTE:

1. Maple syrup is one of the best kinds of survival food.
2. Maple trees should be tapped for sap in early spring.
3. Best sap flow requires cold nights followed by warm days.
4. About 35 gallons of sap makes one gallon of maple syrup.
5. Leaves are very rich in sugar content.

SYCAMORE

INNER BARK:

1. Eat raw strips as good emergency rations.
2. Dry and pound or grind into flour for making bread.

3. Cut in strips and boil as noodles in soup or stew.

4. Boil or steam until tender and eat as vegetable.

TWIGS:

1. A crispy treat when dried and then eaten as snack.

SEEDS:

1. Chew raw or roasted seeds as quick snack when hungry.

2. Dry seeds in sun or over embers and eat for tasty snack.

YOUNG LEAVES:

1. Eat raw by themselves for quick snack.

2. Tear in pieces and use as salad green.

3. Cook (boil or steam) and eat as vegetable (*Chapter 2*).

4. Saute leaves (*Chapter 2*).

SAP:

1. Drink nutritious sap in spring as it comes from tree.

2. Boil sap down to make delicious syrup or sugar (*Chapter 2*).

NOTE:

1. Sycamore sap is one of the best kinds of survival foods.

2. Save cooking liquid and drink or use for soups and stews.

3. Sycamore sap is excellent substitute for drinking and cooking water when no unpolluted water is available.

MORE TREES WITH EDIBLE NUTS

Nuts can be eaten either fresh or dried, cooked or uncooked. They contain much valuable protein. Great food value is derived from a nut's oil content. A single pound of nut meats provides a person with around 3,000 calories. Nuts are unquestionably among the most nutritious of all plant foods. They are extremely useful in a survival situation.

AMERICAN BEECH

NUTS:

1. Eat beechnut kernels raw as nutritious snack.

2. Slowly roast nuts and eat them whole for snack.

3. Add nuts to salad made up of various greens.

4. Use nuts in baking — cakes, pies, breads, etc.

5. Add nut meats to soups and stews.

6. Boil crushed nuts and skim oil from surface of water to make cooking oil and butter substitute.

7. Grind or crush nuts to make peanut butter substitute.

8. Pulverize roasted nuts to make coffee substitute (*Chapter 2*).

9. Use nuts to make candy (*Chapter 2*).

INNER BARK:

1. Eat raw strips as good emergency rations.

2. Dry and pound or grind into flour for making bread.

3. Cut in strips and boil as noodles in soup or stew.

SAWDUST:

1. Boil or roast sawdust and blend with flour for making bread.

LEAVES:

1. Cook (boil or steam) young leaves in the spring.

NOTE:

1. Thin shelled beechnuts are small, sweet and delicious.
2. Heat nuts and briskly rub together to remove shells.
3. Gather nuts just after they have fallen in October.

CHESTNUT

NUTS:

1. Eat chestnuts raw as nutritious snack.
2. Add nuts to salad made up of various greens.
3. Roast nuts on embers of fire or in oven and eat.
4. Boil crushed nuts and skim oil from surface of water.
   a. Chestnut oil makes a great cooking oil.
   b. Let oil cool and use as excellent butter substitute.
5. Grind or crush nuts to make peanut butter substitute.
6. Grind or pulverize dried nuts and eat as grits or gruel.
   a. Grind dried nuts finer and blend with equal amount of cornmeal.
   b. Shape into small cakes or cookies and bake until crispy.
7. Mash boiled nuts and eat as you would mashed potatoes.
8. Use nut meats to make candy (*Chapter 2*).
9. Add nut meats to soups and stews.
10. Put cooked or uncooked nutmeats in various desserts.
11. Grind roasted nuts to make coffee substitute (*Chapter 2*).

NOTE:

1. Pick nuts off ground in late summer and early fall.
2. Both raw and cooked nuts are one of best survival foods.

COOKING POINTERS:

1. Blend acorn nutmeat meal with regular baking flour. Adds agreeable nutty flavor to the cooked or baked items.
2. Put nutmeats with cornmeal to make terrific stuffing.

WHITE OAK

NUTS:

1. Eat acorns raw or cooked for nutritious snack.
2. Add acorn nuts to salads made up of various greens.
3. Roast acorns on embers of fire or in oven and eat.
4. Boil crushed nuts and skim oil from surface of water to make cooking oil and butter substitute.
5. Crush acorn nuts to make peanut butter substitute.
6. Pulverize dried nuts and eat as oatmeal-like cereal.

a. Grind nuts finer and blend with equal amount of cornmeal.

7. Form into small cakes and cookies and bake until crispy.

8. Use acorn nut meats to make candy (*Chapter 2*).

9. Add acorn nut meats to soups and stews.

10. Grind roasted nuts to make coffee substitute (*Chapter 2*).

11. Mash boiled nuts and eat as you would mashed potatoes.

   a. Make into patties and fry as you would potato patties.

NOTE:

1. Most acorns aren't edible because of bitterness caused by an excess of tannin.

2. Only the white oak family of trees have sweet acorns.

3. Put acorn nuts with cornmeal to make stuffing for game.

4. Add acorn meal to baking flour for nutty flavored bread.

5. Leave acorns in their shells to store.

6. Pick acorns off ground in late summer and early fall.

7. Raw and cooked acorns are one of best survival foods.

7

MUSHROOMS TO EAT OR NOT TO EAT

There are many misconceptions about mushrooms and their imagined food value. In reality, they offer absolutely **nothing** in the way of nutrition. And they're terribly difficult to digest. It's hardly worth the effort required to bend over and pick one up! It's certainly ridiculous to waste time and energy hunting for, cooking and eating mushrooms.

Even the world's most widely respected survival experts can make mistakes regarding mushrooms. John "Lofty" Wiseman, for example, makes these erroneous statements in his otherwise excellent video presentation: OUTDOOR SURVIVAL: SAS SURVIVAL TECHNIQUES:

1. "Fungi [mushrooms]: slightly more nutritional value than plants."

2. "Fungi [mushrooms] come above plants on the nutritional ladder. Pound for pound they got more nourishment."

This is simply not the case and John certainly should have known better!

Then why even bother learning to identify edible mushrooms? Why go to the trouble of preparing mushrooms to eat? And better yet, why eat mushrooms in the first place? No logical answers can be given to the above questions.

On the other hand, the U.S. military takes another approach to the identifying, collecting and eating of mushrooms. The official Army survival manual, FM 21-76, gives this warning: **"Do not eat mushrooms in a survival situation! The only way to tell if a mushroom is edible is by positive identification. There is no room for experimentation. Symptoms of the most dangerous mushrooms affecting the central nervous system may show up after several days have passed when it is too late to reverse their effects."**

Need more be said?

8

THE ART OF MAKING NOURISHING TEA

Place broken sprigs or crushed leaves and water in large saucepan or pot. Bring to boil. Cover. Cut heat down. Simmer 15 minutes. Take off heat. Steep 10 minutes. Strain and sweeten to taste.

or

Pour boiling water over crushed dried or fresh leaves. Cover. Set aside to steep 15 minutes. Strain and sweeten to taste.

or

Put dried or fresh leaves in pot boiling water. Cover. Simmer 5 minutes. Steep 10 minutes or more. Strain and sweeten to taste.

NOTE: Use about 1 teaspoon crushed dried leaves or 2 tablespoons of crushed fresh leaves for each cup of water used.

A FEW EASY-TO-FIND PLANTS FOR BREWING NUTRITIOUS TEA IN THE WILDS:

ALFALFA: Dry and crush flower heads and young leaves.

BIRCH: Twigs. Nutritious but bland. Mix with others.

BLACKBERRY: Crumpled dry leaves. Loaded with vitamin C.

CHAMOMILE: Entire fresh or dried plant, flowers and/or leaves.

CLOVER: Dried flower heads. An extremely healthy tea.

COMFREY: Crushed Dried leaves.

MINT: Crushed fresh or dry leaves. Highly aromatic.

PEPPERMINT: Fresh or dried leaves. Highly aromatic.

PERSIMMON: Dried leaves. Loaded with vitamin C.

PINE: Fresh pine needles. Rich in vitamin A and C.

   1. Light green needles in spring make best tea.

   2. Older needles also bring good results.

PLANTAIN: Crushed fresh or dried leaves. One of best backwoods teas.

RED RASPBERRY: Dried leaves. Rich in vitamin C.

ROSE: Fresh or dried rose hips and fresh petals. Very rich in vitamin C.

SLIPPERY ELM: Small chips of slimy inner bark. Pleasant tasting and quite nourishing.

SPEARMINT: Fresh or dried leaves. Highly aromatic.

STINGING NETTLE:Young leaves and shoots. Good source of iron, protein and vitamin A and C.

STRAWBERRY: Dried leaves. Rich in vitamin C. Pleasant tasting.

SWEET VIOLET: Dried leaves. Loaded with vitamin A and C.

WINTERGREEN: Fresh or dried leaves. Delightfully aromatic.

PART III

EDIBLE PLANT IDENTIFICATION
(BY WORDS AND PICTURES)

ALFALFA

Grows to 3" tall. Oval shaped leaves in groups of three. Tiny blue-violet clover-like blossoms (June to August). Widespread throughout U.S. (March to October).

AMARANTH, GREEN

Shallow bright red taproot. Stout, slightly hairy stem. Dull green egg-shaped leaves. Small greenish flower clusters. Commonly found in rich cultivated fields, yards, fence rows and waste ground. Troublesome weed. Widespread throughout U.S.

AMERICAN BEECH

Tree 45' to 60' tall. Bark dark gray with light gray patches. Leaves paper thin with coarse sharp teeth. Nuts in tiny burr with prickles. Ripe (September/October). Found in rich woods, uplands, moist rocky ground, etc. Common in many parts of U.S.

ASPARAGUS, WILD

Grows to 9' high. Lacy, dark green plant. Needle-like branches. Small green bell-shaped flowers followed by bright red berries (June) with black seeds. Found in gardens, fields, along ditches, meadows, orchards, etc. Widespread throughout U.S.

BIRCH

Tree to 60' tall. Simple saw tooth leaves. Sweet, aromatic black or gray papery bark peels in curls. Broken twigs and leaves have strong wintergreen smell. Found in rich woods, on river banks, moist fertile ground, etc. Common in many parts of U.S.

276

BLACKBERRY

Sprawling thorny shrub. Double toothed leaves. Showy white/pale pink 5-petaled flowers (April to July). Juicy purple/black berries (June to September). Found on wood edges, fields, roadsides, along fences, sunny thickets, etc. Widespread throughout U.S.

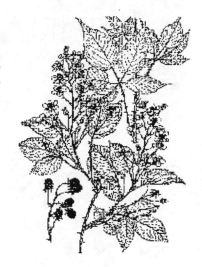

BLUEBERRY (BILBERRY)

Grows to 2' tall. Narrow leaves with tiny teeth. Urn-shaped 5- petaled flowers (May to July). Berries (August to September) covered with white powder. Found in open woods, bogs, fields, thickets. Extremely adaptable. Widespread throughout U.S.

BRACKEN FERN

Grows 1' TO 6' high. Broad triangle-shaped leaves. Long creeping rootstalks. New ferns (fiddleheads) come out in spring. Found in open woods, thickets, waste places. Often found in areas with no other plants. Widespread throughout U.S.

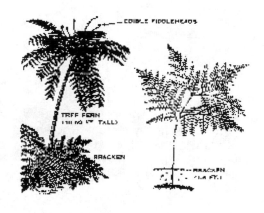

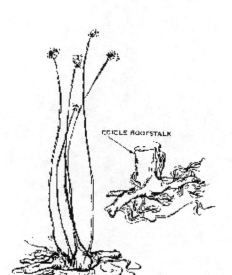

BULRUSH

Tall dark green water plant. Leafless, stiff round stems grow 3' to 9' high. Clusters of brown, bristly flower spikes loaded with seeds and pollen. Thick scaly root stocks. Found growing in thick patches in shallow water or mud. Widespread throughout U.S.

BUTTERNUT TREE

Grows 30' to 45' tall. Broad, spreading medium-size tree. Heavy lower branches. Scraggly limbs, sparse foliage. Very rough bark. Found in pastures, rich woods and on moist hillsides. Widespread throughout U.S.

BURDOCK

Grows 4' to 6' high. Resembles rhubarb. Hollow stems. Thick fleshy taproot. Egg-shaped deep green leaves. Clusters of purple flowers on short stalks (July to October). Found on neglected farm lands, fields, gardens, etc. Widespread throughout U.S.

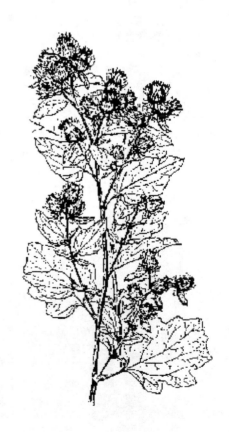

278

CAMAS LILY (WILD HYACINTH)

Grows 1' to 2' high. Showy spikes of bright blue flowers (May through June). No branches or stems. Grass-like leaves. Found along streams, in wet fields, meadows, moist woods. Widespread throughout U.S.

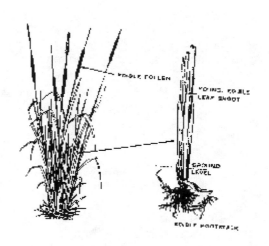

CATTAIL

Grows to 8' high. Spongy, long sword-like leaves. Yellow pollen laden flowers on erect stalks (May to July). Straight stems with no branches have hotdog-like seed heads. Found in marshes, ponds, shallow water, ditches, etc. Widespread throughout U.S.

CHAMOMILE

Grows to 2' high. Bright green stemless leaves. Delicate apple or pineapple smell. Daisy-like, yellow-centered, tight budded flowers with white petals (May through October). Found almost everywhere — meadows, alongside roads, etc. Widespread throughout U.S.

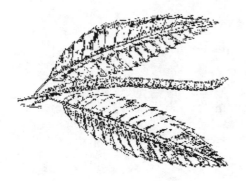

CHESTNUT

Tree grows from 50' to 70' tall. Large spear-shaped, coarsely toothed leaves. Fruit is 2" to 3" bur containing 3 nuts. Yellow-green flowers (July). Found in well-drained forests, rocky woods, pastures, etc. Common in many parts of U.S.

CHICKWEED

Succulent, straggling stems up to 1' long. Hairy leaf stalks creep on ground. Oval, pointed leaves. Small white star-shaped flowers (March to September). Found in fields, gardens, sides of road, waste places and cultivated grounds. Widespread throughout U.S.

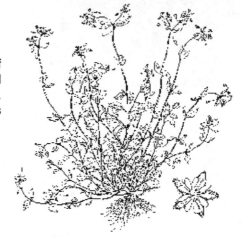

CHICORY, WILD

Grows 1' to 4' high. Deep carrot-like root. Red veined leaves at plant base like the dandelion. Milky sap. Hollow stems rough and hairy. Abundant flowers. Found in fields, on sides of roads, fence rows, lawns, pastures, etc. Widespread throughout U.S.

CLOVERS

Grows to 18" high. Pink to red to purple rounded heads (May to September). Found everywhere — fields, lawns, parks, roadsides, etc. Widespread throughout U.S.

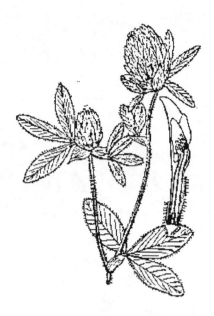

COMFREY

Grows to 3' high. Rough and hairy. Stout, hollow stem. Large hairy leaves cause itching when skin is touched. Bell-shaped flowers (May to June). Found in wet soils — river banks, ditches, roadsides, meadows, etc. Widespread throughout U.S.

CRAB APPLE, WILD

Small tree. Leaves and fruit on short twigs with thorny ends. Short pink-white flowers with 5 petals (April to June). Tiny apples 1" across (September to November). Found in open woodlands and edge of woods, yards, etc. Widespread throughout U.S.

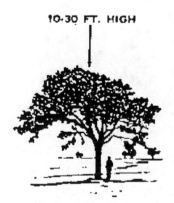

10-30 FT. HIGH

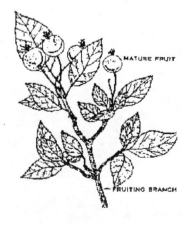

MATURE FRUIT

FRUITING BRANCH

281

DAISY OX-EYE

Grows to 2' high. Numerous flowers with oblong, white petals around yellow pebbly center. Leaves narrow, oblong and deeply scalloped. Found in meadows, waste areas, fields, on side of roads, etc. Widespread throughout U.S.

DANDELION

Grows to 2' high or more. Milky juice in hollow stalk. Thick taproot. Jagged dark green, hairless leaves clustered near ground. Golden yellow flowers on leafless stem. Found everywhere — lawns, parks, gardens, roadside, fields, etc. Widespread throughout U.S.

DAY LILY

Grows from 2' to 5' high. 3" to 5" showy orange flowers on top of naked stem open only one day. Frequently grows in dense patches. Found along roadsides and ditches in large clumps. Widespread throughout U.S.

EDIBLE FERN PARTS
Widespread throughout U.S.

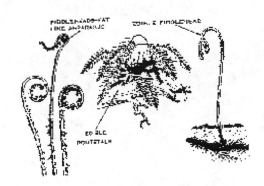

ELDERBERRY

Grows to 12' high. Thick, white pith in stems and twigs. Fragrant white flowers in bowl-shaped clusters (June to July). Tiny purplish berries (July to September). Found in ditches, thickets, along roads, river banks, etc. Widespread throughout U.S.

EVENING PRIMROSE

Grows to 5' high. Long carrot-like root. Reddish stem. Tall leafy spike with large yellow flowers. Subtle fragrance. Flowers open in late afternoon. Found on roadsides, in fields, sand dunes and waste places. Widespread throughout U.S.

FIREWEED (PILEWORT)

Grows 3' TO 8' high. Erect unbranched stems. Narrow willow-like leaves. Showy rose-purple flower spikes. Grows thickly in burned over areas, logging sites and on river banks. Widespread throughout U.S.

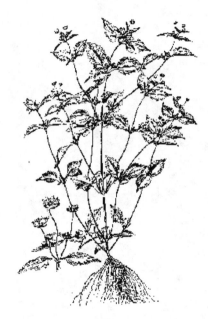

GALINSOGA (QUICKWEED)

Very small white flowers surround small yellow stems with many branches. Leaves pointed at tip. Found in weedy gardens, lowland fields, waste places — especially damp areas with rich soil. Widespread throughout U.S.

GOOSEGRASS (CLEAVERS)

Straggly stems grow to 4' high. Tiny star-shaped greenish-white flowers. Stems and leaves covered with rough bristles. Found in moist woods, thickets, roadsides, waste places, fields, etc. Widespread throughout U.S.

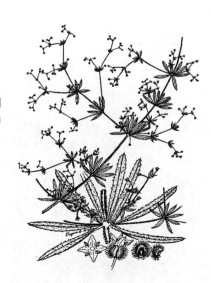

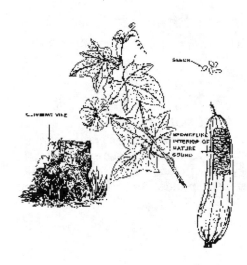

GOURD, WILD (LUFFA SPONGE)

Found on a climbing vine 20' to 30' long. Yellow flowers. Member of squash family. Grows similar to watermelon, cantaloupe and cucumber. Often found in old gardens, clearings and barnyards. Widespread throughout U.S.

GRAPE, WILD

High climbing vine. Shreddy, brown bark. Heart-shaped leaves. Greenish flowers (August to October). Fleshy grapes in clusters. Found in gardens, thickets, edges of woods, etc. Check grape smell since there is a poisonous look-alike. Widespread throughout U.S.

GREENBRIER

Vine with scattered thorns. Triangle-shaped leathery leaves. Greenish flower clusters. Round black berries follow flowers. Found on sides of roads, old fields and open woods and thickets. Widespread throughout Midwest and Eastern U.S.

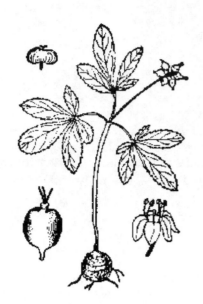

GROUNDNUT

Slender vine often found climbing on other plants. Climbs to height of several feet. Purplish-brown highly fragrant flowers. Milky sap in plant. Found in rich thickets and wet areas, around ponds and along streams. Widespread throughout U.S.

HICKORY

Grows 60'to 90' tall. Deep yellowish green leaves from 8" top 14". Trunk 4' diameter with light colored bark. Egg-shaped nuts with thick husks. Extremely hard wood. Found in rich soil, river bottoms, hillsides, etc. Common in many parts of U.S.

HOG PEANUT

Low growing vine with slender stems. Light green leaves. Clusters of white to pale lilac flowers. Likes rich soil, thickets, wet or moist wooded areas. Widespread throughout U.S.

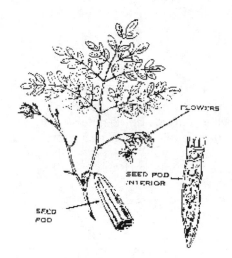

SEED POD INTERIOR

SEED POD

FLOWERS

HORSERADISH TREE

Grows 15' to 45' tall. Flowers on branch ends as are long pendulum fruit resembling giant beans. Leaves fern-like. Found in abandoned fields, gardens, edges of forests, etc. Widespread throughout U.S.

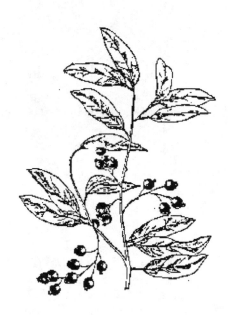

HUCKLEBERRY

Narrow leaves with tiny teeth. Hairy twigs and branches. Bell-like white or pink or green flowers (May to July). Blue-black berries follow flowers. Found in swamps, bogs, moist woods, etc. Often grows near blueberries. Widespread throughout U.S.

HYSSOP HEDGE-NETTLE

Small erect shrub. Square stems grow to 2' long. Dense flowers on top of stem. Found in wet places, usually sandy soil, meadows, fields, shores, etc. Widespread throughout U.S.

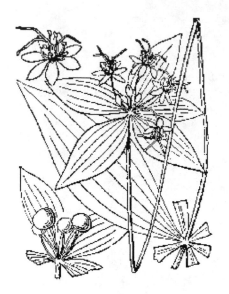

INDIAN CUCUMBER ROOT

Grows to 2' high. Yellow-green flower cluster at stem top. Inedible purple berries follow flowers. Underground swollen white tuber. Found on swamp edges and in moist woods. Widespread throughout Eastern U.S. and as far west as Kansas.

LADY'S THUMB (REDLEG)

Leaves vary — some narrow, others broad. Flowering spikes. Flowers purplish, green and pink. Found in cultivated fields, roadsides, in ditches, damp clearings. Widespread throughout U.S. except for some of southwest.

LAMB'S QUARTER (PIGWEED)

Grows 1' to 5' high. Short taproot. Triangular dark green leaves covered with mealy white powder. Irregular spike clusters at branch ends. Flowers small and green. Commonly found growing in gardens and grain fields. Widespread throughout U.S.

LEEK, WILD (RAMP)

Grows to 18" high. Slim fleshy leaves (2 or 3). Whitish bulbs exude strong onion smell. Spoke-like cluster of creamy white flowers (May to June). Found in rich moist soil, wet woods, alongside roads, thickets, etc. Widespread throughout U.S.

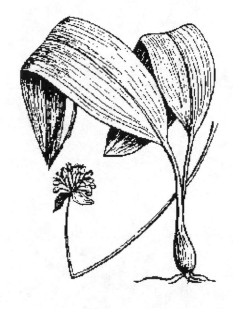

LETTUCE, WILD

Grows 1' to 5' high. Rounded blue-green leaves similar to dandelion. Strong odor. Prickly stem. Small blue or yellow flowers. Broken parts exude milk-like juice. Found on roadsides, banks, vacant lots, waste areas, etc. Widespread throughout U.S.

LOTUS LILY

Grows 5' to 6' high. Aquatic plant. Leaves measure 3' to 5' wide. Thick tubers grow 50' long. Yellow or pink flowers are 4" or more across. Found in quiet pond or lake water, along sluggish rivers, etc. Common in many parts of U.S.

MALLOW, COMMON

Grows to 14" high or more. Deep rooted plant. Low growing trailing stems. Rounded, slightly toothed leaves. Small white or lilac flowers (April to October). Found in yards, gardens, cultivated fields, barnyards, etc. Widespread throughout U.S.

MAPLE TREE

Grows from 50' to 70' high with 5' trunk. Serrated 3-sectional leaves. Light brown/gray bark. Some bark smooth, some rough. Gray lichen moss often found on these trees. Common in rich woods, rocky hillsides, yards, etc. Widespread throughout U.S.

MARSH-MARIGOLD (COWSLIP)

Aquatic plant. Succulent hollow stem. Kidney-shaped leaves. Large buttercup-like flowers (May to July). Grows around swamps, ponds, lakes, in wet ditches, etc. Common in many parts of U.S.

MINT

Grows to 3' high. Fine hairs on stem. Strong mint aroma. Tiny clusters bell-shaped flowers (June to October). Found in wet or damp places such as roadsides, beside streams, around ponds, etc. Widespread throughout U.S.

MULBERRY, RED

Grows 20' to 60' tall. Heart-shaped sandpapery leaves. Milky sap in twigs. Flowers in tight drooping clusters (April to May). Fruit resembles blackberries. Found along roads, in abandoned fields, open woods, thickets, etc. Widespread throughout U.S.

MUSTARD, WILD

Grows to 10' high. Rounded, coarse and bristly leaves. Bright yellow flowers in clusters at branch ends (May to September). Pods contain black seeds. Found in hedges, waste places, cliffs, roadsides, etc. Widespread throughout U.S.

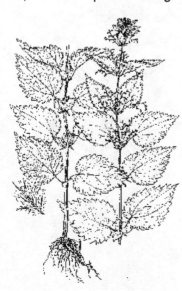

NETTLE, STINGING

Grows to 5' high. Coarse saw-toothed leaves. Erect square stems. Covered with stiff stinging hairs. Green flower clusters on many spikes (June to September). Found in vacant lots, gardens, ditches, along sides of roads, etc. Widespread throughout U.S.

291

NUT GRASS (CHUFA)

Clump of grassy-looking leaves on triangular stem. Flowers in clusters of feathery spikelets (June to October). Found in moist, sandy places, edge of streams and ponds, and moist or wet ground. Widespread throughout U.S.

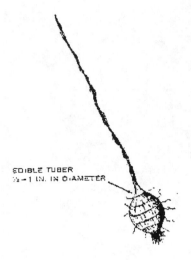

EDIBLE TUBER
½ - 1 IN. IN DIAMETER

OAK, WHITE

Grows to 110' tall. Bright olive-green leaves to 7" long. Light brown, sweet, edible acorns grow in pairs. Flowers (May and June). Found in dry woods, sandy places, gravelly ridges, etc. Common in many parts of U.S.

292

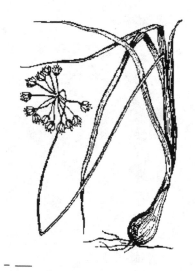

ONION, WILD

Grows to 2' high. Grass-like leaves in early spring. Small pink or purple flowers form clusters on stem end (June to August). Found on slopes, ridges, open woods, hillsides, etc. One of the most abundant food plants. Widespread throughout U.S.

OSTRICH FERN

Grows from 2' to 6' high in graceful bunches. Rich looking, dark green fronds (fern leaves). Found on banks of rivers and streams, around swamps, clear areas of woods, in rich, wet dirt. Common in many parts U.S.

of

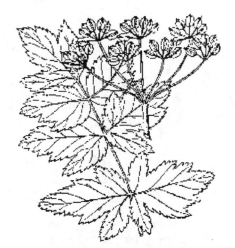

PARSNIP, WILD

Grows to 4' high. Erect flowering stalk. Fruits have flat narrow wings. Commonly found on roadsides, in ditches, grassy waste places, etc. Widespread throughout U.S.

PENNYCRESS, FIELD

Grows 16" to 32" high. White flowers (April to August). Deeply notched seedpods. Found in waste places, gardens, grasslands, fields, roadsides, etc. A troublesome weed in grain fields. Widespread throughout U.S.

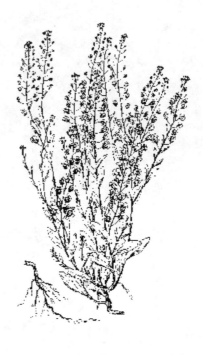

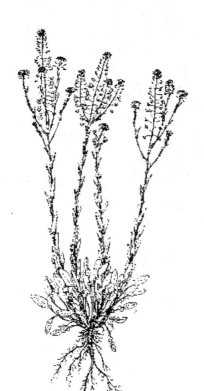

PEPPERGRASS

Grows from 1' to 2' high. Leaves on stem are arrow-shaped. Flower clusters on top of plant. Found on roadsides, fields of clover, alfalfa and winter wheat. Widespread throughout U.S.

PEPPERMINT

Grows to 3' high. Dark green, sharply toothed leaves. Mint smell. Taste initially burns, followed by cool sensation. Small lavender blossoms in spike-like groups (July to September). Found on brook and river banks, wet meadows, etc. Widespread throughout U.S.

PERSIMMON TREE

Grows to 60' tall. Stiff, glossy, dark green leaves. Fruit orange to reddish-purple when ready to eat (August to October or later). Found in old fields, open woods and in dry non-evergreen woods. Common in many parts of U.S.

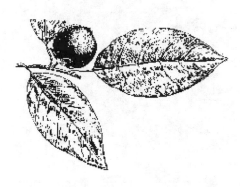

PICKERELWEED

Aquatic plant. Grows to 3' high. Arrowhead-shaped leaves. Dense blue flowers. Found in marshes, on banks of slow moving streams, around lakes and ponds. Common in many parts of U.S.

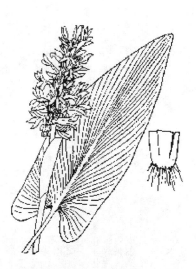

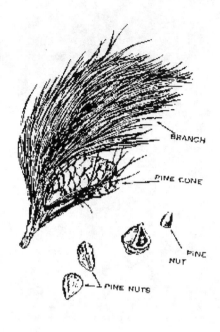

PINE TREE
Grows to 150' tall. Soft blue-green needles, 4" long or longer. Needles in clusters of 2 to 5. Crushed needles have characteristic taste and smell. Dark brown cylindrical pine cones, 4" to 8" long. Common in many parts of U.S.

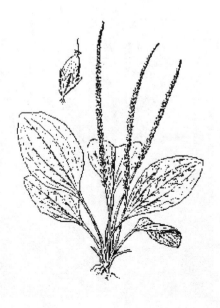

PLANTAIN, BROADLEAF
Grows to 18" high. Broad, oval, toothed leaves. Compact clusters of tiny yellow or greenish flowers on leafless stems (May to August). Found in lawns, fields, pastures, waste areas and gardens. Widespread throughout U.S.

POKEWEED (POKE)

Grows 2' to 3' high. Smooth reddish stems. Plant has disagreeable smell. Small flower clusters greenish to white. Found in pastures, cultivated fields, roadsides, open places in woods, etc. Common in many parts of U.S.

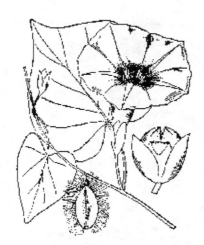

POTATO VINE, WILD

Trailing vine. Grows to 12' long. Heart-shaped leaves. Large white funnel-shaped flowers with pink-purple centers. Large root resembles sweet potato. Found in fields, pastures, on roadsides, etc. Widespread throughout U.S.

PRICKLY PEAR CACTUS

Grows 1' to 8' high. Tufts of bristles. Sharp spines. Showy yellow flowers (May to August). Fruit dull red, pulpy (August to October). Found in dry, sandy soils, rocks, etc. Common in many parts of U.S.

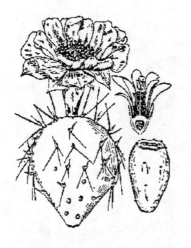

EDIBLE FRUIT

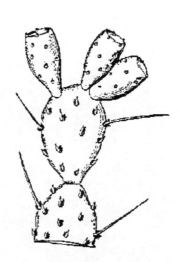

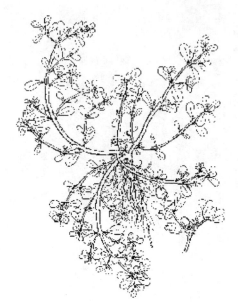

PURSLANE

Grows to 12" high. Creeps on ground. Reddish stems. Thick, succulent, spatula-shaped bright green leaf clusters. Tiny yellow flowers open only on sunny mornings (June until frost). So common that all you have to do is look for it. Widespread throughout U.S.

QUEEN ANNE'S LACE (WILD CARROT)

Grows to 4' high. Hairy oblong leaves and stems. Lacy clusters of purple-pink flowers (June to September). Leaves and flowers smell like carrots or plant may be poisonous look-alike. Found in meadows, fields, pastures, roadsides, etc. Widespread through U.S.

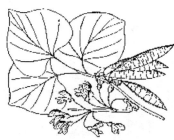

RED BUD TREE

Grows to 40' tall. Heart-shaped leaves. Showy clusters of purple-red, pea-like flowers (March to May). Found on stream borders, roadsides, in woods, yards, etc. Widespread throughout U.S.

298

RED RASPBERRY (CLOUDBERRY, DEWBERRY, ETC.)

Grows to 5' high. Prickly shrub. White powder on stems. Tiny white flower clusters (April to August). Red berries. Found along roadsides, edges of woods, fields, etc. Widespread throughout U.S.

RHUBARB

Grows 1' to 4' high. Large, compact plant. Thick stems. Leaves 2' to 4' long. Peculiar aromatic odor. Greenish-white flower clusters (May to August). Found everywhere — parks, roadsides, fields, yards, etc. Widespread throughout U.S.

RICE, WILD

Grows 3' to 4' high. Aquatic grass. Coarse and graceful. Light green 2" wide leaves. Large plumes of flowers. 2' flower stems. Found in wet areas, edges of ponds, rivers and lakes — in shallow water. Common in many parts of U.S.

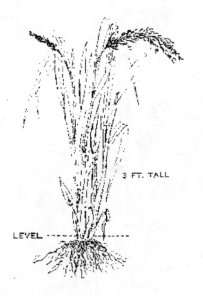

3 FT. TALL

LEVEL

RICE GRAIN INSIDE HUSK

299

ROSE, MULTI-FACETED WILD
Grows to 10' high. Look for fruit — fleshy red hips. All rose varieties have thorns or briars. Found on edges of woods, old pastures, yards, sides of roads, etc. Widespread throughout U.S.

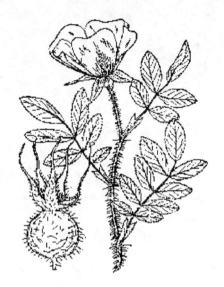

SERVICEBERRY
Small tree or shrub. Oval leaves with blunt tips and tiny teeth. White flowers in drooping clusters (May to July). Fruit dark blue and much like huckleberries (June to September). Found in thickets, open woods, fields, swampy areas, etc. Widespread.

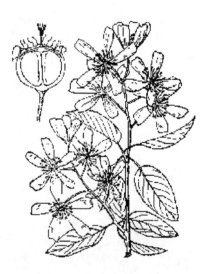

SHEEP SORREL (COMMON SORREL)

Grows to 18" high. Stems erect, slender and branched at top. Hard to eradicate. Arrow-shaped leaves have pleasant, mildly sour taste. Flowers yellow to red. Found in fields, gardens, pastures, roadsides, meadows, lawns, etc. Widespread throughout U.S.

SHEPHERD'S PURSE

Grows to 4' high. Dandelion-like, hairy leaves in clusters. Hairy stems. Triangular seed pods. Small erect spikes of tiny white flowers (January to December.) Found on roadsides, in backyards, gardens, fields, etc. Widespread throughout U.S.

SLIPPERY ELM TREE

Grows to 50' feet. Rough bark. Mildly aromatic. Oval leaves with sandpapery tops. Fruits winged. Ripen (March to May). Flowers bloom (March to April). Found in moist woods, thickets, rocky hillsides, etc. Common in many parts of U.S.

SOLOMON'S SEAL

Grows to 3' high. 4" x 2" dark green leaves. Drooping clusters of fragrant, greenish-white flowers (April to May). Black or blue berries. Found in rocky woods, shady places, dry to moist thickets, fields, etc. Widespread throughout U.S.

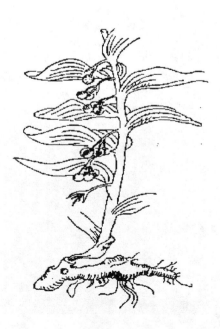

SPEARMINT

Grows to 3' high. Oblong, bright green leaves. Spearmint smell. Spikes of pale pink flowers (June to October). Found in fairly shady, moist places. Extremely common. Widespread throughout U.S.

SPINY LEAVED SOW THISTLE

Stem grows 1' to 5' high. Short taproot. Leaves crowded along reddish stem. Pale yellow flowers. Found in orchards, grain fields, lawns, etc. Widespread throughout U.S.

SPRING BEAUTY

Grows 6" to 12" high. Pair smooth leaves halfway up stem. Flowers white to pale pink with dark pink veins (March to May). Solid bulb-like root buried 3" to 5" inches. Found in moist woods, rich soil, thickets, etc. Widespread throughout U.S.

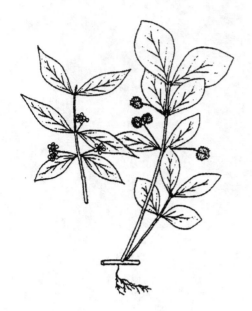

STRAWBERRY, WILD
Similar to cultivated strawberries. Smaller berries. Flat clusters of white flowers. Leaves dark green and hairy. Found in moist, rocky areas, clearings in woods, fields and other open places. Some in swampy areas. Widespread throughout U.S.

SUNFLOWER
Grows 3' to 12' high. Stems are stout, coarse, rough and hairy. Saw-tooth leaves. Showy bright yellow flowers are 3" to 5" across. Found on open plains, cultivated fields, waste places, pastures, grain fields, roadsides, etc. Widespread throughout U.S.

SWEET FLAG (CALAMUS)
Grows to 4' high. Cattail-like leaves have spicy smell when bruised. Stalks covered with tiny green-yellow flowers (May to August). Found in shallow water, swamps, wet fields, stream edges, etc. Widespread throughout U.S.

SYCAMORE TREE

Grows 50' to 80' tall. Massive tree with 12' diameter trunk. Dull brown brittle and flaking bark. Coarsely toothed leaves 6' to 10' long. Small flowers (April to May). Tiny hairy fruit. Found on banks of streams, bottom lands. Widespread throughout U.S.

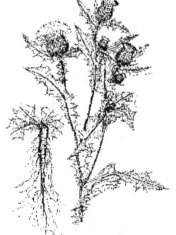

THISTLE, BULL

Grows to 8' high. Dozens of varieties. Long needle-pointed barbs on leaves and stems. Leaves often white and wooly on underside. Many white, red or purple burr-like flowers. An aggressive weed. Found in pastures, roadsides, etc. Widespread.

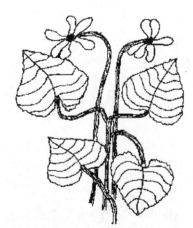

VIOLET, SWEET

Grows 4" to 6" high. Green heart-shaped leaves with scalloped edges. Sweet smelling, pansy-like flowers (May to June). Found on playgrounds, damp shady woods, yards, moist soil on stream edge, etc. Widespread throughout U.S.

WAPATO (DUCK POTATO)

Grows 10' or more. Aquatic plant. Leaves shaped like arrowheads. Waxy white flowers near tops of stalks (July and August). Smooth potato-like tubers on ends of underground runners. Found on stream edges and road borders, swampy areas. Widespread.

WATER CHESTNUT

Aquatic free floating plant. Some leaves on top of water. Others, root-like and feathery are underwater. Pods underwater. Nuts resemble horned steer. Found on lakes, ponds, rivers in quiet water. Common in many parts of U.S.

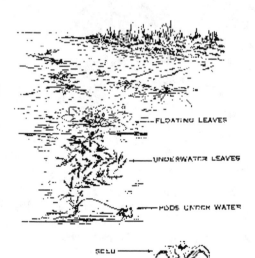

WATERCRESS

Aquatic plant Succulent creeping stems. Grows underwater or partly submerged in water. Shiny dark green leaves. Tiny white flower petals. Found in slow moving streams, springs, ponds, lakes, etc. Widespread throughout U.S.

WATER LILY, FRAGRANT

Aquatic plant with large 8" to 28" floating leaves. Showy sweet smelling pale yellow flowers 5" to 10" across (June to September). Found in quiet or dead water — ponds, slow running brooks, lakes, etc. Widespread throughout U.S.

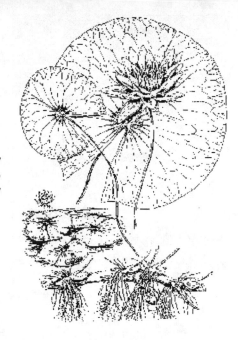

WATER PLANTAIN

Heart-shaped leaves. Thick bulb-like root stocks. White or yellow flowers. Plant stays partially submerged in water. Found in lakes, ponds and slow-moving streams. Common in many parts of U.S.

WINTERGREEN

Grows to 6" high. Low-creeping evergreen. Aromatic. Glossy, oval leaves. Waxy white flowers like drooping bells (June to August). Dry red berries (fall and winter). Found in dry, wooded areas, clearings, base of trees, etc. Widespread throughout U.S.

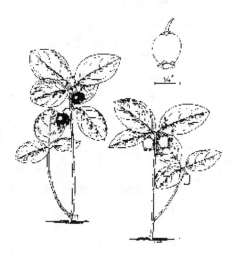

YELLOW DOCK (CURLY DOCK)

Grows to 5' high. Deep yellow root. Lance-shaped leaves. Winged, heart-shaped seeds (June to September). Small greenish flowers densely clustered on stalks (May to September). Found in pastures, meadows, gardens, roadsides, etc. Widespread in U.S.

PART IV

WHAT NOT TO EAT — POISONOUS PLANTS

9

Poisonous Plants and How to Avoid Eating Them

1. Poisonous plants are those containing natural toxic substances.
2. Poisonous plants aren't a serious hazard. Their number is small compared to the number of edible nonpoisonous plants.
3. Under certain conditions a person can be poisoned by:
    a. Eating poisonous plants or plant parts.
    b. Absorbing plant poison through the skin.
    c. Inhaling the plant poison.
4. Mistakenly eating poisonous plants can be extremely dangerous. They can cause:
    a. Serious illnesses.
    b. Death.
5. The critical factor in using plants for food:
    a. Eat only plants known for certain to be safe.

Learn All You Can About Poisonous and Nonpoisonous Plants

1. Preparation for survival situations includes learning to readily identify poisonous plants.
    a. Recognizing poisonous plants is as important as recognizing edible plants.
    b. Successful use of plants for food in a survival situation depends entirely upon positive identification.
2. Recognizing poisonous plants will help avoid unnecessary problems resulting from eating them:
    a. Remember: poison hemlock has killed people who mistook it for wild carrots and wild parsnips.
3. There is no room for experimentation where plants are concerned. There may be no second chance.
4. Many deadly poisonous plants:
    a. Look like their safe to eat relatives
    b. Have other edible look-a-likes in the plant world.
    c. For example, poison hemlock is similar to Queen Anne's Lace (wild carrot).
5. A few plants are safe to eat in certain stages of growth and poisonous in other stages.
    a. For example, pokeweed leaves are edible when first starting to grow, but they soon become poisonous.
6. A person can eat the fruit of some plants only when ripe.
    a. For example, the ripe fruit of mayapple is edible, but the green fruit is poisonous.
7. Some plants contain both edible and poisonous parts.
    a. Tomatoes and potatoes are commonly known plant foods, but their green parts are poisonous.
8. Some plants become poisonous after they wilt.
    a. For example, when the black cherry wilts, hydrocyanic acid is developed.
9. Specific preparation methods make some plants edible that are poisonous when eaten raw.
    a. A person can eat the dried corms of the Jack-in-the pulpit.
    b. The same corms are poisonous if not thoroughly dried.

How Poisonous are Poisonous Plants?

1. Poisons in plants exist in great number and variety.
2. Phytotoxins are the deadliest of poisons. These include:
    a. Abrin found in the Rosary Pea.
    b. Jatrophin found in the Physic Nut.
    c. Ricin found in the Castor Bean.
3. Ricin, for example, is:
    a. 12,000 times more potent than rattlesnake venom.
4. How poisonous is a poisonous plant? It's impossible to tell how potent a poisonous plant may be. It's equally as hard to tell how the particular plant will affect an individual who has eaten it. A number of factors must be considered:
    a. Every person has a different level of resistance to poisons.
    b. Some people are more sensitive than others to the poisons in a particular plant.
    c. The severity of poisoning in a person is influenced by the quantity of the poison ingested and by health and age.
    d. Poisonous plants vary in their amount of toxicity due to climate, habitat, soil, season, age, etc.
    e. A person must eat a lot of some poisonous plants before experiencing anything adverse.
    f. Other poisonous plants will cause a person's death after eating only small amounts.
    g. Poison may be distributed throughout a plant or may be concentrated in seeds, fruit, leaves or the roots.

Common Misconceptions About Poisonous Plants

1. Watch what animals eat and eat what they eat.
    a. This statement is not always accurate. Some animals can eat plants that would be poisonous to humans.
2. Boil the plant and poisons will disappear.
    a. Boiling does remove some poisons, but not always.
3. All red-colored plants are poisonous.
    a. Some red plants are poisonous, but certainly not all of them.
4. The most important point is this:
    a. Make an effort to learn as much about plants as possible.
    b. Never eat a plant unless it's positively identified.

Signs and Symptoms of Plant Poisoning

1. Signs and symptoms of plant poisoning may include:
    a. Severe abdominal cramps.
    b. Depressed heartbeat.
    c. Slower than normal respiration.
    d. Dry mouth.
    e. Nausea and/or vomiting.
    f. Diarrhea.
    g. Headaches.
    h. Hallucinations.
    i. Unconsciousness.
    j. Coma.
    k. Death.

What to do When Someone is Poisoned by Eating a Plant

1. Act quickly and pull poisonous material out of victim's mouth.
2. Try to induce vomiting:
   a. Tickle back of person's throat.
   b. Make person drink warm salt water, if conscious.
3. Dilute poison by making person, if conscious:
   a. Drink large quantities of milk.
   b. Drink large quantities of water.

Plants and the Universal Edibility Test

1. When unsure as to how safe a plant may be, use the Universal Edibility Test. Apply test before eating. This is a reliable way to determine which plants to eat and which plants to avoid.
2. Before testing a plant for edibility, be sure there are enough plants around to make testing worth the time and effort.
   a. Each part of a plant (roots, flowers, leaves, etc) requires more than 24 hours to test.
3. To avoid potential danger, stay away from unidentified plants having:
   a. A bitter or soapy taste.
   b. Discolored or milky sap.
   c. Carrot, dill, parsnip or parsley-like foliage.
   d. Almond smell in woody parts and leaves.
   e. Fine hairs, thorns and spines.
   f. Pods with bulbs, beans or seeds.
   g. Pink, purplish or black spurs on grain heads.
4. Use the above criteria as eliminators when choosing plants for the Universal Edibility Test. These guidelines will help a person avoid eating dangerous toxic plants.

The Universal Edibility Test for Plants Found in the Wilds.

1. Don't eat for eight hours prior to taking the test.
2. Separate the plant into its basic components:
   a. Leaves.
   b. Buds and/or flowers.
   c. Stems.
   d. Roots.
3. Test only one part of a plant at a time.
4. Smell the plant for strong or acid odors.
   a. Keep in mind that smell alone doesn't indicate a plant is safe to eat.
5. Select small portion of a single plant:
   a. Prepare it the way you plan to eat it.
6. During test period:
   a. Drink nothing other than purified water.
   b. Eat nothing more than plant being tested.
7. Before placing the cooked plant part in mouth:
   a. Touch a small portion (a pinch) to outer lip.
   b. Check for any burning or itching sensation.

8. If after 3 minutes there's no reaction on lip:
    a. Place plant part on tongue.
    b. Leave there 15 minutes.
9. If still no reaction:
    a. Thoroughly chew tiny piece.
    b. Hold in mouth for 15 minutes.
    c. Do not, under any circumstances, swallow.
10. If no itching, stinging, burning, numbness, dizziness, etc., occurs:
    a. Swallow the plant piece.
11. Now wait 8 more hours. If ill effects occur:
    a. Induce vomiting.
    b. Drink plenty of water.
12. If no ill effects occur:
    a. Eat 1/4 cup of same plant part, prepared same way.
    b. Wait another 8 hours.
13. If no ill effects occur:
    a. The plant part as prepared is safe for eating.
14. Test every part of plant for edibility.
    a. Some plants have both edible and inedible parts.
    b. Never assume a part found edible when cooked is also edible when raw.
    c. Test the raw part before eating to ensure edibility.
    d. The same part or plant may produce varying reactions in different individuals.

PART V

IDENTIFICATION OF COMMON POISONOUS PLANTS
(BY WORDS AND PICTURES)

AZALEA

TOXIC PART: Entire Plant.
SYMPTOMS: Slow pulse, vomiting, low-blood pressure, watering of the nose, mouth and eyes, paralysis, convulsions, death.
DESCRIPTION: Shrub. Grows to 2'. Showy masses of white, red and other color flowers.
WHERE FOUND: Moist woods, thickets, around homes, schools. Widespread throughout U.S.

BANEBERRY
(DOLL'S EYES)

TOXIC PART: Roots and berries.
SYMPTOMS: Dizziness, increased pulse, vomiting, stomach cramps, circulatory failure, delirium, headache, death.
DESCRIPTION: Grows 1' to 3' high. Toothed leaves 1" to 3" long. Spike-like clusters of small white flowers (May - June).
WHERE FOUND: Woods, moist shaded areas. Not very abundant. Widespread throughout U.S.

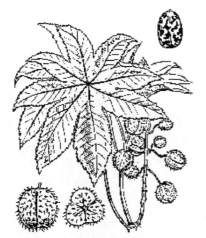

CASTOR BEAN
(CASTOR OIL PLANT)

TOXIC PART: Entire plant.
SYMPTOMS: Diarrhea, convulsions, excessive thirst, nausea, vomiting, stomach pains, dullness of vision, death in 1 to 12 days.
WARNING: Seeds contain more poison than any other plant part. 1 to 3 seeds fatal to child; 2 to 8 seeds fatal to adults. Seeds may be mistaken for bean-like food.
DESCRIPTION: Shrub-like plant 5' to 12' tall. Star-like leaves. Clusters of inconspicuous flowers (July to September) on plant top.
WHERE FOUND: Common in fields, open woods area, pastures, thickets. Widespread throughout U.S.

312

CHINABERRY TREE

TOXIC PART: Entire plant.
SYMPTOMS: irregular breathing, paralysis, signs of suffocation, vomiting, nausea, diarrhea.
DESCRIPTION: To 35' tall. Light purple flowers in ball-like masses. Leaves are a natural insecticide.
WHERE FOUND: Thickets, old fields, pastures, etc. Widespread throughout U.S.

DEATH CAMAS

TOXIC PART Entire plant.
SYMPTOMS: Subnormal body temperature, nausea, vomiting, muscular weakness, slow heart beat, diarrhea, stomach pains, death.
DESCRIPTION: Grass-like leaves grow from base. White flowers clustered on top of stem. Green/yellow heart-shaped structure found on petals near flower base.
WARNING: Bulbs of death camas easily confused with wild onions and camas lily bulbs. Remember — death camas have no onion smell.
WHERE FOUND: Wet or damp meadows, pastures, fields, etc. Dry, rocky slopes. Sunny places. Widespread throughout U.S.

FOXGLOVE

TOXIC PART: Entire plant.
SYMPTOMS: Irregular heartbeat, convulsions, nausea, vomiting, severe headache, tremors, stomach pains, diarrhea, death. Death can occur extremely rapidly.
DESCRIPTION: Tall and erect. Lance-shaped leaves. Stout stem. Showy clusters of thimble-shaped white or colored flowers.
WHERE FOUND: Along roads, logged areas, burned out places. Widespread throughout U.S.

☠

HOLLIES

TOXIC PART: Berries.
SYMPTOMS: Violent vomiting, nausea, stupor, diarrhea.
DESCRIPTION: Evergreen shrub or small tree. Leathery leaves with spine-tipped teeth. Clusters of red or orange berries.
WHERE FOUND: Sandy woods, yards around homes, etc. Widespread throughout U.S.

☠

HORSECHESTNUT

TOXIC PART: Seeds, flowers, leaves, nuts.
SYMPTOMS: Depression, nausea, diarrhea, vomiting, paralysis.
DESCRIPTION: Large tree, palm-shaped leaves. Spikes of showy white flowers (September-October). Thorny husks cover brown nuts.
WHERE FOUND: Often found in yards around homes, parks, etc. Widespread throughout U.S.

☠

IVIES

TOXIC PART: Berries, leaves.
SYMPTOMS: Difficulty breathing, extremely hyper, excitable, profuse sweating, coma.
DESCRIPTION: Hardy creeping plant. Scalloped edged leaves. Violet flowers.
WHERE FOUND: Along sides of roads, lawns. Found climbing on homes, embankments, up trees, etc. Widespread throughout U.S.

JIMSON WEED
(JAMESTOWN WEED, THORNAPPLE)

TOXIC PART: Entire plant.
SYMPTOMS: Hallucination, excessive thirst, dry mouth, vomiting, nausea, delirium, convulsions, coma, death.
DESCRIPTION: Grows 2' to 5'. Purplish branching stems. Trumpet-shaped flowers. Hard, prickly fruit 1" to 2" long. Foul smelling.
WHERE FOUND: Yards, rich farmland, fields, pastures, sides of roads, banks of streams, etc. Widespread throughout U.S.

LANTANA

TOXIC PART: Entire plant.
SYMPTOMS: Collapse of circulatory system. Death.
DESCRIPTION: Shrub-like plant. Strong smelling. Flat-topped colorful flower clusters. Dark blue or black berries.
WHERE FOUND: Along sides of roads, old fields, pastures, etc. Widespread throughout U.S.

LILY-OF-THE-VALLEY

TOXIC PART: Entire plant.
SYMPTOMS: Vomiting, dizziness, stimulation of heart, loss of balance, nausea.
DESCRIPTION: Grows 4' to 8' high. Oval-shaped leaves with smooth edges. Spread by root runners.
WHERE FOUND: Almost everywhere. Widespread throughout U.S.

MONKSHOOD

TOXIC PART: Entire plant.

SYMPTOMS: Convulsions, vomiting, weak pulse, diarrhea, spasms, paralysis of respiratory system, death.

DESCRIPTION: Grows from 2'to 3' high. Buttercup-like leaves. Flowers partly covered by helmet-like growth.

WHERE FOUND: Damp or wet areas. Slopes, woods, thickets, road sides, etc. Widespread throughout U.S.

OLEANDER

TOXIC PART: Entire plant.

SYMPTOMS: Unconsciousness, dizziness, irregular heartbeat, nausea, slowed pulse, vomiting, bloody diarrhea, severe stomach pains, paralysis of lungs, death.

DESCRIPTION: Bush grows to over 18'. Dark green cylindrical leaves. Various flower colors. Brown pod filled with many seeds.

WHERE FOUND: Grown as ornamental plant in yards, playgrounds, parks, etc. Widespread through U.S.

WARNING: Don't cook with oleander wood. Fumes will poison food.

PHYSIC NUT

TOXIC PART: Entire plant.

SYMPTOMS: Stomach cramps, severe diarrhea, convulsions.

DESCRIPTION: Shrub or small tree. Small greenish-yellow flowers. Each apple size fruit has 3 poisonous seeds.

NOTE: Seeds have sweet taste but their oil is violently purgative.

WHERE FOUND: Widespread throughout U.S.

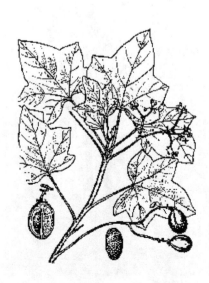

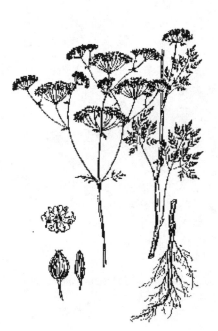

POISON HEMLOCK
(FOOL'S PARSLEY)

TOXIC PART: Entire plant.

SYMPTOMS: Attacks central nervous system. Weakness of muscles, respiratory paralysis, trembling, coldness, coma, death.

DESCRIPTION: Grows to 10'. Smooth, hollow stems with purple spots on lower section. Flat clusters of white flowers (May-August). Turnip-like taproot. Bruised fern-like leaves have obnoxious odor.

WHERE FOUND: Sides of roads, gullies, wet or moist ground, swamps, banks of streams, ditches. Widespread throughout U.S.

WARNING: Don't confuse with Queen Anne's Lace (Wild Carrot). Queen Anne's Lace has carrot smell. Poison Hemlock does not! Queen Anne's Lace has hairy leaves and stems. Poison hemlock does not! A small piece of stem, only 1/2" in diameter, may be fatal.

RHODODENDRON

TOXIC PART: Entire plant.

SYMPTOMS: Blood pressure drops, vomiting, slow pulse, convulsions, watery eyes, runny nose, drooling mouth, paralysis, coma, death.

DESCRIPTION: Small evergreen, 10' to 14' tall. Large, leathery, toothless leaves, rolled under edges. Showy clusters of white or pink spotted flowers (June-July).

WHERE FOUND: Widespread throughout U.S.

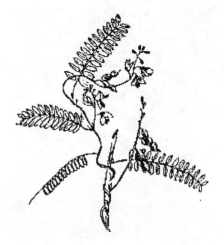

ROSARY PEA

TOXIC PART: Seeds.

SYMPTOMS: Circulatory collapse, vomiting, drowsiness, colic, trembling, pulse weak and fast, diarrhea, nausea, cold sweat, stomach pains, weakness, coma, death in 1 to 3 days.

DESCRIPTION: Vine with light purple flowers. Red and black seeds frequently used to make rosaries.

WHERE FOUND: Widespread throughout U.S.

WATER HEMLOCK
(SPOTTED COWBANE)

TOXIC PART: Entire plant.

SYMPTOMS: Tremors, violent convulsions, stomach pain, delirium, diarrhea, death.

DESCRIPTION: Grows 2' to 7' high. Stout, bamboo-like hollow stalks. Yellowish oil oozes from roots when cut. Flat spreading clusters of small white flowers.

WHERE FOUND: Ditches, wet or moist ground, swamps, along roadsides. Widespread throughout U.S.

WARNING: One of the most poisonous plants known. Its most poisonous part, the roots, are sometimes mistaken for parsnips.

WISTERIA

TOXIC PART: Seeds.

SYMPTOMS: Stomach pains, diarrhea, nausea, repeated vomiting.

DESCRIPTION: Sturdy vines, smooth bark. Showy clusters of drooping white, lilac or purple flowers. Flat, knobby seed pods.

WHERE FOUND: Wet or damp areas of woods, thickets, pastures, fields. River and stream banks. Widespread throughout U.S.

WARNING: Two seeds are enough to cause serious illness.

YEW

TOXIC PART: Seeds, twigs and leaves.

SYMPTOMS: Difficulty in breathing, trembling, nausea, diarrhea, vomiting, muscular weakness, collapse, coma, death.

DESCRIPTION: Straggly evergreen shrub. Short flat pointed green needles. Twigs smooth. Juicy red berries.

WHERE FOUND: Cool, moist woods. Widespread throughout U.S.

NOTE: Although seeds are poisonous, the red fleshy sweet pulp surrounding the seed is edible in small amounts.

SECTION 4

THE

OFFICIAL

URBAN & WILDERNESS

MEDICINAL PLANT

SURVIVAL MANUAL

CONTENTS

PART I

INTRODUCTION TO MEDICINAL PLANT SURVIVAL MATERIAL

PART II

MEDICAL PROBLEMS AND PROPER PLANT REMEDIES

(A)

Abrasions

Abscess (Also see "Boils" and "Gum Boils")

Acne

Allergies

Anemia

Anxiety

Arthritis (Also see "Joints")

Asthma

Athlete's Foot (Also see "Fungus")

(B)

Back Pain (Also see "Pain Relief")

Bites (Also see "Stings")

Bleeding (Also see "Nose Bleed")

Bladder (See "Diuretic")

Blisters

Blood Pressure, High

Blood Purifier

Boils (Also see "Abscess")

Breasts (Also see "Nursing")

Bronchitis

Bruise

Bunion

Burns and Scalds

(C)

Canker Sores (Also see "Cold Sores")

Chest Congestion (Also see "Cold" and "Cough")

Chilblains (Also see "Frostbite")

Chills (Also see "Fever" and "Sweating")

Cold (Also see "Cough" and "Sore Throat)

Cold Sores (Also see Canker Sores")

Constipation (Also see "Laxative")

Corns

Cough (Also see "Chest Congestion")
Cough Syrup
Cramps (Also see "Menstrual Cramps")
Cuts (Also see "Wounds")

(D)

Diarrhea
Diarrhea — Food
Diarrhea — Syrup
Diarrhea — Tea
Digestion
Diuretic
Diuretic — eat
Diuretic — Syrup
Diuretic — Tincture

(E)

Earache
Earache — Compress
Earache — Ear Drops
Earache — Poultice
Eczema
Exhaustion
Eye Problems
Eye Strain

(F)

Felon (finger or toe infection)
Fever (Also see "Chills" and "Sweating")
Flatulence (Also see "Gas")
Flu
Frostbite (Also see "Chilblains")
Fungus (Also see "Athlete's Foot")

(G)

Gas (Also see "Flatulence")
Glands, Swollen
Gout
Gum Boils
Gums, Bleeding
Gums, Sore (Also see "Mouth Sores")

(H)

Headache
Heartburn (Also see "Indigestion")
Hemorrhoids
Hernia
Hiccups

(I)

Indigestion (Also see "Heartburn")
Inflammation
Insomnia (Also see "Sleep Disorders")

(J)

Joints (Also see "Arthritis")

(K)

Kidney Problems (See "Diuretic")

(L)

Laxative (Also see "Constipation")
Lice, Head (Also see "Parasites")
Lungs

(M)

Menstrual Cramps (Also see "Cramps")
Morning Sickness (Also see "Nausea" and "Vomiting")
Mouth Sores (Also see "Gums, Sore")
Muscle Pulled (Also see "Muscle Strain")
Muscle Spasms
Muscle Strain (also see "Muscle Pulled")

(N)

Nausea (Also see "Morning Sickness" and "Vomiting")
Nervousness
Nose Bleed (Also see "Bleeding")
Nursing (Also see "Breasts")

(P)

Pain Relief (Also see "Arthritis" and "Swelling")
Parasites (Also see "Lice, Head" and "Worms")
Pneumonia (Also see "Chest Congestion")
Poison Ivy/Poison Oak (Also see "Rash")

(R)

Rash (Also see "Poison Ivy/Poison Oak")
Restlessness (See "Nervousness")
Ring Worm (Also see "Worms")

PART III

THE PLANTS AND THEIR PICTURES

PART I

INTRODUCTORY PLANT SURVIVAL MATERIAL

1

Medicinal Plants Important to Your Survival

More than 2,000 years ago, Egypt's Queen Cleopatra used freshly cut aloe vera leaves as a soothing burn ointment.

Around the same time period, people sipped a brew made from white willow bark to relieve the pain of gout. Why did it work so well? Because white willow bark contains a natural form of aspirin!

Modern medicine borrowed heavily from natural plant remedies. Of all the prescription drugs sold in the United States, a full one-third are derived from plants.

Approximately 80 percent of the world's people rely on folk medicine for treatment of their illnesses. People in developing areas of the world depend wholly upon the sometimes strange practices of a local healer, medicine man or witch doctor. Many of their unique plant concoctions are as effective as modern medications in the civilized world.

American Indians, by necessity, developed a vast expertise in plant medicines. And early settlers from England and Western Europe brought to the New World their knowledge of medical treatment with plants. Herbal home remedies were handed down in those families over many generations.

In Colonial days, there were no drugstores to be found on street corners and few, if any, trained doctors. People had no choice but to rely on homemade medicines. Plant remedies developed by the Colonists and Indian tribes worked well. It goes without saying that the greatest pharmacy in the world is found in plants scattered throughout the countryside. When properly used, these plants have incredibly effective medicinal properties. Plants can and should be utilized when faced with an emergency medical situation or where survival may be in question. **The Official Urban & Wilderness Medicinal Plant Survival Manual** is designed to show you exactly what to do.

2

Making Plant Medicines for Survival

1. **COMPRESS:** To make a hot compress:

   a. Brew a batch of strong plant tea and strain.

   b. Dip soft cloth in the hot tea and wring out.

   c. Apply to affected area as hot as person can stand it.

2. **DECOCTION:** Simmering berries, roots, barks and seeds for half hour or more is a decoction. Here's how it's done:

   a. Slice fresh plant parts or crush dry plant parts.

   b. Simmer in pot and strain while hot.

   c. Pour into containers and shake when ready to use.

3. **INFUSIONS:** An infusion is identical to a decoction except for the plant parts used.

   a. Soak leaves or flowers in hot water for 30 minutes or

   b. Make cold infusion by soaking leaves or flowers overnight in cold water.

4. **LINIMENTS:** Liniments are concocted for rubbing on the skin at body temperature.

5. **POULTICE:** a moist paste made from mashed plants or parts of plants and spread on affected areas of the body.

6. **TEA:** Bring cold water to a rolling boil. Pour over dried, pulverized root or over leaves (dry or fresh). Steep (infuse) 15 to 20 minutes. Strain and sweeten to taste.

7. **TINCTURE:** A tincture is merely plant parts dissolved in brandy, wine or other alcoholic beverages. Tinctures are made in this way:

   a. Steep leaves, etc., in brandy (or other) for 2 weeks.

   b. Shake mixture daily.

   c. Strain through paper coffee filter before bottling.

8. **WASHES:** Washes are a cooled tea used externally rather than for drinking. They are wiped on skin around affected area. Washes are easily made:

   a. Put plant parts in boiling water.

   b. Simmer until soft.

   c. Leave to steep for at least 10 days.

PART II

MEDICAL PROBLEMS AND PROPER PLANT REMEDIES

3

Medical Problems and How to Treat Them

(A)

ABRASIONS

Dictionary definition: *"Raw place on arm, leg or elsewhere made by rubbing or scraping."*

SWEET VIOLET:

A. Crush handful of roots to pulpy mass.

B. Spread as poultice on abrasion.

    1. Heals sore place and relieves pain.

ABSCESS

(also see "BOILS" and "GUM BOILS")

Dictionary definition: *"Pus collected in tissues of some part of the body."*

DANDELION:

A. Crush tops and leaves to pulpy mass.

B. Spread as poultice on abscess.

    1. Draws pus to head.

ELDERBERRY:

A. Crush handful of fresh leaves to pulpy mass.

B. Spread pulp as poultice on boils.

    1. Alleviates pain and brings abscess to head.

THYME:

A. Put handful of thyme in pint boiling water.

B. Steep for 20 to 30 minutes.

C. Soak folded cloth in hot tea-like mixture.

D. Lay compress over abscess.

E. Repeat when cloth cools.

    1. Draws pus to a head.

YARROW, COMMON:

A. Crush yarrow plant to pulpy mass.

B. Spread poultice on abscess.

    1. Draws pus to head.

ACNE

Dictionary definition: *"Skin disease where oil glands get clogged. often causes pimples."*

GARLIC, WILD:

A. Slice end off a garlic clove.

B. Rub garlic juice on the acne.

1. Clears up acne.

HORSERADISH:

A. Grate small handful of horseradish.

B. Add pinch of grated nutmeg.

C. Blend in grated peel of a bitter orange.

D. Blend and put mixture in pint of vinegar.

E. Shake daily while steeping for 2 weeks.

F. Dip Q-tip in tincture and touch to each pimple.

G. Repeat a number of times daily.

    1. Quickly dries pimples.

SOAPWORT:

A. Crush handful of leaves to pulpy mass.

B. Blend with warm water and make a lather.

C. Spread poultice on all acne spots.

    1. Dries and heals acne.

    2. Crushed roots can be substituted for crushed leaves.

ALLERGIES

Dictionary definition: *"Overly sensitive to certain substances such as pollens or dust."*

GINSENG, AMERICAN:

A. Crush piece of dry root to a powder.

B. Add 1/4 teaspoon to any herbal tea.

C. Drink the tea.

    1. Overcomes person's inclination for allergy attacks.

    2. Especially good for sinus problems and hay fever.

MULLEIN, COMMON:

A. Put handful of crushed whole plant in quart boiling water.

B. Steep 15 to 20 minutes.

C. Strain.

D. Drink 1/2 to 1 cup every 3 to 4 hours.

    1. Clears up respiratory allergies.

    2. Relieves hay fever and aids swollen membranes.

THYME:

A. Put handful leaves in pint boiling water.

B. Steep 30 minutes.

C. Strain.

D. Allow to cool.

E. Take 1 tablespoonful every 3 to 4 hours.

    1. Stops post nasal drip.

YELLOW DOCK:

A. Pulverize small piece of dried root to fine powder.

B. Sniff this powder up the nostrils.

    1. Excellent remedy for hay fever.

ANEMIA

Dictionary definition: *"A blood deficiency; not enough red corpuscles or hemoglobin in the blood."*

DANDELION:

A. Put handful of crushed leaves quart boiling water.

B. Steep 15 to 20 minutes.

C. Strain.

D. Drink 1/2 to 1 cup every 3 to 4 hours.

    1. Helps build and purify the blood.

    2. Crushed root can be substituted for crushed leaves.

    3. Purifies blood.

GRAPE, WILD:

A. Make a drink from the following:

    1. 2 pints grape juice.

    2. 1 pint beet juice.

B. Put in tightly closed container.

C. Keep in cool place.

D. Take 1 teaspoon 3 to 4 times daily.

    1. Overcomes anemia.

ROSE HIPS:

A. Put 2 teaspoon of rose hips in large cup water.

B. Simmer 10 minutes to brew strong tea.

C. Strain.

D. Drink on a regular basis.

    1. Overcomes anemia.

STINGING NETTLE:

A. Put handful of crushed whole plant in quart boiling water.

B. Steep 15 to 20 minutes.

C. Strain.

D. Drink 1/2 to 1 cup every 3 to 4 hours.

    1. Crushed root can be substituted for crushed whole plant.

    2. Overcomes anemia and builds blood.

ANXIETY

Dictionary definition: *"Uneasy feeling, troubled, worried. An anxious state of feeling."*

FIELD HORSETAIL:

A. Put handful of crushed whole plant in quart boiling water.

B. Steep 20 to 25 minutes.

C. Strain.

D. Drink 1/2 to 1 cup every 3 to 4 hours.

    1. Overcomes feelings of anxiety.

RED CLOVER:

A. Put large handful of blossoms in quart boiling water.

B. Steep 20 to 25 minutes.

C. Strain.

D. Drink 1/2 to 1 cup with each meal and at bedtime.

    1. Alleviates anxiety.

ROSEMARY:

A. Put 2 teaspoons of crushed leaves in quart white wine.

B. Add 2 tablespoons crushed skullcap leaves.

C. Shake thoroughly to blend.

D. Set aside to steep for 3 to 4 weeks.

E. Strain.

F. Take 1 tablespoon 3 times a day plus 1 on retiring.

    1. An excellent sedative.

    2. Overcomes feelings of anxiety.

VALERIAN:

A. Put handful of crushed root in quart boiling water.

B. Steep 20 to 25 minutes.

C. Strain.

D. Take 1 to 2 tablespoons before each meal.

E. Drink 1/2 cup before going to bed at night.

    1. Valerian is Europe's best selling over the counter tranquilizer.

ARTHRITIS

(Also see "JOINTS")

Dictionary definition: *"Inflammation of a joint or joints."*

CHRISTMAS FERN:

A. Stuff fresh ferns into 1/2 gallon container.

B. Fill with rubbing alcohol.

C. Leave container outside in sun for 4 weeks.

D. Strain.

E. Rub lotion on sore or aching areas of body.

    1. An old Amish folk medicine remedy.

POTATO, WILD:

A. Bake potato until soft.

B. Scoop out entire inside and mash.

C. Put potato pulp on sore joints.

D. Lay potato skin over this.

E. Hold in place with elastic bandage.

    1. An old American Indian remedy.

WHITE WILLOW:

A. Fill a pot with inner bark from a white willow tree.

B. Add enough water to cover bark.

C. Bring to boil.

D. Add more water and simmer 3 to 4 hours.

E. Strain.

F. Bottle and cap this tea.

G. Drink good-sized glass when needed.

    1. Alleviates pain.

    2. White willow bark has natural form of aspirin.

ASTHMA

Dictionary definition: *"Chronic disease causing breathing difficulty, coughing and suffocating feelings."*

ANISE ROOT (SWEET CICELY):

A. Crush handful anise seeds.

B. Put seeds in pint brandy.

C. Steep 3 to 4 weeks.

D. Strain.

E. Take 1 teaspoon by itself.

or

F. Put 1 teaspoon in cup tea.

    1. Alleviates asthma attack.

    2. Also works to stop hacking cough.

CRANBERRY:

A. Take pot and fill half way with fresh cranberries.

B. Add water until pan is full and bring to boil.

C. Simmer until water level in at top of berries.

D. Pour off water and set aside for tea.

E. Run cranberries through strainer.

F. Discard skins.

G. Put cranberry pulp in jar.

H. Set pulp outside in cool place.

I. Take 2 teaspoons pulp in 1 cup hot water.

    1. Thwarts asthma attack.

    2. Normal breathing restored after couple sips.

LOBELIA:

A. Put handful of crushed root in quart rum.

B. Steep 7 days.

C. Strain.

D. Take 3 drops 4 times daily to thwart attacks.

   or

E. Take 1/2 teaspoon during asthma attack.

    1. Crushed leaves can be substituted for crushed roots.

    2. An old Amish folk medicine remedy.

SMOOTH SUMAC:

A. Put handful of leaves in quart boiling water.

B. Steep 20 to 25 minutes.

C. Strain.

D. Drink 1/2 to 1 cup every 3 to 4 hours.

    1. Alleviates asthma.

ATHLETE'S FOOT
(Also see "FUNGUS")

Dictionary definition: *"A contagious skin disease caused by fungus. Found on the feet, generally between toes."*

CELANDINE:

A. Crush handful of fresh stems.

B. Wipe plant juice on each infected part of foot.

    1. Eliminates athlete's foot problem.

ELECAMPANE:

A. Put handful of crushed root in pint boiling water.

B. Steep 30 to 35 minutes.

C. Strain.

D. Dip cloth in this tea.

E. Wring out so it doesn't drip.

F. Wash lightly over fungus on feet.

G. Repeat 3 or 4 times each day.

    1. Crushed leaves can be substituted for crushed root.

    2. Good remedy for athlete's foot.

HORSERADISH:

A. Pound handful of roots to pulpy mass.

B. Spread as poultice on infected parts of feet.

C. Wrap towel on feet to keep poultice in place.

    1. Excellent remedy for athlete's foot.

(B)

BACK PAIN
(Also see "PAIN RELIEF")

HORSEMINT:

A. Put large handful of leaves in pint boiling water.

B. Steep 1 hour.

C. Strain.

D. Drink 1 cup prior to going to bed.

    1. An old American Indian remedy.

QUEEN-OF-THE-PRAIRIE:

A. Chew a piece of this root.

B. Swallow the juice.

    1. Sure fire cure for backache.

    2. Wisconsin's Fox Indian remedy of days gone by.

BITES
(Also see "STINGS")

ALOE VERA:

A. Break or cut leaves into pieces.

B. Smear gel-like juice on insect bites.

    1. Relieves pain of ant bites, etc.

    2. Also good for bee, caterpillar and other stings.

GARLIC, WILD:

A. Slice raw garlic cloves lengthwise.

B. Rub juice from clove slices on bites.

C. After short wait, slice cucumber.

D. Rub cucumber juice over bites.

    1. Excellent for use on various stings.

GOAT'S RUE:

A. Crush entire plant to pulpy mass.

B. Spread as poultice on bites.

C. Or make compress by soaking up oil in cloth.

D. Lay compress over bite.

    1. Alleviates pain and discomfort of bites.

    2. Good for scorpion, spider, ant bites, etc.

    3. Also good for jelly fish and all insect stings.

MARSH-MARIGOLD:

A. Wet flower petals.

B. Press on ant bites and others.

    1. Excellent for bee stings and others.

BLEEDING

(Also see "NOSE BLEED")

BUGLEWEED:

A. Pound entire plant to pulpy mass.

B. Spread as poultice on cut or wound.

    1. Stops bleeding.

FRAGRANT WATER LILY:

A. Soak leaves in warm water for few minutes.

B. Plaster wet leaves on cuts or wounds.

    1. Stops bleeding.

JUNIPER, COMMON:

A. Pick a pint or more of the berries.

B. Mash thoroughly to a pulp.

C. Pack on bleeding cut or wound.

    1. Quickly stops bleeding.

MARSH-MARIGOLD:

A. Pulverize large handful flowers.

B. Extract juice and strain.

C. Rub juice directly on cut or wound.

    1. Good for stopping bleeding.

PUFF BALL (WHITE ONLY—*BLACK PUFF BALLS ARE POISONOUS*):

A. Rub ripe puff ball mushroom into fine powder.

B. Sprinkle or rub powder on small cuts and wounds.

C. Place entire puff ball over larger cuts and wounds.

D. Leave on until healed.

    1. White puff balls were collected and kept on hand for emergencies in England in the 1800s.

SNAKEROOT, WHITE:

A. Crush entire plant to pulpy mass.

B. Spread as poultice on cut or wound.

    1. Works quickly to stop bleeding.

    2. An old American Indian remedy.

BLADDER

(See "DIURETIC")

BLISTERS

FIELD HORSETAIL:

A. Pulverize handful of whole plant to pulpy mass.

B. Spread as poultice over blistered area of body.

C. Lay cloth over poultice to keep in place.

    1. Promotes healing of blistered area.

2. An old American Indian remedy.

HEMLOCK:

A. Half fill large pot or bucket with bark and leaves.

B. Fill with water and bring to boil.

C. Boil 2 hours while adding water as needed.

D. Strain.

E. Soak blistered feet in this until it's cool.

    1. Heals blisters on feet.

    2. American Indians of long ago used this remedy.

MULLEIN, COMMON:

A. Put fresh leaves in shoes.

B. Wear when walking to prevent blisters.

    1. Plantain leaves can also be used.

PLANTAIN, COMMON:

A. Pulverize handful of leaves to pulpy mass.

B. Spread as poultice on blisters.

C. Lay small cloth over poultice to keep in place.

    1. Alleviates discomfort of blisters.

BLOOD PRESSURE, HIGH

ANGELICA:

A. Take some Angelic root and grind to fine powder.

B. Take some Mistletoe root and grind to fine powder.

C. Take 2 tablespoons each and put in quart boiling water.

D. Steep for time it takes to cool.

E. Strain.

F. Drink 2 to 3 cups daily.

    1. Keeps blood pressure down.

    2. Used by American Indians of old.

GARLIC, WILD:

A. Peel cloves of 1 whole garlic.

B. Crush cloves to pulpy mass.

C. Add 1 large raw grated carrot and blend.

D. Spread on bread daily and eat as sandwich.

    1. Keeps blood pressure down.

RED ALDER:

A. Pound handful of bark to powder and crumbs.

B. Put 1/2 cup in quart boiling water.

C. Steep 25 to 30 minutes.

D. Strain.

E. Drink 1/2 cup as needed.

1. Brings down high blood pressure.
2. Used by American Indians of old.

BLOOD PURIFIER

BURDOCK, COMMON:

A. Take 1 teaspoon seeds.
B. Eat them with each meal.
    1. Cleanses blood and eliminates boils.
    2. An old Amish folk medicine remedy.
    3. Commonly used by American Indians of old.

SARSAPARILLA, WILD:

A. Put handful of crushed root in quart boiling water.
B. Steep 20 to 25 minutes.
C. Strain.
D. Drink 1/2 to 1 cup with each meal.
    1. Used long ago as blood purifier by American Indians.

YARROW, COMMON:

A. Put handful of crushed whole plant in quart boiling water.
B. Steep 15 to 20 minutes.
C. Strain.
D. Drink 1/2 to 1 cup every 2 to 3 hours.
    1. Builds and cleanses the blood.

BOILS

(Also see "ABSCESS")

Dictionary definition: *"A painful, red swelling on the skin, formed by pus around a hard core."*

CATTAIL:

A. Pound handful of dried roots to powder.
B. Make paste by adding a little tepid water.
C. Apply paste as poultice and cover boils.
    1. Draws boils to head.
    2. An old American Indian remedy.

PASSION FLOWER:

A. Crush a few fresh roots to pulpy mass.
B. Spread as poultice directly on boil.
    1. Quickly draws boils to head.
    2. Used by American Indians many years ago.

SWEET VIOLET:

A. Crush some roots to pulpy mass.
B. Spread poultice on boils.
    1. Draws pus to head.

2. American Indians used this long ago.

WHITE PINE:

A. Gather some pitch from the tree.

B. Spread over boils as poultice.

    1. Used long ago by American Indians.

WILD HYDRANGEA:

A. Soak a few fresh leaves in hot water.

B. Lay softened leaves directly on boils.

C. Replace leaves when cool.

    1. American Indians used this for boils.

    2. An old Amish folk medicine remedy.

BREASTS
(Also see "NURSING")

COMFREY:

A. Put handful of crushed root in quart boiling water.

B. Steep 25 to 30 minutes.

C. Strain.

D. Soak folded cloth in this tea.

E. Lay hot compress on breasts.

F. Reapply hot compress when other cools.

    1. Crushed leaves can be substituted for crushed root.

    2. Alleviates soreness and swelling in breasts.

SHEPHERD'S PURSE:

A. Put handful of leaves and seeds in quart boiling water.

B. Steep 25 to 30 minutes.

C. Strain.

D. Soak folded cloth in this tea.

E. Lay hot compress on breasts.

F. Put on fresh compress when first feels cool.

    1. Soothes and heals sore or swollen breasts.

    2. Promotes healing of cracked nipples.

YARROW, COMMON:

A: Put handful of crushed whole plant in quart boiling water.

B. Steep 20 to 25 minutes.

C. Strain.

D. Soak folded cloth in this tea.

E. Lay hot compresses on breasts.

F. Expose breasts to sunshine.

    1. Alleviates sore nipples.

BRONCHITIS

Dictionary definition: *"Inflammation of the lining of the bronchial tubes."*

COLT'S FOOT:

A. Put handful of crushed leaves in quart boiling water.

B. Add tablespoon crushed wild licorice leaves.

C. Steep 25 to 30 minutes.

D. Strain.

E. Stir in 2 tablespoons honey.

F. Drink 1 cup 4 times a day.

    1. Alleviates bronchitis.

FLAX:

A. Put large handful of seeds in cooking pot.

B. Rinse with cold water.

C. Drain and add quart cold water.

D. Toss in rind of whole lemon.

E. Simmer 20 to 25 minutes.

F. Strain.

G. Squeeze juice of 1 lemon in tea.

H. Stir in tablespoon honey.

I. Take 1 one or 2 tablespoons as needed.

    1. Relieves chest tightness associated with bronchitis.

GINSENG, AMERICAN:

A. Chew on small piece of peeled ginseng root.

B. Swallow root pieces and juices.

    1. Alleviates deep bronchial cough.

    2. An old Amish folk medicine remedy.

HOREHOUND, COMMON:

A. Put 2 tablespoons of fresh leaves in pint boiling water.

B. Steep 20 to 25 minutes.

C. Strain.

D. Sweeten tea to suit with honey.

E. Drink 1/4 to 1/2 cup as needed.

    1. Overcomes bronchitis.

BRUISE

AGRIMONY:

A. Crush handful of leaves to pulpy mass.

B. Spread thickly as poultice on bruise.

C. Lay cloth over poultice to keep in place.

    1. Roots can be substituted for leaves.

    2. Alleviates soreness and quickens healing.

ONION, WILD:

A. Peel and slice a raw onion.

B. Lay onion slices on bruises.

C. If large bruised area, cover onion slices.

D. Use plastic wrap held in place with elastic bandages.

    1. Alleviates bruises.

PARSLEY:

A. Take 2 tablespoons of crushed parsley leaves.

B. Blend crushed leaves with 2 tablespoons butter.

C. Use as poultice to spread on bruises.

    1. Good remedy for bruises.

SOLOMON'S SEAL:

A. Take handful roots and crush to pulpy mass.

B. Spread as poultice on bruised area.

C. Cover with damp cloth to hold in place.

    1. Takes away black and blue marks.

    2. An old American Indian remedy.

BUNION

Dictionary definition: *"Painful, inflamed swelling on the joint of the big toe."*

CATNIP:

A. Put 1/2 cup of shredded leaves in bowl.

B. Break up and add 4 inch square yeast bread.

C. Stir in 1/2 cup milk.

D. Blend and boil until it thickens.

E. Let cool slightly.

F. Use as poultice on bunion 2 successive nights.

G. Reapply poultice for 2 mornings.

    1. Reduces pain of bunion.

BURNS AND SCALDS

ALOE VERA:

A. Break or cut succulent leaves in pieces.

B. Smear gel-like juice on minor burns.

    1. Relieves pain and promotes healing.

AMERICAN BEECH:

A. Put handful of crushed leaves in pint boiling water.

B. Steep 35 to 40 minutes.

C. Strain.

D. Put 2 tablespoons in pint salt water and blend.

E. Wet cloth in this mixture.

F. Lightly wash over burn areas or pat them wet.

1. Soothes burned places.
2. American Indians of old used this remedy.

CATTAIL:

A. Scrape the fuzz from flower heads.

B. Lightly pat this fuzz on burn or scald area.

    1. American Indians long ago used this remedy.

MALLOW, COMMON:

A. Put handful crushed of leaves in quart boiling water.

B. Steep 15 to 20 minutes.

C. Strain.

D. Set aside to cool.

E. Wet cloth in cooled concoction.

F. Wipe lightly over or pat on burn or scald area.

    1. Soothes and heals.

    2. Crushed root can be substituted for crushed leaves.

    3. An old Amish folk medicine remedy.

POTATO, WILD:

A. Grate a few raw potatoes as needed.

B. Apply grated potato mash to burn or scald area.

C. Cover with towel to keep in place.

    1. Excellent burn remedy.

QUININE, WILD:

A. Crush handful of fresh leaves to pulpy mass.

B. Spread as poultice on burn and scald areas.

    1. Promotes healing of burn area.

    2. Used years ago by Catawba Indians.

(C)

CANKER SORES

Dictionary definition: *"A sore that spreads, especially one on or around the mouth."*

ALUMROOT:

A. Pulverize a little dried root to fine powder.

B. Put on canker sores with Q-tip.

    1. Relieves discomfort of sores.

    2. Heals sores around mouth.

GOLDENSEAL:

A. Pulverize a little dried root to fine powder.

B. Put on canker sores with Q-tip.

    1. Relieves pain associated with sores.

    2. Sometimes heals sores overnight.

ONION, WILD:

A. Peel and slice a raw onion.

B. Lay onion slice directly over canker sore.

C. Cover with gauze and tape securely to hold in place.

    1. Excellent for healing sores around mouth.

RED CEDAR:

A. Chew a few pieces of the fruit.

B. Spit juice in soft cloth and pat on canker sore.

    1. Helps clear up sores on mouth.

    2. An old American Indian remedy.

CHAPPED SKIN

Dictionary definition: "To crack open, become rough. Lips often chap or crack open in cold weather."

ALOE VERA:

A. Cut succulent leaves of plant in pieces.

B. Gently rub plant gel on chapped hands and face.

    1. Alleviates burning and cracking of skin.

LEEK, WILD:

A. Crush handful fresh leeks.

B. Blend leek juice with any kind face cream.

C. Gently rub on hands and other chapped areas.

    1. An old Cherokee Indian remedy.

RED CLOVER:

A. Put pound clover blossoms in big pot.

B. Add pound vaseline.

C. Set in sun or place pot in oven to heat.

D. Blend thoroughly by stirring.

E. Strain out blossoms through cheese cloth.

F. Store in tightly closed jar until needed.

G. Rub on hands and face when chapped.

    1. Wonderful remedy for chapped skin.

CHEST CONGESTION
(Also see "COLD," "COUGH","SORE THROAT")

COLT'S FOOT:

A. Put handful of fresh leaves in quart boiling water.

B. Add tablespoon wild licorice leaves.

C. Stir in 2 tablespoons honey.

D. Steep 25 to 30 minutes.

E. Strain.

F. Drink 1 cup 4 times daily.

    1. Clears up lung congestion.

    2. Can use dried leaves in lieu of fresh leaves.

FLAX:

A. Put 1/2 cup seeds in small bowl.

B. Pour cup of boiling water over seeds.

C. Dump this seed mixture on clean cloth.

D. Apply as poultice on chest area.

    1. Simple but effective remedy.

MULLEIN, COMMON:

A. Pour pint milk in pot and heat.

B. Stir in 2 tablespoons dried leaves.

C. Simmer 10 to 15 minutes.

D. Steep 30 minutes more.

E. Strain.

F. Drink while warm 2 times daily.

    1. Alleviates chest congestion.

CHILBLAINS
(Also see "FROSTBITE")

Dictionary definition: "Sore spots or itching redness areas on hands, face and feet caused by extreme cold weather."

ONION, WILD:

A. Peel and roast a whole onion.

B. Crush and use as poultice.

C. Spread on affected areas of cracked or broken skin.

    1. Alleviates itching and burning areas on skin.

PEPPERMINT:

A. Blend 2 tablespoons each of the following:

    1. Peppermint oil.

    2. Olive oil.

    3. Ammonia.

B. Lightly rub on frostbitten hands, face and feet.

C. Reapply frequently as needed.

    1. Alleviates pain and discomfort.

    2. An old Amish folk medicine remedy.

POTATO, WILD:

A. Bake white potato until soft inside.

B. Cut in half lengthwise.

C. Scoop out the inside and add a little vegetable oil.

D. Spread on frostbitten areas.

    1. Soothes and heals frostbite.

CHILLS

(Also see "FEVER" and "SWEATING")

Dictionary definition: *"Uncontrollable shivering caused by sudden coldness of the body."*

FLOWERING DOGWOOD:

A. Put handful of pulverized root in quart boiling water.

B. Add small piece sassafras root for flavor.

C. Steep 20 to 25 minutes.

D. Strain.

E. Drink 1/2 to 1 cup as needed.

    1. Alleviates chills and fever.

    2. An old American Indian remedy.

    3. Pulverized bark can be substituted for pulverized root.

QUININE, WILD:

A. Put handful of dried leaves in pint brandy.

B. Steep 4 to 6 weeks.

C. Strain.

D. Take 1 to 2 tablespoons as needed.

    1. Stops chills and brings down fever.

REDBUD TREE:

A. Put handful of crushed inner bark in quart boiling water.

B. Steep 20 to 25 minutes.

C. Strain.

D. Drink 1/2 to 1 cup as needed for chills.

    1. Overcomes chills.

SMARTWEED:

A. Put handful of crushed leaves in quart boiling water.

B. Steep 20 to 25 minutes.

C. Strain.

D. Drink 1/2 to 1 cup as needed.

    1. Alleviates chills and shakes.

    2. American Indians of old commonly used this remedy.

COLD
(also see "COUGH" and "SORE THROAT")

BONESET:

A. Put handful of leaves in quart boiling water.

B. Steep 20 to 25 minutes.

C. Strain.

D. Drink 1/2 to 1 cup as hot as possible when needed.

    1. Wards off colds.

    2. American Indians of old used this remedy.

    3. Stalks can be substituted for leaves.

HEMLOCK:

A. Put large handful of leaves in quart boiling water.

B. Steep 30 to 35 minutes.

C. Strain.

D. Drink cup as hot as possible when needed.

    1. An old American Indian cold remedy.

PRICKLY ASH:

A. Put handful of crushed bark in quart boiling water.

B. Steep 20 to 25 minutes.

C. Strain.

D. Drink 1/2 to 1 cup hot every 2 to 3 hours.

    1. Quickly alleviates symptoms of cold.

    2. American Indians used this remedy many years ago.

SILVER SPRUCE:

A. Put handful of spruce tips in quart boiling water.

B. Steep 25 to 30 minutes.

C. Strain.

D. Drink 1 cup hot as possible when needed.

    1. An old American Indian cold remedy.

SWEET EVERLASTING:

A. Put handful of leaves and stalks in quart boiling water.

B. Steep 20 to 25 minutes.

C. Strain.

D. Drink 1/2 to 1 cup every 3 to 4 hours.

    1. An old Appalachian Mountain folk cold remedy.

COLD SORES

(Also see "CANKER SORES")

Dictionary definition: *"Spreading sore, in or on mouth, usually accompanied by cold or fever."*

GERANIUM:

A. Pound small amount of dried root to fine powder.

B. Make thick paste by adding little tepid water.

C. Apply as poultice on cold sores.

D. Cover with cloth and tape to hold in place.

E. Repeat as needed until sore heals.

    1. Works well in getting cold sores to heal.

SOAPWORT:

A. Put handful of crushed leaves in pint boiling water.

B. Steep 15 to 20 minutes.

C. Strain.

D. Put lather from this directly on cold sores.

E. Reapply healing lather as needed.

    1. Crushed roots can be substituted for crushed leaves.

    2. American Indians used this many years ago.

YELLOWROOT:

A. Pound a few dried roots to fine powder.

B. Make salve by adding little tepid water.

C. Dab directly on cold sores.

    1. Same salve used by American Indians years ago.

CONSTIPATION

(Also see "LAXATIVE")

Dictionary definition: *"When the bowels are found to be in sluggish condition."*

DANDELION:

A. Put large handful of dried roots in quart boiling water.

B. Steep 25 to 30 minutes.

C. Strain.

D. Drink 1/2 to 1 cup every 2 hours until problem solved.

    1. Excellent aid in ending constipation.

FLAX:

A. Take 1 teaspoon seeds.

B. Chew thoroughly and swallow.

    1. Helps alleviate constipation.

    2. An old Amish folk medicine remedy.

SENNA, WILD 1:

A. Chew on a few leaves

B. Swallow the juice.

    1. A simple yet effective remedy for constipation.

SENNA, WILD 11:

A. Put large handful of leaves in quart boiling water.

B. Steep 20 to 25 minutes.

C. Strain.

D. Drink 1/2 to 1 cup every 1 to 2 hours as needed.

    1. Acts as strong laxative.

WAHOO TREE:

A. Put handful of crushed bark in quart boiling water.

B. Steep 15 to 20 minutes.

C. Strain.

D. Flavor and sweeten with molasses to taste.

E. Drink 1 cup every 2 hours as needed.

    1. Excellent for overcoming constipation.

CORNS

Dictionary definition: *"The hardening of a specific area of skin, usually found on a toe."*

CELANDINE:

A. Pound handful of stems to pulpy mass.

B. Pare corns slightly.

C. Saturate corns with juice from stems.

D. Repeat as necessary.

    1. An excellent corn remedy.

CHICKWEED, COMMON:

A. Pound handful of stems to pulpy mass.

B. Pare corns slightly.

C. Saturate corns with juice from stems.

D. Repeat as necessary.

    1. A good corn remedy.

GROUND IVY:

A. Put handful of ivy leaves in vinegar.

B. Let soak 4 to 6 hours.

C. Lay 1 freshly soaked leaf on each corn.

D. This must be done each morning and evening.

E. Tie these leaves with thread to hold in place.

F. Continue applying leaves for 2 to 3 days.

G. Corn will painlessly lift off in a few days.

    1. Gets rid of any corn remnants.

LEEK, WILD:

A. Crush fleshy leek leaves to pulpy mass.

B. Pare corns slightly.

C. Saturate corns with leek juice.

D. Reapply as necessary.

    1. Excellent remedy for corns.

    2. An old Cherokee Indian remedy.

PRICKLY PEAR CACTUS:

A. Crush handful of prickly pear cactus chunks.

B. Saturate each corn with cactus juice.

    1. Quickly gets rid of unwanted corns.

    2. An old American Indian remedy.

COUGH

(Also see "CHEST CONGESTION")

GINSENG, AMERICAN:

A. Pulverize dry root to fine powder.

B. Put pinch in cup tea or water.

C. Drink this to eradicate tenacious cough.

    1. Simple yet excellent cold remedy.

RATTLESNAKE WEED:

A. Put handful of crushed root in quart boiling water.

B. Steep 20 to 25 minutes.

C. Strain.

D. Drink 1/2 to 1 cup every 2 to 3 hours.

    1. Overcomes coughing.

    2. Leaves can be substituted for crushed root.

SLIPPERY ELM:

A. Chew a piece of inner bark.

B. Swallow the juice.

    1. Relieves bad cough.

WHITE PINE:

A. Put handful of pine needles in quart boiling water.

B. Steep 20 to 25 minutes.

C. Strain.

D. Drink 1/2 to 1 cup every 3 to 4 hours.

    1. Stops coughing.

    2. An old American Indian remedy.

COUGH SYRUP

ELECAMPANE:

A. Place large handful of crushed root in quart boiling water.

B. Steep 2 hours.

C. Bring to boil and cook down to 1 pint.

D. Strain.

E. Pour into bottles and tightly cap.

F. Take 1 or 2 tablespoons every 3 to 4 hours.

    1. Stops persistent old cough.

FLAX:

A. Put 1 cup flaxseed in quart boiling water.

B. Add 1 thinly sliced lemon.

C. Simmer 4 hours.

D. Strain.

E. Stir in honey to suit taste.

F. Take 1 teaspoon 3 times a day.

    1. Take more following severe coughing session.

    2. An old Amish cough syrup concoction.

GARLIC, WILD:

A. Bruise 1 tablespoon fennel seeds.

B. Bruise 1 tablespoon caraway seeds.

C. Put seeds in 1 cup boiling vinegar.

D. Steep 15 minutes.

E. Strain and set aside.

F. Slice up 1 pound garlic.

G. Put garlic in quart boiling water.

H. Cover and steep 12 hours.

I. Strain.

J. Blend vinegar mixture with this.

K. Add honey to suit for flavor.

L. Take 1 to 2 teaspoons as needed.

    1. Alleviates persistent cough.

    2. An old Amish cough syrup blend.

HOREHOUND, COMMON:

A. Put 1-1/2 handfuls of leaves in quart boiling water.

B. Boil 10 minutes.

C. Steep 20 minutes more.

D. Strain.

E. Stir in 6 cups sugar.

F. Boil until mixture thickens.

G. Add 1/2 cup whiskey.

H. Store in closed jar or bottle.

I. Take 1 teaspoon every 3 to 4 hours.

    1. Quickly alleviates a cough.

SLIPPERY ELM:

A. Put handful of chopped inner bark in quart boiling water.

B. Simmer until this starts to thicken.

C. Strain.

D. Stir in 2 tablespoon honey.

E. Stir in juice of 1 lemon.

F. Stir in 1 crushed garlic clove.

G. Continue boiling until quite thick.

H. Blend with 1/2 pint brandy.

I. Take 1 to 2 teaspoons as needed.

    1. Excellent for the harshest of coughs.

CRAMPS

(Also see "MENSTRUAL CRAMPS")

Dictionary definition: *"A sudden, painful contracting of muscles in legs, arms, back, chest, feet or elsewhere."*

FENNEL:

A. Put 2 teaspoons of seeds in pint boiling water.

B. Steep 10 to 15 minutes.

C. Strain.

D. Drink 1/2 to 1 cup as needed.

    1. Stops cramping.

    2. An old Amish folk medicine remedy.

PIPSISSEWA 1:

A. Chew a few leaves.

B. Swallow juice and leaves.

    1. Stops cramping.

PIPSISSEWA 11:

A. Put handful of leaves in quart boiling water.

B. Steep 25 to 30 minutes.

C. Strain.

D. Drink 1/2 to 1 cup as needed.

    1. Stops severe cramping.

SQUAW-WEED 1:

A. Put 2 handfuls of crushed leaves in pint boiling water.

B. Steep 30 to 35 minutes.

C. Strain.

D. Sip off and on during the day.

    1. Excellent for stopping cramping.

    2. An old American Indian remedy.

SQUAW-WEED 11:

A. Chew on small piece of root.

B. Swallow juice and pieces of chewed root.

    1. Good for cramping legs.

2. An old American Indian remedy.

SWEET FLAG:

A. Chew on small piece of root.

B. Swallow juice and pieces of chewed root.

    1. Excellent for stopping cramps.

    2. Long ago used by American Indians.

CUTS
(Also see "WOUNDS")

BLACKBERRY:

A. Put handful of crushed leaves in strainer.

B. Pour boiling water over leaves.

C. Lay soft wet leaves as poultice on cut.

    1. Heals sores.

HORSERADISH:

A. Grate 1 tablespoon of horseradish root.

B. Put grated horseradish in quart — apply cider.

C. Add 1/2 cup garlic juice.

D. Shake well and set in warm place 12 hours.

E. Shake container every 2 hours while set aside.

F. Let stand in warm place another 12 hours.

G. Strain and store in cool place.

H. Dip folded cloth in this mixture.

I. Lay cloth over cut area.

J. Repeat as necessary.

    1. Soothes and heals.

PRICKLY PEAR CACTUS:

A. Crush handful of peeled cactus pods.

B. Spread as poultice on cuts.

C. Cover with cloth and tape securely.

D. Change poultice and cloth as needed.

    1. Alleviates pain and hastens healing.

    2. Used many years ago by American Indians.

PUFF BALL (WHITE ONLY—*BLACK PUFF BALLS ARE POISONOUS*):

A. Gather up numerous puff balls and let them dry.

B. Crush and lay directly on cuts as needed.

    1. Stops bleeding and prevents infection.

(D)

DIARRHEA

Dictionary definition: *"Bowels have too many movements and movements of bowels too loose."*

DIARRHEA — FOOD

BARLEY, LITTLE:

A. Eat 1 cup cooked barley.

B. Do this with every meal.

    1. Controls diarrhea.

CATTAIL:

A. Eat half a handful of fuzz from flowerheads.

    1. Overcomes diarrhea.

    2. Used by American Indians many years ago.

DIARRHEA — SYRUP

BARLEY, LITTLE:

A. Put 1 to 1-1/2 cups of barley in quart boiling water.

B. Add peeling from 3 lemons.

C. Allow to simmer until barley is soft.

D. Steep for 2 to 3 hours.

E. Stir in 1/2 cup honey.

F. Take 2 to 3 teaspoons every 3 to 4 hours.

G. Repeat as often as needed.

    1. Stops diarrhea.

BLACKBERRY:

A. Add 2 tablespoons crushed bark to 3 cups boiling water.

B. Let simmer until only 2 cups liquid are left.

C. Take 1 to 2 tablespoons 4 times a day.

D. Increase dosage if necessary.

    1. Stops diarrhea.

    2. Can substitute roots for bark if necessary.

BLUEBERRY, WILD:

A. Put handful of berries in pint of brandy.

B. Let soak for 21 days.

C. Strain.

D. Blend 1 tablespoon extract with small glass water.

E. Take every 2 hours until diarrhea ceases.

    1. American Indians of old used this remedy.

    2. An old Amish folk medicine remedy.

HUCKLEBERRY:

A. Make the same as blueberry above.

DIARRHEA—TEA

AGRIMONY:

A. Put entire plant in pint boiling water.

B. Steep 20 minutes.

C. Strain.

D. Drink 1/2 to 1 cup.

    1. Instant loose-bowel remedy.

FLOWERING DOGWOOD:

A. Crush handful of dry root and put in quart boiling water.

B. Add small piece sassafras root for flavoring.

C. Steep 20 to 25 minutes.

D. Strain.

E. Drink 1/2 to 1 cup as needed.

    1. Alleviates diarrhea.

    2. Bark can be substituted for root if necessary.

GERANIUM:

A. Put handful of crushed leaves in quart boiling water.

B. Steep 20 to 25 minutes.

C. Strain.

D. Drink 1/2 to 1 cup every hour as needed.

    1. Stops diarrhea.

RATTLESNAKE WEED:

A. Put handful of crushed roots in quart boiling water.

B. Steep 20 to 25 minutes.

C. Strain.

D. Drink 1/2 to 1 cup every 2 hours as needed.

    1. Alleviates diarrhea.

    2. Crushed leaves can be substituted for crushed roots.

REDBUD TREE:

A. Put handful of crushed inner bark in quart boiling water.

B. Steep 25 to 30 minutes.

C. Strain.

D. Drink 1/2 to 1 cup every 2 hours as needed.

    1. Alleviates diarrhea.

SMOOTH SUMAC:

A. Put handful of crushed bark in quart boiling water.

B. Steep 20 to 25 minutes.

C. Strain.

D. Drink 1/2 to 1 cup every 2 to 3 hours.

    1. Clears up diarrhea.

    2. Leaves can be substituted for crushed bark.

    3. An old American Indian remedy.

DIGESTION

CARAWAY:

A. Bruise 2 tablespoons caraway seeds.

B. Put in pint cold water.

C. Steep 6 to 8 hours.

D. Strain.

E. Take one tablespoon at a time.

    1. Aids in digestion.

FLOWERING DOGWOOD:

A. Take 2 tablespoons dried flowering dogwood berries.

B. Take 2 medium nutmegs.

C. Take 3 small cinnamon sticks.

D. Crush all 3 of the above to powder.

E. Blend and keep in airtight container.

F. Put 1/4 teaspoon in cup hot tea each day.

    1. Aids digestive process.

DIURETIC

Dictionary definition: *"Causing the flow of urine to increase considerably."*

DIURETIC — EAT

GRAPE, WILD:

A. Eat grapes regularly and in quantity.

    1. Induces urine flow from body.

    2. Used by American Indians of days gone by.

JUNIPER, COMMON:

A. Eat dried juniper berries regularly and in quantity.

    1. Induces urine flow from body.

PARSLEY I:

A. Put 4 bunches of parsley in pint of boiling water.

B. Peel and cut up 3 onions and add to water.

C. Stir in handful juniper berries.

D. Simmer until done and eat as soup.

    1. Induces urine flow from body.

PARSLEY II:

A. Take some parsley root and chew thoroughly.

B. Swallow the juice.

    1. Increases urine flow from body.

DIURETIC — SYRUP

DANDELION:

A. Put tablespoon of crushed root in quart boiling water.

B. Add 1 tablespoon broom top leaves.

C. Stir in 1 tablespoon juniper berries.

D. Simmer until liquid left is 1 pint.

E. Strain.

F. Take 2 tablespoons every 4 hours.

    1. Stimulates kidneys.

FLAX:

A. Put handful of bruised seeds in pot.

B. Rinse with cold water.

C. Drain water off seeds and add quart cold water.

D. Add rind of whole lemon.

E. Bring to boil and let simmer 20 to 25 minutes.

F. Strain.

G. Squeeze in juice of 1 lemon.

H. Stir in honey to suit taste.

I. Take 1 to 2 tablespoons as needed.

    1. Induces kidney activity.

DIURETIC — TINCTURE

ASPARAGUS:

A. Steam 6 pounds fresh asparagus in quart hot water.

B. Crush stalks to extract juice when ready.

C. Strain.

D. Simmer liquid until reduced to 1 pint.

E. Blend with pint brandy.

F. Set aside in dark place for 2 to 3 weeks.

G. Take 1/2 to 1 teaspoon of this tincture.

    1. Alleviates problems with kidney activity.

CELERY, WILD:

A. Bruise 1/2 cup celery seeds.

B. Stir seeds into pint brandy.

C. Steep 1 to 2 weeks.

D. Strain.

E. Blend 1 tablespoon mixture with 2 tablespoons water.

F. Take 3 times daily when needed.

    1. Red wine can be substituted for brandy.

    2. Increases urine flow.

JUNIPER, COMMON:

A. Put handful of juniper berries in pint wine.

B. Steep 3 to 4 weeks.

C. Strain.

D. Take 1 to 2 tablespoons 3 times each day.

    1. Induces kidney activity.

EARACHE

COMPRESS

CHAMOMILE:

A. Put handful of crushed leaves in pint boiling water.

B. Steep 20 to 25 minutes.

C. Remove leaves from water.

D. Wrap hot leaves in gauze or this cloth.

E. Apply this compress over and behind ear.

    1. Alleviates earache

MARJORAM:

A. Put handful of marjoram in pint boiling water.

B. Bruise handful of flaxseed.

C. Add flaxseed to boiling water as well.

D. Steep 20 to 25 minutes.

E. Strain.

F. Soak folded cloth in this liquid.

G. Lay hot compress over aching ear.

    1. Works quickly and well.

EAR DROPS

MULLEIN, COMMON:

A. Put handful of crushed leaves in pint jar.

B. Fill jar with olive oil or mineral oil.

C. Steep in direct sunlight 4 to 6 weeks.

D. Rotate jar twice daily.

E. Strain.

F. Put 3 drops in ear when aching.

    1. Alleviates pain of earache.

YARROW, COMMON:

A. Put whole plant in pan of boiling water.

B. Steep 25 to 30 minutes.

C. Strain.

D. Push cotton ball into ear.

E. Pour warm liquid into ear.

    1. An old Winnebago Indian recipe.

POULTICE

CARAWAY:

A. Crush handful of caraway seeds.

B. Heat small loaf unsliced bread in oven.

C. Cut away crust and discard.

D. Knead bread and seeds together.

E. Add enough hot brandy to make paste.

F. Apply as poultice behind and over ear.

    1. Alleviates earache.

GARLIC, WILD:

A. Peel outer skin from garlic clove.

B. Trim clove to fit outer ear canal.

C. Wrap clove in gauze to prevent blistering.

D. Fit in outer ear canal.

    1. An old Appalachian mountain remedy.

GINGER, WILD:

A. Put handful of crushed ginger root in some boiling water.

B. Simmer 20 to 30 minutes.

C. Drain water from crushed root and throw away.

D. Spread root as poultice all around ear.

    1. Stops ear aches.

    2. The Meskwaki Indians long ago used this remedy.

PASSION FLOWER:

A. Pulverize handful of roots to pulpy mass.

B. Spread as poultice in and around ear.

    1. Alleviates ear ache.

    2. An old American Indian remedy.

ECZEMA

Dictionary definition: "Skin inflammation characterized by formation of red scaly patches and itching."

ALOE VERA:

A. Slice leaves.

B. Take gel from leaves.

C. Spread on skin where needed.

    1. Effective remedy for eczema.

BLUEBERRY, WILD:

A. Put handful of berries in pint brandy.

B. Soak 2 to 3 weeks.

C. Strain.

D. Put blueberry poultice on dry patches.

E. Cover with gauze to keep in place.

F. Do daily until eczema in under control.

    1. Works well to get rid of skin problem.

    2. Used years ago by American Indians.

SOAPWORT:

A. Pulverize handful of soapwort leaves to pulpy mass.

B. Mix in enough water to make a lather.

C. Spread as poultice on red scaly patches.

D. Repeat as necessary.

    1. Excellent remedy for eczema.

    2. Soapwort roots can be substituted for leaves.

EXHAUSTION

Dictionary definition: *"Extreme fatigue, weariness."*

ALFALFA:

A. Put handful of leaves in quart boiling water.

B Steep 15 to 20 minutes.

C. Strain.

D. Drink 1/2 to 1 cup every 4 to 3 hours.

    1. Quickly increases energy level.

    2. Flowers can be substituted for leaves.

DANDELION:

A. Put handful of leaves in pint boiling water.

B. Steep 20 to 25 minutes.

C. Strain.

D. Drink as much as possible.

    1. Increases energy.

    2. Crushed roots can be substituted for leaves.

MULLEIN, COMMON:

A. Put handful of leaves in quart boiling water.

B. Steep 15 to 20 minutes.

C. Strain.

D. Drink 1/2 cup 3 to 4 times daily.

    1. Increases energy level.

    2. Flowering tops can be substituted for leaves.

EYE PROBLEMS

BURDOCK, COMMON:

A. Chew 1 teaspoon seeds with every meal.

    1. Helps get rid of styes.

    2. Remedy used long ago by American Indians.

    3. An old Amish folk medicine remedy.

ELDERBERRY:

A. Put large handful of blossoms in quart boiling water.

B. Steep 20 minutes.

C. Strain.

D. Use warm liquid as an eyewash.
   1. Helps cure pink eye.
   2. American Indians of long ago used this.
   3. An old Amish folk medicine remedy.

LOBELIA:

A. Put handful of fresh plant in pint brandy.
B. Steep 2 to 3 weeks.
C. Strain.
D. Blend 15 drops with 2 tablespoons water.
E. Put on small pads and lay on eyes.
   1. Works well for most eye problems with mucous discharge.

EYE STRAIN

CHAMOMILE:

A. Put handful of flowers in quart boiling water.
B. Simmer 15 minutes.
C. Strain.
D. Soak folded cloth in this tea.
E. Lay hot compress on eyes.
F. Repeat if necessary when cloth cools.
   1. Relieves eye strain.

LOBELIA:

A. Put handful of whole plant in quart boiling water.
B. Steep 3 to 4 days.
C. Strain.
D. Stir 1 tablespoon in cup boiled water.
E. Use warm as eyewash.
   1. Strengthens eyes.

(F)

FELON

Dictionary definition: *"Extremely painful infection on finger or toe, usually close to the nail."*

ELECAMPANE:

A. Put handful of crushed root in pint boiling water.

B. Steep 30 to 35 minutes.

C. Strain.

D. Soak finger or toe in this for 30 minutes.

E. Reheat and soak again.

F. Reheat and soak a third time if necessary.

G. All soreness should be gone after 2 or 3 soakings.

    1. Clears up infections.

PLANTAIN, COMMON:

A. Crush 1 tablespoon of leaves and seeds to pulpy mass.

B. Put pulp on infected finger or toe.

C. Wrap bandage around pulp to hold in place.

D. Change pulp when hot.

    1. Clears up infection.

FEVER:

(Also see "CHILLS" and "SWEATING")

Dictionary definition: "Unhealthy body condition when temperature is higher than normal."

BONESET:

A. Put handful of leaves in quart boiling water.

B. Steep 20 to 25 minutes.

C. Strain.

D. Drink 1/2 cup as needed.

    1. Alleviates fever.

    2. Stalks can be substituted for leaves.

    3. An old American Indian remedy.

HAZELNUT:

A. Put large handful of bark in quart boiling water.

B. Steep 30 to 35 minutes.

C. Strain.

D. Drink 1/2 to 1 cup when necessary.

    1. Brings fever down.

    2. Used by American Indians long ago.

RED MULBERRY:

A. Chew on some of the fruit.

B. Swallow the juice.

1. Lowers the temperature.

RED RASPBERRY:

A. Pulverize quart of fresh raspberries.

B. Add berries in 1/2 gallon apple cider vinegar.

C. Dilute with water to taste.

D. Drink 1/2 to 1 cup every 1 to 2 hours.

    1. Reduces fever.

SMOOTH SUMAC:

A. Put handful of leaves in quart boiling water.

B. Steep 20 to 25 minutes.

C. Strain.

D. Drink 1/2 to 1 cup every 3 to 4 hours.

    1. Brings down fever.

    2. Used by American Indians years ago.

WAHOO TREE:

A. Put handful crushed bark in quart boiling water.

B. Steep 15 to 20 minutes.

C. Strain.

D. Drink 1 cup every 3 to 4 hours as needed.

    1. Brings down fever.

FLATULENCE

(Also see "GAS")

Dictionary definition: *"Having gas in the intestines or the stomach. ... passing gas."*

CARAWAY:

A. Chew and swallow 1 teaspoon seeds.

    1. Quickly relieves flatulence.

    2. An old Amish folk medicine remedy.

CELERY, WILD:

A. Put 1/2 cup bruised seeds in pint brandy.

B. Steep 1 week.

C. Strain.

D. Blend 1 tablespoon mixture with 2 tablespoon water.

E. Take 3 times daily as needed.

    1. Alleviates flatulence.

FENNEL:

A. Bruise teaspoonful seeds.

B. Pour pint boiling water over seeds.

C. Steep 25 to 30 minutes.

D. Strain.

E. Drink 1/2 to 1 cup as needed.

1. Excellent for relief of flatulence.

SPEARMINT:

A. Chew a few leaves.

B. Swallow juice.

    1. Simple but effective means of flatulence relief.

FLU

BONESET:

A. Put handful of leaves in pint boiling water.

B. Steep 25 to 30 minutes.

C. Strain.

D. Add lemon juice to suit for flavoring.

E. Add sugar to suit for sweetening.

F. Drink 1/2 to 1 cup as needed while hot.

    1. Remedy used by early settlers in Colonies.

    2. American Indians of long ago used this.

MOTHERWORT:

A. Put handful of leaves in pint boiling water.

B. Boil down until 1 cup is left.

C. Strain.

D. Sweeten to taste with honey.

E. Drink hot when going to bed.

    1. Overcomes flu symptoms.

SAGE:

A. Put handful of leaves in pint boiling water.

B. Add a few pine needles.

C. Steep 15 to 20 minutes.

D. Strain.

E. Sweeten to taste.

F. Drink as hot as possible.

    1. Use to create sweat bath.

    2. Favorite of American Indians many years ago.

SWEET EVERLASTING:

A. Put double handful of leaves in quart boiling water.

B. Boil down until 1 pint is left.

C. Strain.

D. Take 1 tablespoon every 3 to 4 hours.

    1. This flu remedy used long ago by American Indians.

FROSTBITE

(Also see "CHILBLAINS")

Dictionary definition: *"Body tissue injury caused by too much exposure to extreme cold weather."*

FIELD HORSETAIL:

A. Put handful of crushed root in quart boiling water.

B. Steep 25 to 30 minutes.

C. Strain.

D. Drink as often as wanted while rewarming.

E. Other teas for frostbite victims can be made with:

    1. Catnip — leaves or flowering tops.

    2. Mint — leaves.

ONION, WILD:

A. Peel skin from raw onion.

B. Cut onion into thin slices.

C. Cover each slice with salt.

D. Crush together to make pulpy mass.

E. Spread over affected area if skin is unbroken.

    1. Effective remedy for frostbite.

POTATO, WILD:

A. Skin and finely grate fresh potato.

B. Spread thickly on gauze.

C. Lay as poultice on frostbitten area of skin.

    1. Relieves burning sensation and associated pain.

SWEET FLAG:

A. Put handful of crushed roots in gallon boiling water.

B. Steep overnight.

C. Strain.

D. Rewarm liquid until comfortable to put hand in.

E. Soak frostbitten hands or feet at least 1 hour.

F. Dip cloth in liquid.

G. Hold to face if frostbitten.

FUNGUS

Dictionary definition: *"A diseased growth on the skin; spongy feeling and unattractive."*

JEWELWEED:

A. Pound handful of leaves to pulpy mass.

B. Rub paste thickly on infected areas of body.

C. Cover with cloth where necessary to hold in place.

    1. Gets rid of unwanted fungus growth.

    2. Excellent for rubbing on infected areas of feet.

MILKWEED, COMMON:

A. Crush all parts of this plant to pulpy mass.

B. Apply the milky latex to infected areas.

    1. Alleviates troublesome fungus.

RED MULBERRY:

A. Take handful of roots and pulverize.

B. Dip cloth in the sap.

C. Apply sap on infected area.

    1. Stops spread of fungus.

STINGING NETTLE:

A. Put handful of crushed whole plant in quart boiling water.

B. Steep 20 to 25 minutes.

C. Strain.

D. Drink 1/2 to 1 cup every 3 to 4 hours.

       and

E. Dip cloth in this tea.

F. Wring out lightly so it doesn't drip.

G. Lay wet cloth over fungus as a compress.

       and

H. Soak fingernails in this tea.

I. Soak toenails in this tea.

    1. Gets rid of fungus.

(G)

GAS
(Also see FLATULENCE)

CORIANDER:

A. Chew and swallow 1 teaspoon of seeds.

    1. Quickly relieves gas.

    2. An old Amish folk medicine remedy.

DILL WEED:

A. Crush 1 tablespoon of seeds.

B. Stir seeds into point boiling water.

C. Steep 20 to 30 minutes.

D. Strain.

E. Drink when needed.

    1. Excellent way to get rid of unwanted gas.

GINGER, WILD:

A. Put 2 tablespoons of crushed root in pint boiling water.

B. Steep 25 to 30 minutes.

C. Strain.

D. Drink 1/2 cup as needed.

    1. American Indians of old used this remedy.

MINT:

A. Put handful of leaves in quart boiling water.

B. Steep 15 to 20 minutes.

C. Strain.

D. Drink 1/2 to 1 cup warm with meals.

    1. Relieves gas.

    2. Remedy long ago used by American Indians.

    3. An old Amish folk medicine remedy.

GLANDS, SWOLLEN

BURDOCK, COMMON:

A. Bruise a few leaves.

B. Wet these leaves.

C. Apply as poultice by laying on area of swollen gland.

    1. Reduces glandular swelling.

FLAX:

A. Crush 1/2 cup of flax seeds.

B. Put crushed seeds in small bowl.

C. Pour cup boiling water over crushed seeds.

D. Put this seed mash on clean cloth.

E. Apply as poultice by laying on area of swollen gland.

    1. Reduces glandular swelling.

MULLEIN, COMMON:

A. Pound handful of leaves to pulpy mass.

B. Spread as poultice on skin where gland is swollen.

C. Cover with small towel to keep pulp in place.

    1. Alleviates swelling and associated discomfort.

PLANTAIN, COMMON:

A. Put handful of leaves in boiling water.

B. Steep 15 to 20 minutes.

C. Strain.

D. Drink 3 to 4 cups daily.

E. Continue drinking until swelling goes down.

    1. Stops swelling.

YARROW, COMMON:

A. Put handful of crushed plant in quart boiling water.

B. Steep 15 to 20 minutes.

C. Strain.

D. Drink 1/2 to 1 cup 3 to 4 times a day.

    1. Relieves swollen glands.

GOUT

Dictionary definition: *"Painful joint disease, often characterized by swelling of big toe."*

STINGING NETTLE:

A. Put handful of crushed root in quart boiling water.

B. Steep 20 to 25 minutes.

C. Strain.

D. Drink 1/2 to 1 cup every 3 to 4 hours as needed.

    1. Relieves pain of gout.

GUM BOILS

MARSHMALLOW:

A. Put handful of crushed leaves in pint boiling water.

B. Steep 20 to 30 minutes.

C. Strain.

D. Use to rinse mouth every 2 to 3 hours.

    1. Alleviates pain of gum boils and heals gums.

SASSAFRAS:

A. Chew small piece sassafras root.

B. Chew this root at least every 2 to 3 hours.

1. Clears up gum boils.

GUMS, BLEEDING

AMERICAN BASSWOOD (LINDEN):

A. Chew on small piece of inner bark.

B. Roll saliva around mouth before swallowing.

   1. Alleviates bleeding gums.

   2. Used long ago by American Indians.

BAYBERRY:

A. Pulverize the following dried leaves to powder:

   1. I handful bayberry.

   2. 1 handful witch-hazel.

   3. 1 handful yellow dock.

B. Blend and use as tooth powder.

   1. Excellent method of overcoming bleeding gums.

QUAKING ASPEN:

A. Chew on small piece of inner bark.

B. Roll saliva around in mouth before swallowing.

   1. Alleviates bleeding gums.

   2. An old American Indian remedy.

SOURWOOD (SORREL-TREE):

A. Chew on small piece of inner bark.

B. Roll saliva around in mouth before swallowing.

   1. Excellent remedy for bleeding gums.

   2. American Indians used this many years ago.

SWEET (BLACK) BIRCH:

A. Chew on small piece of inner bark.

B. Roll saliva around in mouth before swallowing.

   1. Clears up bleeding gums.

   2. Used long ago by American Indians.

TAMARACK:

A. Chew on small piece of inner bark.

B. Roll saliva around in mouth before swallowing.

   1. Alleviates bleeding gums.

   2. Used long ago by American Indians.

GUMS, SORE
(Also see "MOUTH SORES")

AMERICAN HOLLY:

A. Take half handful dried leaves and burn them.

B. Set aside to cool.

C. After cooled, blend the ashes with honey.

D. Dip finger in and rub on infected gums.

    1. Excellent for curing early stages of pyorrhea.

    2. American Indians used this remedy years ago.

GERANIUM:

A. Put large handful of leaves in quart boiling water.

B. Steep 25 to 30 minutes.

C. Strain.

D. Gargle 1/2 cup every hour as needed.

    1. Alleviates gum soreness.

GOLDENSEAL:

A. Pound handful of dried roots to fine powder.

B. Add enough water to make thick paste.

C. Spread paste thickly on gums.

    1. Alleviates sore gums.

WHITE OAK:

A. Pound handful dried oak bark to fine powder.

B. Put powder in pint boiling water.

C. Steep 30 minutes.

D. Strain.

E. Use as gargle and rinse mouth 3 times daily.

    1. Heals gums and combats infections.

(H)

HEADACHE

GINGER, WILD:

A. Put handful of crushed root in quart boiling water.

B. Steep 15 minutes.

C. Strain.

D. Soak folded cloth in this hot concoction.

E. Lay compress on forehead.

F. Repeat if necessary when cloth cools.

    1. Alleviates headache.

    2. American Indians long ago used this remedy.

SPEARMINT:

A. Chew a few spearmint leaves.

B. Swallow the leaves and juice.

    1. Effective in stopping headaches.

HEARTBURN

(Also see "INDIGESTION")

Dictionary definition: "A disconcerting burning sensation felt in the stomach, often moving up to chest and throat."

ANGELICA:

A. Put tablespoon of crushed root in pint boiling water.

B. Simmer 15 minutes.

C. Strain.

D. Take 1 tablespoon when needed.

    1. Relieves heartburn.

CARAWAY:

A. Chew a few caraway seeds.

B. Swallow the juice.

    1. Alleviates heartburn.

    2. An old Amish folk medicine remedy.

MINT:

A. Put handful of leaves in quart boiling water.

B. Steep 15 to 20 minutes.

C. Strain.

D. Drink 1 cup warm tea with meals.

    1. Alleviates heartburn.

    2. American Indians used this remedy many years ago.

    3. An old Amish folk medicine remedy.

HEMORRHOIDS

Dictionary definition: *"Painful swellings formed by the dilation of blood vessels."*

BALSAM POPLAR:

A. Put handful of crushed inner bark in quart boiling water.

B. Steep 25 to 30 minutes.

C. Strain.

D. Drink 1/2 to 1 cup every 3 to 4 hours.

     and

E. Dip cloth in tea.

F. Wring out lightly so it doesn't drip.

G. Lay wet cloth against hemorrhoids.

    1. Excellent hemorrhoid treatment.

HONEYSUCKLE, JAPANESE I:

A. Put handful of crushed honeysuckle flowers in bowl or jar.

B. Add handful of crushed elderberry flowers.

C. Pour 1 pint almost boiling milk over flowers.

D. Steep 30 minutes.

E. Strain.

F. Stir in another pint almost boiling milk.

G. Soak absorbent cloth in liquid.

H. Wring out lightly so it doesn't drip.

I. Hold cloth on area of hemorrhoid pain.

    1. Water can be substituted for the milk.

    2. Excellent hemorrhoid remedy.

HONEYSUCKLE, JAPANESE II:

A. Put handful of green honeysuckle berries in bowl or jar.

B. Blend in handful green elderberry berries.

C. Add pint boiling water.

D. Steep 15 minutes.

E. Strain.

F. Crush berries to pulpy mass.

G. Put ointment directly on hemorrhoids.

    1. Leaves can be substituted for berries.

SWEET (BLACK) BIRCH:

A. Put handful of crushed inner bark in quart boiling water.

B. Steep 25 to 30 minutes.

C. Strain.

D. Drink 1/2 to 1 cup every 3 to 4 hours.

     and

E. Dip cloth in tea.

F. Wring out lightly so it doesn't drip.

G. Lay wet cloth against hemorrhoids.

   1. Quickly relieves discomfort.

TAMARACK:

A. Put handful of crushed inner bark in quart boiling water.

B. Steep 25 to 30 minutes.

C. Strain.

D. Drink 1/2 to 1 cup every 3 to 4 hours.

      and

E. Dip cloth in tea.

F. Wring out lightly so it doesn't drip.

G. Lay wet cloth against hemorrhoids.

   1. Excellent hemorrhoid treatment.

   2. American Indians used this remedy many years ago.

WHITE OAK:

A. Put handful of crushed inner bark in quart boiling water.

B. Steep 25 to 30 minutes.

C. Strain.

D. Drink 1/2 to 1 cup every 3 to 4 hours.

      and

E. Dip cloth in tea.

F. Wring out lightly so it doesn't drip.

G. Lay wet cloth against hemorrhoids.

   1. Reduces swelling of hemorrhoids.

HERNIA

Dictionary definition: *"A rupture; protrusion of part of the intestine or some other organ through a break in its surrounding wall."*

FIELD HORSETAIL:

A. Put handful crushed plant in quart boiling water.

B. Steep 25 to 30 minutes.

C. Strain.

D. Drink 1 cup every 2 to 3 hours.

      and

E. Soak small folded cloth in hot tea.

F. Wring out cloth so it doesn't drip.

G. Lay wet cloth directly on hernia.

H. Cover with piece of plastic wrap and secure with tape.

I. Reapply when cloth is no longer warm.

   1. Relieves discomfort of hernia.

SHEPHERD'S PURSE:

A. Put handful of leaves and seeds in quart boiling water.

B. Steep 25 to 30 minutes.

C. Strain.

D. Soak folded cloth in hot tea.

E. Wring out lightly so it doesn't drip.

F. Lay wet cloth on area of hernia.

G. Cover with piece of plastic wrap and tape securely.

H. Reapply when cloth is no longer warm.

    1. Alleviates hernia problem.

HICCUPS

DILL WEED:

A. Put handful of plant in pint boiling water.

B. Steep 30 to 35 minutes.

C. Strain.

D. Set aside to cool and keep in bottles.

E. Take 1 teaspoon doses as needed.

    1. Stops hiccups almost immediately.

HIVES

Dictionary definition: "One of a variety of diseases in which the skin itches and becomes inflamed in patches."

CATNIP:

A. Put handful of crushed leaves in quart boiling water.

B. Steep 10 to 15 minutes.

C. Strain.

D. Dip cloth in this and wash over hives.

    1. Alleviates itching and swelling of hives.

    2. An old Amish folk medicine remedy.

GROUND IVY:

A. Put handful of leaves and stems in quart boiling water.

B. Steep 25 to 30 minutes.

C. Strain out leaves and stems.

D. Drink 1/2 to 1 cup as needed.

    1. Alleviates itching and swelling.

HAZELNUT:

A. Put handful of crushed bark in quart boiling water.

B. Steep 30 to 35 minutes.

C. Strain.

D. Stir in honey or sugar to suit taste.

E. Drink 1/2 to 1 cup every 3 to 4 hours.

    1. Stops itching and brings down swelling.

(I)

INDIGESTION
(Also see "HEARTBURN")

Dictionary definition: *"The inability to properly digest food; having difficulty digesting food."*

CARAWAY:

A. Chew a few caraway seeds.
B. Swallow the juice and seeds.
   1. Effective in thwarting indigestion.
   2. An old Amish folk medicine remedy.

CELERY, WILD:

A. Bruise 1/2 cup celery seeds.
B. Stir seeds into pint of brandy.
C. Steep 2 to 3 weeks.
D. Strain.
E. Blend 1 tablespoon mixture with 2 tablespoons water.
F. Take 3 times daily when needed.
   1. Overcomes discomfort of indigestion.

MINT:

A. Put handful of leaves in quart boiling water.
B. Steep 15 to 20 minutes.
C. Strain.
D. Drink 1 cup warm tea with meals.
   1. Alleviates indigestion.
   2. Commonly used by American Indians in years past.
   3. An old Amish folk medicine remedy.

SPEARMINT:

A. Chew a few spearmint leaves.
B. Swallow the leaves and juice.
   1. Quickly relieves indigestion.

INFLAMMATION

Dictionary definition: *"Diseased condition somewhere in body; signs are pain, redness, swelling and heat."*

COMFREY:

A. Crush handful of dry root to powder.
B. Add enough water to make thick paste.
C. Place paste on small piece of cloth.
D. Lay this on inflamed area.
   1. Clears up redness and pain.

PASSION FLOWER:

A. Crush handful of roots to pulpy mass.

B. Spread as poultice on inflamed area.

    1. Reduces soreness and inflammation.

    2. American Indians long ago used this remedy.

PLANTAIN, COMMON:

A. Bruise a large wet plantain leaf.

B. Wrap around inflamed area.

C. Cover with plastic wrap to hold in moisture.

    1. Reduces pain and swelling.

INSOMNIA
(Also see "SLEEP DISORDERS")

Dictionary definition: *"Inability to sleep; sleeplessness."*

ANGELICA:

A. Put 1 tablespoon of pulverized root in quart boiling water.

B. Add 1 tablespoon ground skullcap leaves.

C Stir in 1 tablespoon ground catnip leaves.

D. Cover tightly and steep 30 minutes.

E. Strain.

F. Drink 1/2 to 1 cup during day.

G. Drink another 1/2 to 1 cup before going to bed.

    1. Overcomes inability to fall asleep.

HOPS:

A. Put handful of hops in quart boiling water.

B. Steep 20 to 25 minutes.

C. Strain.

D. Drink 1 cup upon going to bed at night.

    1. Person falls asleep easily.

    2. Using hop-stuffed pillow brings uneventful sleep.

SKULLCAP (MAD-DOG SKULLCAP):

A. Put 2 tablespoons of crushed leaves in quart white wine.

B. Add 2 tablespoons of rosemary leaves.

C. Shake thoroughly.

D. Steep 3 to 4 weeks.

E. Strain.

F. Take tablespoon 3 times daily and at bedtime.

    1. Wonderful sedative for those with insomnia.

(J)

JOINTS
(Also see "ARTHRITIS")

EUCALYPTUS

A. Put 1/4 cup crushed leaves in 3 quarts water.

B. Bring to boil.

C. Simmer 20 minutes.

D. Strain.

E. Add this tea to bath water.

F. Soak in bathtub at least 20 to 30 minutes.

    1. Quickly relieves joint pain.

ONION, WILD:

A. Skin 4 to 6 large onions.

B. Slice onions in half.

C. Put in oven and roast.

D. Let cool a little before using.

E. Place halves directly on swollen or painful joints.

F. Keep securely in place with cloth strips and tape.

    1. Excellent for painful and swollen joints.

(K)

KIDNEY PROBLEMS
(See "DIURETIC")

(L)

LAXATIVE
(Also see "CONSTIPATION")

Dictionary definition: *"Medicine taken to make the bowels move."*

SENNA, WILD:

A. Crush 1 teaspoon of dry leaves to powder.

B. Stir into 1 cup boiling water.

C. Add 1 teaspoon of ground coriander seeds.

D. Steep 20 to 25 minutes.

E. Strain.

F. Drink as needed for constipation.

   1. Pinch of fennel can be substituted for coriander.

   2. Pinch of ginger can be substituted for coriander.

   3. Excellent laxative.

SWEET FLAG:

A. Crush a little dried root to powder.

B. Put 1 teaspoonful in cereal bowl.

C. Add 1 cup wheat flour.

D. Stir in enough hot milk to make gruel.

E. Sweeten to taste with honey.

F. Eat for breakfast every morning as needed.

   1. One of best remedies for constipation.

   2. Popular years ago with American Indians.

TURTLEHEAD:

A. Put handful of crushed leaves in pint boiling water.

B. Steep 15 to 20 minutes.

C. Strain.

D. Set aside to cool.

E. Drink 1/2 to 1 cup as needed.

   1. An effective laxative.

LICE, HEAD
(Also see "PARASITES")

ANISE ROOT (SWEET CICELY):

A. Bruise a handful of anise seeds.

B. Add these seeds to 1 pint avocado oil.

C. Bake slowly for 2 hours.

D. Strain.

E. Thoroughly rub ointment on lice-infested head.

   1. Any of these can be substituted for avocado oil:

a. Wheat germ oil.

b. Vitamin E oil.

c. Lanolin.

d. Lard.

2. Extremely good method of eliminating head lice.

GOAT'S RUE:

A. Pulverize 6 tablespoons of dry roots and leaves.

B. Put in quart of vinegar or white wine.

C. Steep in closed bottle 2 weeks or longer.

D. Strain.

E. Use to wash head and hair.

1. Good way to get rid of head lice.

LUNGS

THYME, WILD:

A. Put handful of leaves in pint boiling water.

B. Steep 30 minutes.

C. Strain and allow to cool.

D. Drink 1/2 cup every 3 to 4 hours.

1. Gets rid of phlegm in lungs.

2. Overcomes shortness of breath.

(M)

MENSTRUAL CRAMPS
(Also see "CRAMPS")

CATNIP:

A. Crush handful of leaves and flowery tops to pulpy mass.

B. Spread as poultice on abdomen while lying down.

    1. Alleviates painful monthly cramping.

    2. An old Amish folk medicine remedy.

ELECAMPANE:

A. Put 1 tablespoon of pulverized root in pint water.

B. Soak 2 to 3 hours.

C. Bring to boil and simmer 15 minutes.

D. Strain.

E. Set aside to cool.

F. Take 1 to 2 tablespoons every 3 to 4 hours.

    1. Regulates menstruation.

FENNEL:

A. Chew 1 tablespoon of seeds.

B. Swallow seeds and juice.

    1. Simple remedy reduces pain while cramping.

YELLOWROOT:

A. Put handful of crushed root in quart boiling water.

B. Steep 15 to 20 minutes.

C. Strain.

D. Drink 1/2 to 1 cup every 2 to 3 hours.

    1. Stops pain associated with cramping.

MORNING SICKNESS
(Also see "NAUSEA" and "VOMITING")

LEMON BALM:

A. Put 1 teaspoon each of following in quart boiling water:

    1. Crushed lemon balm leaves.

    2. Crushed stinging nettle leaves.

    3. Crushed red raspberry leaves.

    4. Crushed alfalfa leaves.

    5. Crushed spearmint leaves.

    6. Fennel seeds.

B. Steep 15 minutes.

C. Strain.

D. Drink 1/2 to 1 cup every 3 to 4 hours.

1. Overcomes morning sickness.
2. An old Amish folk medicine remedy.

RED RASPBERRY:

A. Put handful of leaves in quart boiling water.

B. Steep 20 to 25 minutes.

C. Strain.

D. Drink 1/2 to 1 cup every 3 to 4 hours.
1. Excellent in alleviating morning sickness.
2. An old Amish folk medicine remedy.

MOUTH SORES
(Also see "GUMS, SORE")

AGRIMONY:

A. Put entire crushed plant in pint boiling water.

B. Steep 25 to 30 minutes.

C. Strain.

D. Use as mouth and throat gargle.
1. Gets rid of ulcers in mouth.

RED RASPBERRY:

A. Put handful of crushed leaves in pint boiling water.

B. Steep 30 minutes.

C. Strain.

D. Set aside and let tea solution cool.

E. Rinse mouth thoroughly every 2 hours.
1. Gets rid of mouth blisters and sores.
2. Shoots can be substituted for leaves.

FRAGRANT WATER LILY:

A. Put handful of crushed roots in quart boiling water.

B. Steep 15 to 20 minutes.

C. Strain.

D. Gargle 1/2 cup every hour as needed.
1. Clears up mouth sores.
2. An old remedy used by American Indians.

YARROW, COMMON:

A. Put handful of dried whole plant in pint boiling water.

B. Steep 2 to 3 hours.

C. Strain.

D. Let liquid cool.

E. Use as mouthwash.
1. Alleviates sores on gums and inside mouth.

MUSCLE PULLED (Also see "MUSCLE STRAINIELD HORSETAIL:

A. Put handful of crushed whole plant in quart boiling water.

B. Steep 25 to 30 minutes.

C. Strain.

D. Drink 1/2 to 1 cup every 3 to 4 hours.

      and

E. Dip folded cloth in this tea.

F. Wring out lightly so it doesn't drip.

G. Lay wet cloth over pulled muscle.

H. Cover with plastic sheet and secure with tape.

    1. Excellent for overcoming pulled muscles.

SHEPHERD'S PURSE:

A. Put handful of crushed leaves in quart boiling water.

B. Steep 25 to 30 minutes.

C. Strain.

D. Dip cloth in tea.

E. Wring out lightly so it doesn't drip.

F. Lay hot wet cloth over pulled muscle.

G. Cover with plastic sheet and secure with tape.

    1. Seeds can be substituted for leaves.

    2. Relieves pain associated with pulled muscles.

MUSCLE SPASMS

CHAMOMILE:

A. Put 4 tablespoons of flowers in pint boiling water.

B. Steep 20 minutes.

C. Strain.

D. Drink 1/2 cup every 3 to 4 hours.

    1. Stops muscle spasms.

PEPPERMINT:

A. Take large handful of crushed leaves.

B. Put in less than 1/2 cup boiling water.

C. Stir to make wet mash.

D. Steep 20 to 30 minutes.

E. Spread as poultice on area of muscle spasm.

F. Cover poultice with cloth.

G. Tape securely in place.

    1. Overcomes muscle spasms and cramps.

    2. Alleviates associated discomfort.

MUSCLE STRAIN
(Also see "MUSCLE PULLED")

STINGING NETTLE:

A. Put handful of crushed plant in quart boiling water.

B. Steep 25 to 30 minutes.

C. Strain.

D. Drink 3 to 4 cups daily.

    1. Overcomes discomfort from muscle strain.

WITCH-HAZEL:

A. Put handful of crushed bark in quart boiling water.

B. Steep 30 minutes.

C. Strain.

D. Rub tea on strained muscles.

    1. Leaves can be substituted for crushed bark.

    2. Acts to heal muscle strains.

(N)

NAUSEA

(Also see "MORNING SICKNESS" and "VOMITING")

Dictionary definition: *"The terrible feeling a person has they are about to vomit."*

AMERICAN BEECH:

A. Put handful of crushed bark in quart boiling water.

B. Steep 20 to 25 minutes.

C. Strain.

D. Drink 1/2 to 1 cup as needed.

    1. Overcomes nausea and stops vomiting.

    2. American Indians long ago used this remedy.

SPEARMINT:

A. Chew a few spearmint leaves.

B. Swallow leaves and juice.

    1. A simple method of overcoming nausea.

NERVOUSNESS

Dictionary definition: *"Highly restless, easily excited, high strung, uneasy, jittery, excitable."*

BLACK COHOSH:

A. Put handful of crushed root in quart boiling water.

B. Steep 20 to 25 minutes.

C. Strain.

D. Drink 1/2 to 1 cup before going to bed at night.

    1. Quickly alleviates nervousness.

    2. An old Amish folk medicine remedy.

BLUE VERVAIN:

A. Put handful of crushed roots in quart boiling water.

B. Steep 20 to 25 minutes.

C. Strain.

D. Drink 1/2 to 1 cup before going to bed at night.

    1. Leaves can be substituted for crushed roots.

    2. Helps greatly in relaxing at bedtime.

    3. An old Amish folk medicine remedy.

    4. American Indians of old used this for calming effect.

SKULLCAP (MAD-DOG SKULLCAP):

A. Put handful of leaves in quart boiling water.

B. Steep 30 minutes.

C. Strain.

D. Drink 1 cup just before going to bed.

    1. A rather strong sedative.

2. An old Amish folk medicine remedy.

NOSE BLEED
(Also see "BLEEDING")

HAZELNUT:

A. Crumble tablespoon of dry leaves.

B. Gently stuff up nose.

   1. Stops nose from bleeding.

   2. American Indians of old used this remedy.

NURSING
(Also see "BREASTS)

Dictionary definition: "To give milk to a baby from a mother's breast."

ALFALFA:

A. Put handful of crushed plant in quart boiling water.

B. Steep 20 to 25 minutes.

C. Strain.

D. Drink 1/2 to 1 cup every 3 to 4 hours.

   1. Increases and enriches milk supply.

ELDERBERRY:

A. Crush quart of berries to pulp.

B. Stir into quart boiling water.

C. Boil down to where 1 pint is left.

D. Drink 1/2 cup daily.

   1. Increases milk supply.

   2. Used long ago by American Indians.

STINGING NETTLE:

A. Put handful of crushed plant in quart boiling water.

B. Steep 20 to 25 minutes.

C. Strain.

D. Drink 1/2 to 1 cup every 3 to 4 hours.

   1. Increases and enriches milk supply.

(P)

PAIN RELIEF
(Also see "ARTHRITIS" and "SWELLING")

CELANDINE:

A. Crush stems to pulpy mass.

B. Dab juice on cut or wound.

    1. Stops pain around wound.

WAHOO TREE:

A. Put handful of crushed bark in quart boiling water.

B. Steep 15 to 20 minutes.

C. Strain.

D. Drink 1/2 to 1 cup every 3 to 4 hours.

    1. Relieves aches and pains.

WHITE WILLOW:

A. Remove inner bark from white willow tree.

B. Put in pot and add enough water to cover.

C. Bring to boil.

D. Add more water.

E. Simmer 3 to 4 hours.

F. Strain.

G. Cover and set aside.

H. Drink good size glass as needed.

    1. Alleviates aches and pains.

PARASITES
(Also see "LICE, HEAD" and "WORMS")

Dictionary definition: *"Animal or plant that lives on, with, or in another from which it gets its food."*

GOAT'S RUE:

A. Dry and crumble roots and leaves.

B. Put 3 tablespoons in pint vinegar or white wine.

C. Steep in closed jar or bottle at least 2 weeks.

D. Rub on skin affected by parasites or body lice.

    1. American Indians used this remedy long ago.

PNEUMONIA
(Also see "CHEST CONGESTION")

Dictionary definition: *"Serious disease in which a person's lungs are inflamed."*

BALSAM FIR:

A. Put handful of crushed leaves in quart boiling water.

B. Steep 25 to 30 minutes.

C. Strain.

D. Drink 1/2 to 1 cup every 3 to 4 hours.

    1. Relieves chest congestion.

    2. American Indians used this remedy many years ago.

BUTTERFLY WEED:

A. Put handful of crushed root in quart boiling water.

B. Steep 20 to 25 minutes.

C. Strain.

D. Sweeten to taste with honey.

E. Drink 1/2 cup every 2 hours.

    1. Greatly relieves chest congestion.

MUSTARD, BLACK:

A. Pulverize pint of dry seeds to a powder.

B. Add enough water to make thick paste.

C. Spread paste all over chest.

D. Cover chest with towel.

    1. Alleviates chest congestion.

POISON IVY/POISON OAK
(Also see "RASH")

AMERICAN BEECH:

A. Put handful of crushed leaves in pint boiling water.

B. Steep 25 to 30 minutes.

C. Strain.

D. Put 2 tablespoons of this in pint salt water.

E. Wet cloth in mixture and wash over rash.

    1. Stops spread of poison ivy or oak.

    2. American Indians of old used this remedy.

BROOMSEDGE:

A. Put handful of crushed leaves in quart boiling water.

B. Steep 20 to 25 minutes.

C. Strain.

D. Dip clean cloth in this tea.

E. Wash affected areas of body.

    1. Combats poison ivy and poison oak.

    2. An old Amish folk medicine remedy.

    3. Used many years ago by American Indians.

TANSY, COMMON:

A. Put handful of whole plant in quart boiling water.

B Steep 15 to 20 minutes.

C. Strain.

D. Dip clean wash cloth in tea.

E. Wash all rash and blistered areas of body.

    1. Warning: Never drink as 1/2 ounce kills within hours.

    2. Use only as "wash" for rash and blisters.

    3. An old Amish folk medicine remedy.

(R)

RASH
(Also see "HIVES" and "POISON IVY/POISON OAK")

GOLDENSEAL:

A. Put handful of chopped root in pint boiling water.

B. Steep 20 minutes.

C. Strain.

D. Wash rash and blisters often with this concoction.

    1. Any sudden skin eruptions respond well to this.

JEWELWEED:

A. Crush handful of stems and leaves to pulpy mass.

B. Spread as poultice on rash area.

    1. Stops itching and heals affected area.

    2. An old Amish folk medicine remedy.

SASSAFRAS:

A. Put handful of crushed root bark in quart boiling water.

B. Steep 20 to 25 minutes.

C. Strain.

D. Soak clean cloth in this tea.

E. Wring out slightly so it doesn't drip.

F. Wash affected areas of body.

    1. Stops itching and heals rash.

    2. An old Amish folk medicine remedy.

RESTLESSNESS
(See "NERVOUSNESS")

RHEUMATISM

Dictionary definition: *"A disease with stiffness, swelling and inflammation of the joints."*

ALOE VERA:

A. Break or cut a few succulent leaves into pieces.

B. Smear gel-like juice all over painful joints.

    1. Alleviates pain.

CHRISTMAS FERN:

A. Pulverize handful of roots to pulpy mass.

B. Spread as poultice over painful areas.

    1. Alleviates aches and pains in joints.

    2. Used long ago by American Indians.

GARLIC, WILD:

A. Peel clove of fresh garlic every day.

B. Cut up into small pieces.

C. Chew and swallow.

D. Continue for 30 days.

    1. Highly recommended old-time remedy.

PRICKLY ASH:

A. Put handful of crushed bark in quart of boiling water.

B. Steep 20 to 25 minutes.

C. Strain.

D. Drink 1/2 to 1 cup as needed.

    1. Alleviates pain of rheumatism.

    2. An old American Indian remedy.

RINGWORM

(Also see "WORMS")

Dictionary definition: *"A parasite caused contagious skin disease characterized by ring-shaped patches."*

GARLIC, WILD:

A. Remove skin from 4 garlic cloves.

B. Chop the cloves into small pieces.

C. Sprinkle garlic pieces between 2 bread slices.

D. Eat as a sandwich.

    1. Rids body of pin worms.

GOLDTHREAD:

A. Put handful of crushed root in quart boiling water.

B. Steep 1 hour to make extremely strong tea.

C. Strain.

D. Dip cloth in this liquid.

E. Wash affected areas 3 to 4 times daily.

    1. Alleviates ring worm problem.

RED MULBERRY:

A. Collect a small quantity of sap.

B. Rub sap on areas indicating ring worm.

    1. Good remedy to rid body of ring worm infestation.

    2. American Indians of long ago used this method.

YELLOW DOCK:

A. Put handful of crushed root in pint vinegar.

B. Steep 3 to 4 weeks.

C. Strain.

D. Soak cloth in mixture.

E. Wash affected areas 3 to 4 times daily.

    1. Old but effective treatment for ring worm.

(S)

SCURVY

Dictionary definition: *"Caused by lack of vitamin C in diet; disease characterized by swollen and bleeding gums."*

HEMLOCK:

A. Put handful of bark or leaves in quart boiling water.

B. Steep 30 to 35 minutes.

C. Strain.

D. Drink 1/2 to 1 cup as needed.

    1. Prevents scurvy.

    2. Leaves can be substituted for bark.

ROSE HIPS:

A. Put 2 teaspoons of rose hips in large cup water.

B. Simmer 10 minutes.

C. Strain.

D. Drink 1/2 to 1 cup at least once daily.

    1. Prevents scurvy.

SHINGLES

Dictionary definition: *"A virus causing itching spots and blisters to break out on skin."*

RED MULBERRY:

A. Put handful of leaves in quart boiling water.

B. Add handful of pine inner bark.

C. Steep 30 minutes.

D. Strain.

E. Dip cloth in this concoction.

F. Lightly wash over blisters and itching spots.

G. Repeat every 3 hours until shingles are gone.

    1. American Indians of old used this remedy.

SINUS CONGESTION

GOLDENSEAL:

A. Crush small piece of dried root to powder.

B. Put pinch on tongue as needed.

    1. Clears up sinus congestion and stuffy nose.

    2. An old Amish folk medicine remedy.

SPEARMINT

A. Put handful of crushed leaves in quart of boiling water.

B. Simmer 25 to 30 minutes.

C. Strain.

D. Drink 1/2 to 1 cup as needed.

1. Clears up sinus congestion.

SLEEP DISORDER
(Also see "INSOMNIA")

Dictionary definition: *"Inability to sleep. Sleeplessness."*

AMERICAN BASSWOOD:

A. Put handful of flowers and buds in quart of boiling water.

B. Steep 20 to 30 minutes.

C. Strain.

D. Drink as tea before going to bed.

    1. Promotes more restful sleep.

    2. American Indians of old used this remedy.

CATNIP:

A. Put handful of leaves and flowers in pint of boiling water.

B. Steep 15 to 20 minutes.

C. Strain.

D. Take teaspoonful just before going to bed.

    1. Induces sleep even for the most restless.

SNAKE BITE
(Also see "PAIN RELIEF" and "SWELLING")

JEWELWEED:

A. Put handful of crushed leaves in pint of boiling water.

B. Steep 10 to 15 minutes.

C. Strain.

D. Drink all the tea as hot as possible.

E. Spread leaves as poultice directly on snake bite.

    1. Use jewelweed poultice and tea for best results.

QUEEN ANN'S LACE:

A. Crush root to pulpy mass.

B. Spread as poultice directly on and around snake bite.

    1. Alleviates pain and swelling.

SWEET VIOLET:

A. Crush handful of sweet violets to pulpy mass.

B. Spread thickly as poultice on snake bite.

C. Wrap cloth over poultice to keep in place.

    1. Draws poison out and alleviates pain.

    2. Remedy used by American Indians of old.

SORE MUSCLES
(Also see "MUSCLE PULLED")

MARJORAM:

A. Crush handful of dried marjoram leaves to fine powder.

B. Crush handful of dried thyme leaves to fine powder.

C. Blend powders in pint olive oil.

D. Steep in closed container 7 days.

E. Rub this lotion on sore muscles to stop pain.

    1. Stops muscular aches and pains.

WITCH-HAZEL:

A. Put handful of crushed leaves in quart of boiling water.

B. Steep 30 minutes.

C. Strain.

D. Rub hot tea on sore or aching muscles.

    1. Alleviates muscular aches and pains.

    2. Crushed bark can be substituted for crushed leaves.

    3. American Indians used this remedy many years ago.

SORES

COMFREY:

A. Mash handful of fresh leaves.

B. Wrap mashed leaves in gauze.

C. Lay compress on sore.

D. Repeat as often as needed.

    1. Hastens healing and relieves pain.

RED CLOVER:

A. Put handful of fresh clover blossoms in canning jar.

B. Add pint of brandy to jar.

C. Put on lid and set aside 3 to 4 weeks.

D. Shake mixture 2 to 3 times daily.

E. Strain through clean cloth or paper coffee filter.

F. Cover and set aside.

G. Use liquid as needed to bathe sores.

    1. A good means of inducing healing.

SAGE:

A. Put handful of leaves in quart of boiling water.

B. Steep 45 to 60 minutes.

C. Strain.

D. Soak folded cloth in this tea.

E. Lay hot compress on sores.

1. Sores can also be washed with this solution.
2. Helps sores heal more readily.
3. Remedy long ago used by American Indians.

WITCH-HAZEL:

A. Put heaping handful of bark in quart of boiling water.

B. Continue boiling until only half the water remains.

C. Strain.

D. Stir in wheat bran until mixture thickens.

E. Spread as poultice over sores.

F. Cover with cloth or bandage if necessary.

    1. This old American Indian remedy induces healing.

SORE THROAT

BALSAM FIR:

A. Chew on chunk of pitch from this tree.

B. Let juice drizzle down throat.

    1. American Indians used to sooth sore throat.

RHUBARB:

A. Thinly slice a few roots.

B. Set aside to dry.

C. When dry, store in closed jar or container.

D. Chew piece of dried root when needed.

E. Swallow juice while chewing.

    1. Soothes and heals sore throat.

    2. American colonists used this simple remedy.

SLIPPERY ELM:

A. Chew a small piece of inner bark.

B. Swallow juice while chewing.

    1. Relieves sore throat.

SORE THROAT COMPRESS

FLAX:

A. Put 1/2 cup of flaxseed in small bowl.

B. Pour cup of boiling water over seeds.

C. Pour this seed mash on clean cloth.

D. Fold cloth and use as neck compress.

    1. Good compress for getting rid of sore throat.

MARJORAM:

A. Put handful of crushed leaves in quart of boiling water.

B. Steep 4 to 6 hours.

C. Strain.

D. Soak folded cloth in this water.

E. Wring out compress so it doesn't drip.

F. Wrap compress around throat.

G. Wrap towel or dry cloth around compress.

H. Secure in place with safety pin.

I. Reapply compress as necessary.

    1. Relieves discomfort of sore throat.

SORE THROAT GARGLE

SAGE:

A. Put 2 tablespoons of leaves in pint of boiling water.

B. Stir in 1/4 cup salt.

C. Add 1/4 cup apple cider vinegar.

D. Stir in 2 tablespoons cayenne (red) pepper.

E. Add 2 tablespoons honey.

F. Blend all ingredients thoroughly.

G. Steep 20 minutes.

H. Strain.

I. Gargle mixture as needed.

    1. Clears up sore throat and helps avoid colds.

SPIKENARD, AMERICAN:

A. Put handful of crushed leaves in quart of boiling water.

B. Simmer 20 to 30 minutes.

C. Strain.

D. Gargle, then swallow 1/2 cup every 1 to 2 hours.

    1. Alleviates sore throat.

    2. This is an old pioneer remedy.

    3. Also used by American Indians of old.

YELLOWROOT:

A. Put handful of crushed root in quart boiling water.

B. Steep 20 to 25 minutes.

C. Strain.

D. Gargle 1/2 cup every 2 hours.

    1. Overcomes sore throat.

    2. American Indians of old used this remedy.

SORE THROAT TEAS

COLT'S FOOT:

A. Put handful of fresh leaves in quart of boiling water.

B. Add 1 tablespoon wild licorice leaves.

C. Stir in 2 tablespoons honey.

D. Steep 25 to 30 minutes.

E. Strain.

F. Drink 1 cup 4 times daily.

1. Excellent remedy for easing sore throat.

CURRANT, WILD BLACK:

A. Crush 1 pint of currants in pot.

B. Strain.

C. Heat juice slowly.

D. Drink when warm.

    1. American Indians of old used this remedy.

HOREHOUND, COMMON:

A. Put 2 tablespoons of fresh leaves in pint of boiling water.

B. Steep 20 to 30 minutes.

C. Strain.

D. Sweeten tea with 1 tablespoon honey.

E. Drink 1/4 to 1/2 cup as needed.

    1. Alleviates sore throat.

RED RASPBERRY:

A. Crush quart of red raspberries to pulpy mass.

B. Put crushed berries in 2 quarts apple cider vinegar.

C. Dilute with water to taste.

D. Drink cup every 2 to 3 hours daily as needed.

    1. Excellent for overcoming sore throat.

SPLINTERS

AMERICAN BASSWOOD (LINDEN):

A. Put handful of crushed inner bark in pint boiling water.

B. Boil down to jelly consistency.

C. Apply as poultice directly on splinter.

D. Wrap with bandage to hold poultice in place.

    1. Wrap in moss if bandage not available.

    2. Draws splinters to surface of skin.

    3. Used long ago by American Indians.

WHITE PINE:

A. Put some white pine sap in pan.

B. Heat until warm.

C. Spread warmed sap directly on splinter.

D. Splinter will be drawn to surface.

    1. Excellent method of splinter removal.

    2. An old American Indian remedy.

SPRAINS

Dictionary definition: *"Injury to a muscle or joint by sudden wrench or twist."*

BURDOCK, COMMON:

A. Crush handful leaves to pulpy mass.

B. Apply as poultice directly on sprain.

C. Wrap with gauze to hold poultice in place.

    1. Effective remedy for sprains.

POTATO, WILD:

A. Peel and grate 2 to 3 raw potatoes.

B. Spread grated potato thickly on sprain.

C. Cover with towel or cloth to hold in place.

    1. Relieves discomfort and speeds healing.

WINTERGREEN:

A. Put handful of leaves in pint of apple cider vinegar.

B. Blend thoroughly.

C. Steep 3 days.

D. Strain.

E. Soak folded cloth in mixture.

F. Wring out so it doesn't drip.

G. Lay compress on sprain.

H. Repeat an needed.

    1. Speeds healing of sprains.

STINGS
(Also see "BITES")

MINT:

A. Rub area of nettle sting with mint leaves.

    1. Will almost immediately relieve discomfort.

    2. An old American Indian remedy.

PURSLANE, COMMON:

A. Crush plant parts to pulpy mass.

B. Soak up juice on small piece of cloth.

C. Lay cloth on sting.

D. Replace with fresh cloth when needed.

    1. Quickly relieves pain.

ROSEMARY:

A. Crush a few rosemary leaves.

B. Rub these leaves over the sting area.

    1. Almost immediately relieves discomfort.

    2. American Indians used this remedy years ago.

STOMACH PROBLEMS
(Also see "ULCERS IN STOMACH")

BLACKBERRY:

A. Chop up handful of root and put in quart of boiling water.

B. Steep 20 to 25 minutes.

C. Strain.

D. Drink 1/2 to 1 cup when needed.

    1. Overcomes stomach aches.

BLUE VERVAIN:

A. Put handful of crushed leaves in quart boiling water.

B. Steep 30 minutes.

C. Strain.

D. Set aside to cool.

E. Drink 1 cup every 2 to 3 hours.

    1. Stops stomachaches.

    2. American Indians long ago used this remedy.

CHRISTMAS FERN:

A. Put handful of leaves in quart of boiling water.

B. Steep 25 to 30 minutes.

C. Strain.

D. Let cool.

E. Drink 1 cup every 1 to 3 hours as needed.

    1. Alleviates stomachache.

    2. Stems can be substituted for leaves.

    3. American Indians of old used this remedy.

SUNBURN

ALOE VERA:

A. Break or cut succulent leaves into pieces.

B. Smear gel-like juice on sunburned area.

    1. Relieves discomfort and pain of sunburn.

PLANTAIN, COMMON:

A. Pound handful of fresh leaves to pulpy mass.

B. Spread as poultice over entire blistered area.

C. Wash off in 1 hour.

D. Wait 30 minutes and reapply poultice.

    1. Soothes pain associated with sunburn.

    2. Heals blisters.

SWEATING
(Also see "CHILLS" and "FEVER")

RED CEDAR:

A. Put handful of fruit in quart boiling water.

B. Steep 20 to 25 minutes.

C. Strain.

D. Drink 1/2 to 1 cup every 2 to 3 hours.

1. Induces sweating.
2. American Indians used this remedy many years ago.

SWELLING

CURRANT, WILD BLACK:

A. Pound handful of dry roots to a powder.

B. Add enough water to make thick paste.

C. Spread as poultice on swollen area.

D. Lay cloth over poultice to keep in place.

    1. Brings down swelling.

    2. American Indians of long ago used this remedy.

    3. Bark can be substituted for roots.

ELDERBERRY:

A. Pulverize fresh leaves to pulpy mass.

B. Spread as poultice on swollen areas.

C. Cover with towel or cloth to keep in place.

    1. Reduces swelling.

    2. Used by American Indians many years ago.

WINTERGREEN:

A. Put handful of crushed leaves in pint of apple cider vinegar.

B. Steep 3 to 4 weeks in closed container.

C. Strain.

D. Soak folded cloth in mixture.

E. Wring out so it doesn't drip.

F. Lay this compress directly on swollen joint.

G. Repeat as needed.

    1. Relieves pain and swelling.

(T)

TAPEWORM
(Also see "RINGWORM" and "WORMS")

Dictionary definition: *"A long flat worm that lives as a parasite in the intestine."*

RED MULBERRY:

A. Put handful of crushed root in quart of boiling water.

B. Steep 15 to 20 minutes.

C. Strain.

D. Drink 1/2 to 1 cup every 3 to 4 hours as needed.

    1. Gets rid of tapeworms.

    2. American Indians of old used this remedy.

TICKS

Dictionary definition: *"A tiny insect that attaches itself to a person's skin and sucks their blood."*

PLANTAIN, COMMON:

A. Crush handful of leaves to pulpy mass.

B. Crush equal amount of yarrow leaves and stems.

C. Blend and spread thickly over tick.

D. Leave on 30 minutes.

    1. Tick should drop off when poultice is removed.

    2. If not, remove with tweezers or scrape off with knife.

E. Wash area when tick is gone.

F. Crush more plantain leaves to pulpy mass.

G. Spread thickly over bite area.

    1. Promotes healing and draws out remaining poison.

TONSILLITIS
(Also see "SORE THROAT GARGLE")

Dictionary definition: *"When a person's tonsils become inflamed and swollen and painful."*

GINGER, WILD:

A. Pulverize handful of wild ginger roots.

B. Crush 2 lemons and put in with roots.

C. Stir in 2 tablespoons honey.

D. Pour pint of boiling water over this.

E. Blend thoroughly.

F. Strain.

G. Gargle and rinse out back of throat.

    1. Soothes throat.

TOOTHACHE

CATNIP:

A. Chew a few fresh catnip leaves.

B. Roll juice around in mouth before swallowing.

    1. Stops toothache.

FENNEL:

A. Crush fennel plant to pulpy mass.

B. Spread on gums and on aching tooth.

    1. Effective remedy for toothache.

PRICKLY ASH:

A. Chew small piece of inner bark.

B. Rub softened innerbark on gums.

    1. Relieves toothache quickly.

    2. American Indians of old used this remedy.

(U)

ULCER ON SKIN

Dictionary definition: *"A nasty looking open sore on the skin that discharges pus"*

BAY, SWEET:

A. Put 1/2 dozen bay leaves in cup boiling water.

B. Steep 30 to 35 minutes.

C. Strain.

D. Bathe open sores.

E. Then lay soaked leaves on runny sores.

F. Cover with cloth and tape in place.

G. Leave on 24 hours.

H. Repeat daily until sores are healed.

    1. Cures skin ulcers.

    2. Pioneers of days gone by used this.

PLANTAIN, COMMON:

A. Pulverize handful of leaves to pulpy mass.

B. Spread thickly as poultice on running skin ulcers.

C. Apply fresh poultice as required.

    1. Readily heals skin ulcers.

ULCERS IN STOMACH

(also see "STOMACH PROBLEMS")

SHEPHERD'S PURSE:

A. Put handful of crushed whole plant in quart boiling water.

B. Steep 35 to 40 minutes.

C. Strain.

D. Drink cup 4 times a day.

    1. Clears up stomach ulcers.

YELLOWROOT:

A. Put handful of crushed root in quart boiling water.

B. Steep 20 to 25 minutes.

C. Strain.

D. Drink 1/2 to 1 cup every 3 to 4 hours.

    1. Heals stomach ulcers.

    2. Used years ago by American Indians.

UPSET STOMACH

HOPS:

A. Put handful of fruit in quart boiling water.

B. Steep 25 to 30 minutes.

C. Strain.

D. Drink 1/2 to 1 cup 3 times daily and before retiring.

    1. Settles upset stomach.

PEPPERMINT:

A. Put handful of crushed leaves in quart boiling water.

B. Soak for 25 to 30 minutes.

C. Do not strain but let cool.

D. Drink glass when needed.

    1. Quickly settles upset stomach.

YELLOW DOCK:

A. Put handful of leaves in quart boiling water.

B. Steep 20 to 25 minutes.

C. Strain.

D. Drink 1/2 to 1 cup before meals and when retiring.

    1. Roots can be substituted for leaves.

    2. Excellent for upset stomach.

(V)

VARICOSE VEINS

Dictionary definition: *"Veins that have become enlarged or swollen."*

MARSH-MARIGOLD:

A. Put handful of flower petals in pint of boiling water.

B. Steep 20 to 25 minutes.

C. Strain.

D. Soak folded cloth in this tea.

E. Lay hot compress over varicose vein area.

F. Repeat when compress cools.

    1. Alleviates discomfort from varicose veins.

WHITE OAK:

A. Put handful of inner bark in pint of boiling water.

B. Steep 20 to 25 minutes.

C. Strain.

D. Soak folded cloth in this concoction.

E. Lay hot compress over varicose vein area.

F. Repeat procedure when compress cools.

    1. Alleviates discomfort from varicose veins.

WOOD SORREL, CREEPING:

A. Get handful of fresh leaves.

B. Lay leaves on varicose veins.

C. Wrap in elastic bandage to keep in place.

    1. Eases discomfort from varicose veins.

VOMITING

(Also see "MORNING SICKNESS" AND "NAUSEA")

Dictionary definition: *"The involuntary act of throwing up what has been eaten."*

HORSEMINT:

A. Put handful of crushed leaves in quart boiling water.

B. Steep 20 to 25 minutes.

C. Strain.

D. Drink 1/2 to 1 cup every 3 to 4 hours.

    1. Makes person start vomiting.

    2. An old American Indian remedy.

SWEET VIOLET:

A. Put handful of crushed leaves in quart of boiling water.

B. Steep 15 to 20 minutes.

C. Strain.

D. Drink 1 cup as needed.

    1. Induces vomiting.

    2. Crushed root can be substituted for crushed leaves.

(W)

WARTS

Dictionary definition: *"A small hard lump on a person's skin."*

BLOODROOT:

A. Pare warts slightly.

B. Crush root to pulpy mass.

C. Rub juice and pulp on warts.

D. Reapply juice and pulp as necessary.

    1. Warts will eventually disappear.

CHICKWEED, COMMON:

A. Pare warts slightly.

B. Crush plant stems and extract juice.

C. Rub juice directly on warts.

D. Reapply juice as necessary.

    1. Rids a person of warts.

GROUND IVY:

A. Pare warts slightly.

B. Soak ivy leaves in vinegar 4 to 6 hours.

C. Lay 1 leaf on wart and tie with thread.

D. Tie on freshly soaked leaf each morning and evening.

E. Wart can be painlessly lifted off in a few days.

F. Continue applying ivy leaves 2 more days.

    1. This will get rid of wart remnants

LEEK, WILD:

A. Pare warts slightly.

B. Crush fleshy leaves and extract juice.

C. Apply juice directly on warts.

    1. Used by Cherokee Indians of old.

MILKWEED, COMMON:

A. Pare warts slightly.

B. Pulverize entire plant and extract juice.

C. Rub this milky latex directly on warts.

D. Continue on daily basis.

    1. An old Amish folk medicine remedy.

WORMS
(Also see "RINGWORMS" and "TAPEWORMS")

CURRANT, WILD BLACK:

A. Put handful of crushed root in quart boiling water.

B. Steep 25 to 30 minutes.

C. Strain.

D. Drink 1/2 to 1 cup every 2 to 3 hours.

    1. Expels worms.

    2. Crushed bark can be substituted for crushed root.

    3. American Indians of long ago used this remedy.

RED CEDAR:

A. Put handful of fruit in quart of boiling water.

B. Steep 20 to 25 minutes.

C. Strain.

D. Drink 1 cup every 3 to 4 hours as needed.

    1. Good for getting rid of worms.

    2. Used long ago by American Indians.

THYME:

A. Eat thyme twigs freely.

B. Eat dried thyme leaves in salads, sandwiches, etc.

    1. Simple but good worm remedy.

WOUNDS
(Also see "CUTS")

CATTAIL:

A. Pound handful of dried roots to fine powder.

B. Make paste by adding small amount of water.

C. Apply as poultice on wounds.

D. Repeat procedure as necessary.

    1. Wounds heal more readily.

    2. American Indians used cattail for this purpose.

PASSION FLOWER:

A. Crush handful of roots to pulpy mass.

B. Spread as poultice on wounds.

C. Reapply poultice as needed.

    1. Induces wounds to heal faster.

    2. Used by American Indians many years ago.

PRICKLY PEAR CACTUS:

A. Pulverize handful of peeled cactus pods.

B. Spread as poultice on wounds.

C. Reapply poultice as needed.

    1. Speeds up healing.

    2. American Indians of days gone by used this remedy often.

PART III

THE PLANTS AND THEIR PICTURES
4

Identifying Plants with Medical Properties

Plants with medicinal qualities are available in stores today. For example, basil, garlic, rosemary, etc., come in powder form and can be purchased in small cans. Dill weed, oregano, and others can be found in small chopped pieces.

Seeds are procurable from such plants as anise, caraway, celery, fennel, ginger root, etc.

Cloves can be purchased whole or powdered as can juniper berries, nutmeg and sage.

Ginger root comes as a powder and in small chunks.

Cinnamon is obtainable as a powder and sticks to 18 inches.

Then there are extracts made from anise and peppermint, etc.

Herbal teas are available — chamomile, sage and lemon, etc. Check your local grocery store, supermarket or pharmacy. Pick up at least one handy little can or bottle or jar every time you shop. Take it home and set it aside in a safe place. Each makes an important addition to an emergency medical bag.

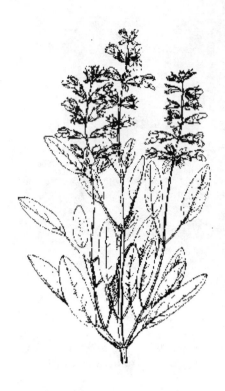

AGRIMONY:

Grows to 7' tall. Usually in clumps. Serrated deep green leaves with apple smell. Tiny aromatic yellow flowers on reddish spikes (June to September). Found on sides of roads, fields, pastures, damp woods, etc.

ALFALFA:

Grows to 3' tall. Leaves in groups of three. Oval-shaped. Tiny blue/violet clover-like blossoms (June to August). Found everywhere (March to October).

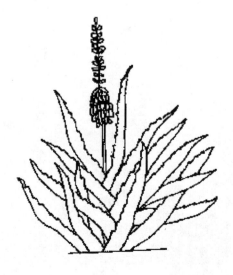

ALOE VERA:

8" to 16" tall. Succulent. Fleshy, lance-shaped leaves. Orangish red/yellow flowers in clusters on stalk ends. Found in nurseries. Easy to grow in pots and garden.

ALUMROOT:

Grows to 2' tall. Single stem. Leaves 12" to 24" long, 3" to 6" across. Toothed. Maple shaped. Greenish-white flowers. 3 to 5 on short stem (April to June). Found in rich woods, meadows, fields, shaded rocks, etc.

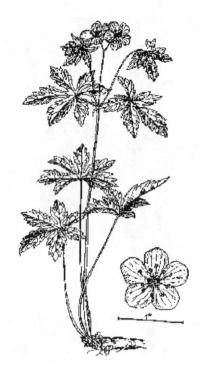

AMERICAN BASSWOOD (LINDEN):

Long heart-shaped leaves. Coarsely double-toothed. Pea-sized fruit, single and clusters. Sweet, scented, creamy white/yellow flowers (May to August). Found in rich woods, fertile soils, etc.

AMERICAN BEECH:

Tree 45' to 60' tall. Bark dark gray/light gray patches. Leaves paper thin with coarse sharp teeth. Nuts in tiny burr with prickles. Ripe (September/October). Found in rich woods, uplands, moist rocky ground, etc.

AMERICAN HOLLY:

To 90' tall. Smooth, leather-like leaves. Needle-like tips. Berries red/orange (August through September). Found in moist areas, woods, thickets, etc.

ANGELICA:

Grows to 7' tall. Three saw-toothed leaves on tip of each stem. White/greenish flower clusters on stalk ends (May to August). Found in gardens, rich low grounds, near streams, swamps, etc.

ANISE ROOT (SWEET CICELY):

Grows to 20" high. Fern-like leaves. Seeds stick to clothing. Tiny white inconspicuous flowers (May to July). Found in woods and thickets, moist areas, etc.

ASPARAGUS:

Grows to 9' tall. Needle-like leaves are really branches. Small greenish-white bell-shaped flowers seldom seen. Reddish fruits (June). Found in gardens, fields, meadows, yards, etc.

BALSAM FIR:

Spire-shaped tree. To 60' tall. Resin pockets in bark. Cones 1" to 4" long, purple/green. Flat needles. Found on moist mountain slopes, woods, thickets, etc.

BALSAM POPLAR:

25' to 85' tall. Gummy buds. Onion-sandalwood fragrance. Thick, firm, smooth leaves. Silver-white underside. Flowers on long drooping spikes (May to June). Found on river banks, swamp borders, moist soils, etc.

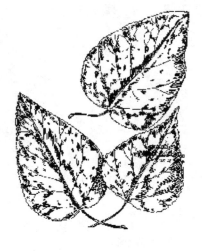

BARLEY, LITTLE:

Stiff bristles. Spikes rough to the touch. Shallow roots. Found in uncultivated places, pastures, plains, etc.

410

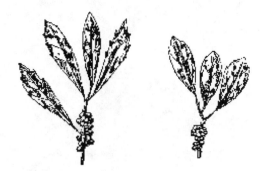

BAYBERRY:

Grows to 12' tall. Compact evergreen. Smooth bark. Fragrant. Spatula shaped leaves. Crowded berry clusters. Very hairy. Yellowish flower clusters under leaf tips (April to June). Found in coastal marshes, stream banks, woods, fields, etc.

BAY (SWEET BAY):

Grows to 6' tall. Evergreen shrub. Dark green leathery leaves. Aromatic when crushed. Small cup-shaped, fragrant yellow flowers (April to July). Dark purple berries follow flowers. Found in low, damp woods, sunny valleys with rich soil, etc.

BLACKBERRY:

Sprawling shrub. Densely prickled stems. Double-toothed leaves. White/pale pink flowers (April to July). Purple/black berries (July to September). Widespread on wood edges, roadsides, thickets, fence rows, etc.

411

BLACK COHOSH:

Grows to 8' tall. Sharply toothed leaves. Small white flowers in long fluffy spikes. Foul smelling. Found in rich woodlands and hillsides, etc.

BLOODROOT:

Fibrous root when cut gives off orange-colored juice. Single, white, cup-shaped flowers, 8 to 10 petals (March to June). Found in woods, moist rich soils, and on shaded slopes, etc.

BLUEBERRY

Grows to 2' high. Narrow, lance-shaped leaves with tiny teeth. Flowers white, urn-shaped, 5 petals (May to July). Berries (August to September). Mostly sandy soils, woods, fields, thickets, etc.

BLUE VERVAIN:

Grows to 4' high. Rough stems. Sharp-toothed leaves. Spikes of blue-violet flowers (June to September). Found in rough fields, thickets, pastures, roadsides, etc.

BONESET:

Grows to 5' high. Thick hairy stems. Clusters of white/purple-tinged flowers (June to September). Moist ground, damp pastures, woods, by streams, river banks, etc.

BROOMSEDGE:

2' to 5' tall grass. Grows erect in small clumps. Very hairy green leaves. Bluish/green stem. Found in dry soil, old woods, open fields, pastures, etc.

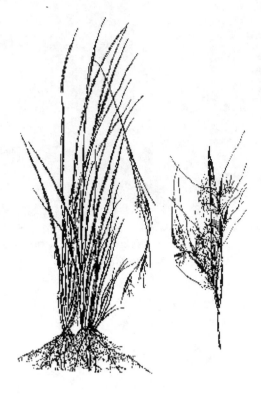

BUGLEWEED:

Leaves green/purple-tinged with coarse teeth. Plant has mint odor, bitter taste. Tubular purple flowers (July to October). Found in rich wet places, fields, forests, thickets, etc.

BURDOCK:

Stems hollow. Numerous red/violet flowers on short stalks (July to October). Found on neglected farm lands, fields, gardens, waste places, etc.

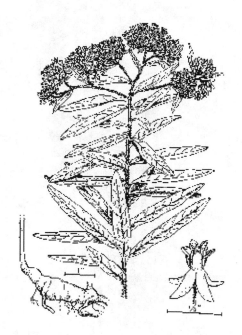

BUTTERFLYWEED:
Grows to 3' high. Leaves long, slim, slightly hairy. Flowers showy orange-yellow, some yellow (May to August). Found in pastures, fields and dry, sandy roadsides, etc.

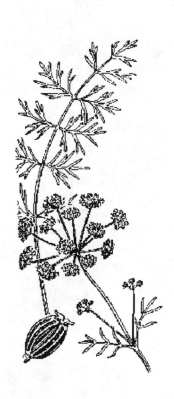

CARAWAY:
Grows to 3' high. Hollow stem. Carrot-like leaves. Flowers very small pink/white (May to June). Found growing as a weed in most soils — fields, pastures, etc.

CATNIP:
Grows 3' tall. Heart-shaped leaves, coarsely toothed. Flower clusters. White with purple dots (July to September). Found in woods, fields, sides of roads, waste places, etc.

CATTAIL:

Grows to 8' tall. Sword-like leaves. Yellow pollen laden flowers on erect stalks (May to July). Found in fresh marshes, ponds, shallow water, ditches, etc.

CELANDINE:

Grows to 2' high. Evergreen. Fleshy. Brittle stems with yellow juice. Bright yellow flowers on hairy stocks (March to September). Found in fields, pastures, other waste places, etc.

CELERY, WILD:

Grows to 2' high. Leaves similar to garden celery. Strong pungent smell. Tastes bitter compared to garden celery. Sparse greenish-white flowers throughout summer. Found in marshy ground near water, wet woodlands, etc.

CHAMOMILE:

Grows to 2' tall. 2' stems. Bright green stemless leaves. Delicate apple scent. Yellow-centered flowers with white petals (May through October). Found almost everywhere – meadows, along sides of roads, etc.

CHICKWEED, COMMON:

Succulent, straggling stems. Hairy leaf stalks. Small white flowers resembling stars (March to September). Found throughout U.S. in waste as well as cultivated places.

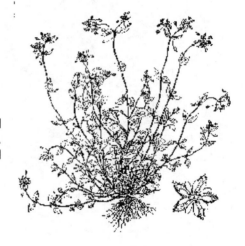

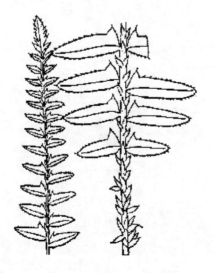

CHRISTMAS FERN:

Evergreen. Tapering leaves are lustrous and leathery. Found on rich wooded slopes, woods, etc. Not uncommon.

COLT'S FOOT:

Grows to 8" high. Fragrant. Jagged dandelion-like leaves covered with white hairs. Large yellow flowers on reddish scaly stalk (March to April). Found in fields, railroad embankments, sides of brooks, etc.

COMFREY:

Grows to 3' high. Rough and hairy. Stout, hollow stem. Large hairy leaves cause itching when touching skin. Bell-shaped flowers, cherry-yellow/pale pink (May to June). Found in wet soils -- river banks, ditches, damp meadows, etc.

CORIANDER:

Grows to 2' high. Flat parsley-like lower leaves. Feathery, thread-like upper leaves. Unpleasant smell if crushed. White/pale mauve flowers (June to August). Found in dry, light soils. Sunny, sheltered locations.

CRANBERRY:

Shrub with odorless and hairless evergreen leaves. Crimson berries edible (September to October). Small pink bell-like flowers (May to July). Found in mossy evergreen woods, exposed rocky or dry soil.

CURRANT:

Thornless shrub. Maple-like leaves. Strong smell. Fruit is smooth and purple, red or black. Drooping clusters of bell-shaped flowers (April to July). Found in woods, thickets and fields. Some grow in swamps, etc.

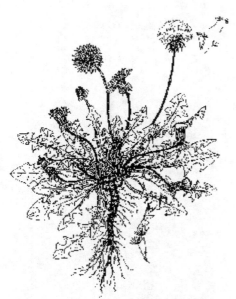

DANDELION:

Flowering stalk hollow with milky juice. Jagged leaves, dark green and hairless. Start at ground. Golden yellow flowers throughout year in warm places. Found everywhere. Lawns, parks, roadsides, fields, etc.

419

DILL WEED:

Grows to 3' high. Smooth and shiny hollow stems. Pale green, feathery, aromatic leaves. Small yellow flowers, petals turned inward (June to July). Found everywhere from waste places to sunny sheltered spots.

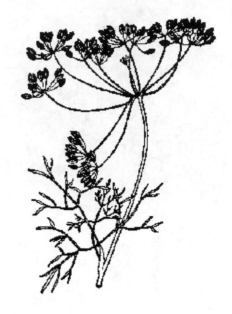

ELCAMPANE:

Grows to 8' tall. Large leaves with wooly underside. Thick aromatic root. Large bright yellow flowers to 4" across (July to September). Found in damp soils — fields, roadsides, pastures, etc.

ELDERBERRY:

Grows to 12' tall. Thick, white pith in stems and twigs. Fragrant white flowers in umbrella-like clusters (June to July). Berries tiny, purplish-black (July to September). Found in wet, rich soils. Ditches, thickets, roadsides, etc.

EUCALYPTUS:

Grows to 190'. Bark peels off in narrow strips. Narrow, fragrant leathery leaves studded with oil glands. Creamy white flowers on short, flat stalks. Fruit follows concealed in woody, camphor scented cup. Found in wet areas. Relatively common.

FENNEL:

Grows to 7' tall. Smells like anise or licorice. Blue-green, fan-shaped, thread-like leaves. Mustard yellow flowers (June to September). Egg-shaped, odorous fruits follow flowers. Found growing as weed all over U.S.

FIELD HORSETAIL:

Grows to 18" high. Stiff-stemmed, leafless herb. Hollow stems with conical spikes. Feathery branches. Found almost everywhere in damp soil — fields, roadsides, etc.

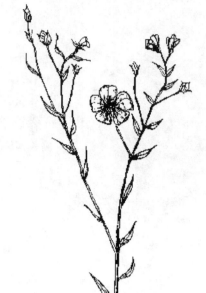

FLAX:
Grows to 22" tall. Lance-shaped stalkless leaves. Delicate blue flowers with 5 overlapping petals (June to August). Found on waste ground everywhere — roadsides, fields, etc.

FLOWERING DOGWOOD:
Grows to 25' tall. Nearly horizontal branches. White/pink flowers in clusters of 20 to 30 each (April to May). Found on sides of roads, fields, etc. Widely cultivated.

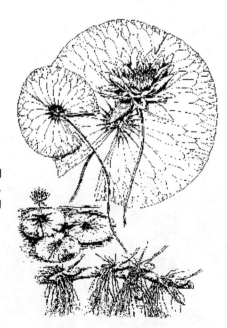

FRAGRANT WATER LILY:
Aquatic. Large floating leaves. Notched at base. Strong sweet smelling flowers. 5" across. (June to September). Found in quiet or dead water —ponds, slow running brooks, etc.

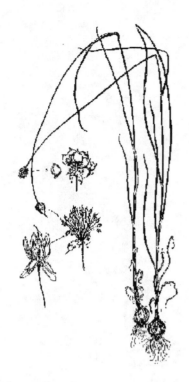

GARLIC, WILD:

Grows to 12" high. 3" leaves. White/pink flowers on stem tip (April to May). Found in well drained places — fields, roadsides, etc.

GERANIUM:

Grows to 2' high. Broad leaves with 5-toothed segments. Pinkish blossoms in pairs, wooly at base (April to August). Found in open woods and low grounds — fields, meadows, etc.

GINGER, WILD:

Grows to 4' tall. Swollen finger-like joints. Heart-shaped leaves on hairy stems. Flowers purple-streaked, urn-shaped, fragrant (April to June). Found in rich, shaded, well drained soil. Woods, thickets, etc.

GINSENG, AMERICAN:

Grows to 2' high. Slow growing. Fleshy root. Pinkish/whitish/yellowish scented flowers (June to July). Fruits are red berries. Found in rich woods, cool wooded areas, thickets, etc.

GOAT'S RUE:

Grows to 2' high. Silky haired. Disagreeable odor when bruised. Unpleasant bitter taste. Flowers yellow base with pink wings (May to August) Found on places with sandy soil — prairies, fields, etc.

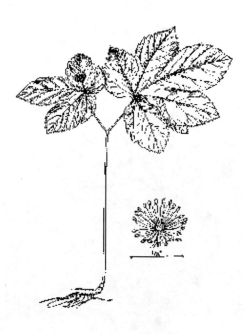

GOLDENSEAL:

Grows to 1' high. Usually 2 leaves on forked branch. Hairy plant. Berries look like raspberries but not edible. Topped with single greenish-white flower (April to May). Found in rich, moist soils, shady woods, meadows, etc.

GOLDTHREAD:

Grows to 3" high. Bright yellow thread-like roots. Shiny, evergreen, strawberry-like leaves. Showy, solitary white flowers. 5 to 7 petals (March to July). Found in cool forests, dark swamps, damp woods, bogs, etc.

GRAPE, WILD:

High climbing vine. No thorns. Shreddy, dark brown bark. Greenish flowers (August to October). Grapes fleshy, purple/black/amber. 20 or more in clusters. Found in thickets, edges of woods, gardens, etc.

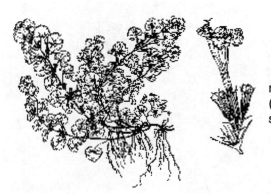

GROUND IVY:

Long, creeping stems. Scalloped edged leaves have mint smell. Small tubular violet flowers with purple spots (March to July). Found in orchards, gardens, rich shaded areas, etc.

HAZELNUT:

Grows to 10' tall. Stems have stiff hairs. Nuts edible. Flowers (April and May). Found in thickets, wooded areas, etc.

HEMLOCK:

Grows to 90' tall. Evergreen. Needles silver-white underneath. Tree produces small cone. Found in rocky woods, mountain slopes, cold swamps, etc.

HOPS:

Rough, prickly plant. Climbs to 20'. Greenish flowers (July and August). Commonly found in woods, thickets, fields, etc. Widely cultivated.

HONEYSUCKLE, JAPANESE:

Evergreen. Climbing shrub. Oval leaves. Purplish-black berries (September to November). Flowers are white/buff (April to July). Found in gardens, fence rows, thickets, roadsides, etc.

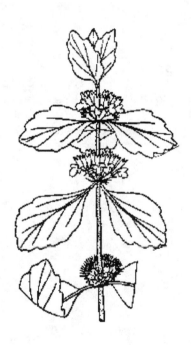

HOREHOUND, COMMON:

Grows to 20" tall. Resembles catnip. Rank smelling. Leaves wrinkled. Plant covered with whitish hairs. Numerous tubular white flowers (May to September). Found almost everywhere — roadsides, field edges, etc.

HORSEMINT:

Grows to 4' high. Leaves lance-shaped, sharp toothed.Pungent smelling. Strongly aromatic. Flowers pale yellow with purple dots (July to October). Found in dry places, sandy soils, coastal plains, etc.

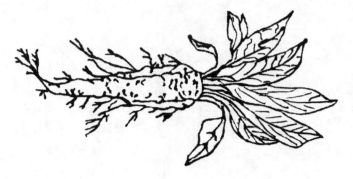

HORSERADISH:

Grows to 4' high. Large, shiny, light green leaves. Hairless plant. Strong odor when thick taproot scraped. Flowers white and tiny (June to August). Found in moist fields, around buildings, etc. Widely cultivated.

HUCKLEBERRY:

Low to tall shrubs. Hairy twigs and branches. Bell-like white/pink/green flowers (April to June). Berries follow flowers (June to September). Found in swamps, dry thickets, moist woods, etc.

JEWELWEED:

Grows to 5' tall. Succulent stems. Oval leaves have red spots and hang like pendant. Flowers pale yellow (June to October). Found in wet soil, shady places, etc.

JUNIPER, COMMON:

Grows to 10' tall. Small evergreen shrub. Bark peels off. Dense blue-green foliage. Needles extremely sharp pointed. Berries bluish black, oval, the size of a pea. Found on dry hillsides, rocky pastures, etc.

LEEK, WILD:

Grows to 18" high. 2 or 3 slim fleshy leaves. Whitish bulbs exude strong onion smell. Spoke-like cluster of creamy white flowers (May to June). Found in rich moist soil, wet woods, thickets, etc.

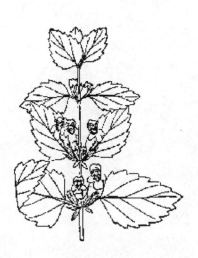

LEMON BALM:

Grows to 20' feet tall. Strong lemon scented leaves. Flowers whitish/pale yellow in whorls (May to August). Found on old building sites, barnyards, fields, etc. Common.

LOBELIA:

Grows to 18" high. Hairy stems contain acrid milk-like juice. Light green leaves with hairy underside. Numerous spikes of tubular pale blue flowers (June to September). Found on sides of road, open woodlands, fields, dry pastures, etc.

MALLOW, COMMON:

Deep rooted herb. Trailing stems. Rounded, slightly toothed leaves. Small white/pale lilac flowers (April to October). Found in yards, cultivated fields, gardens, barnyards, etc.

MARSH-MARIGOLD:

Aquatic plant. Succulent hollow stem. Kidney-shaped leaves. Large buttercup-like flowers (May to July). Found around swamps, wet ditches, etc.

MARJORAM, WILD:

Grows to 2' high. Aromatic. Frequently bushy. Hairy stems. Clusters of pinkish-purple/white flowers (July to September). Flowers bloom on short spikes. Clusters often 2' across. Found in fields, roadsides and waste places, etc.

MARSH-MALLOW:

Grows to 4' high. Wooly stems. Serrated leaves 1" to 3" long. Covered with soft hairs. Flowers purple/pale pink/white in clusters (June to August). Found in sunny wet places, damp inland soils, salt marshes, etc.

MILKWEED, COMMON:

Grows to 4' tall. Milky fluid exudes from stem when broken. Clusters of sweet smelling white flowers (June to August). Found almost everywhere – fields, pastures, roadsides, etc.

MINT:

   Grows to 2' high. Fine hairs on stem. Strong mint odor. Tiny clusters bell-shaped flowers (June to October). Found on stream banks, damp soil, etc. Widely cultivated.

MOTHERWORT:

   Grows to 5' tall. Long, hairy, dark green leaves. Sharp teeth. Purple spotted, hairy flowers on spike (June to September). Found almost everywhere -- roadsides, fields, gardens, etc.

MULLEIN, COMMON:

   Grows to 6' tall. Branches covered with woolly hairs. Large oval leaves with dense woolly hairs. Long spikes of lemon-yellow flowers (June to August). Found in pastures, old fields, roadsides, railroad yards, etc.

432

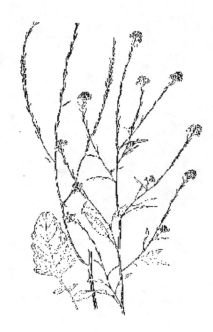

MUSTARD, BLACK:
    Grows to 6' tall. Lower leaves coarse and bristly. Bright yellow flowers on twig-like stem (May to September). Found in hedges, waste places, roadsides, cliffs, etc.

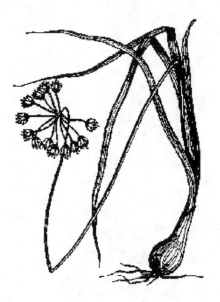

ONION, WILD:
    Grows to 2' high. Long, slender grass-like leaves. Small pink-white flowers (July to August). Found on slopes, open woods, rocky soil, etc. Widespread.

PARSLEY:
    Grows to 2' high. Bright green, feathery leaves. Stout taproot. Tiny, greenish-yellow flowers (April to July). Found in rich moist soil, wild rocky places, shaded areas, etc.

PASSION FLOWER:

Hairy vine climbs to 25' high. Smooth egg-shaped yellow fruit to 3" long (August to November). Large, sweet-smelling, flesh color flowers (May to September). Found in sandy soil, shady areas, edge of woods, along fences, etc.

PEPPERMINT:

Grows to 3' high. Dark green, sharply toothed leaves. Mint smell. Taste initially burns, followed by cool sensation. Small, lavender blossoms in spike-like groups (July to September). Found on brook banks, wet meadows, other moist areas, etc.

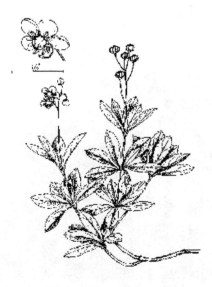

PIPSISSEWA:

Evergreen. Grows to 10" high. Dark green, wedge-shaped leaves. Flowers flesh-colored, fragrant (May to July). Found in hardwood, coniferous forests, other rich woodlands, etc.

PLANTAIN, COMMON:

Grows to 18" high. Broad, oval leaves, toothed. Tiny yellow flowers on cylindrical spikes (May to August). Found on waste and cultivated land, lawns, gardens, etc.

POTATO, WILD:

Leaves 2" to 3" long. Large tapering root. Large, white flowers open every morning (June to August). Found in light, moist, sandy soils, etc. Quite common.

PRICKLY ASH:

Aromatic shrub. Leaves give off lemon smell when crushed. Reddish-green berries with yellow dots (August to October). Tiny yellow flowers (April to May). Found in moist woods, wet thickets, river banks, etc.

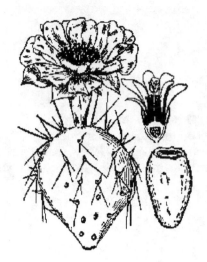

PRICKLY PEAR CACTUS:

Grows to 1' high. Tufts of bristles. Sharp spines. Fruit dull red, pulpy (August to October). Large, showy yellow flowers (May to August). Found in dry, sandy soils, rocks, etc.

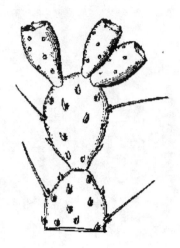

PUFF BALL:

Grows directly from ground. Easy to recognize. White when young. Dingy looking when older. Found in open places — pastures, fields, barnyards, etc.

(CHALK-WHITE INSIDE)

(1-12 IN. IN DIAMETER)

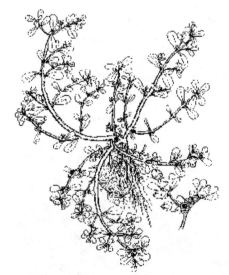

PURSLANE, COMMON:

Grows to 12" high. Fleshy, sprawling herb. Reddish stems. succulent, spatula-shaped leaves. Bright green in clusters. Tiny yellow flowers open only on sunny mornings (June until frost). Found everywhere — lawns, gardens, woods, cultivated lands, etc.

436

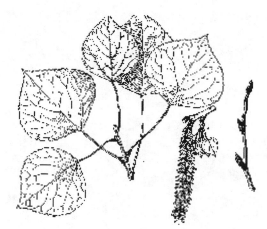

QUAKING ASPEN:
Grows to 40' tall. Smooth gray-white to greenish bark. Broad, heart-shaped leaves. Whitish to dark blue-green. Found on hillsides, stream borders, etc. Prefers dry ground.

QUEEN ANNE'S LACE:
Grows to 4' high. Bristly-hairy stems. Delicate leaves — oblong, somewhat hairy. Flat clusters of purple-pink flowers (June to September). Found in old meadows, dry fields, pastures, roadsides, etc.

QUEEN OF THE PRAIRIE:
Grows to 8' tall. Smooth stems. Deeply divided leaves. Pink-red flowers in spreading clusters (June to August). Found in moist meadows, bogs, fields, pastures, etc.

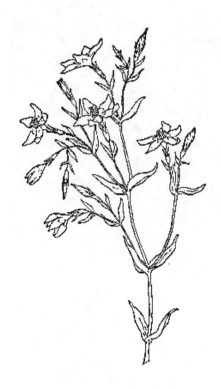

QUININE, WILD:
Grows to 5' tall. Large, oval, lance-shaped leaves. White flowers in loose umbrella-like clusters (May to July) Found on roadsides, rock outcroppings, prairies, etc.

RATTLESNAKE WEED:
Grows to 2' high. Lance-shaped leaves. Dandelion-like flowers on stalks (May to October). Found in clearings, pastures, fields, etc Quite common.

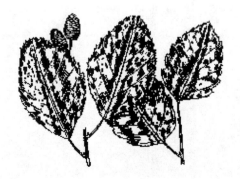

RED ALDER:
Grows to 90' tall. Bark smooth. Leaves rusty haired. Cones small — up to 1" long. Flowers bloom (September). Found on stream borders, riverbanks, moist ground, etc.

REDBUD TREE:

Grows to 40'. Heart-shaped leaves. Showy clusters of purple-red, pea-like flowers (March to May). Found in rich bottom lands, stream borders, woods, roadsides. etc.

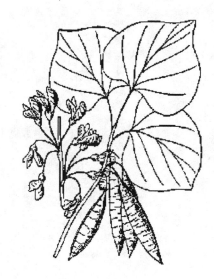

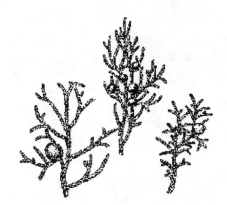

RED CEDAR:

Grows to 40'. Spire-shaped. Dark evergreen. Shreddy bark. Pea-size fruit, purplish covered with gray-white bloom. Flowers appear (April and May). Found on edges of lakes and streams, pastures, etc.

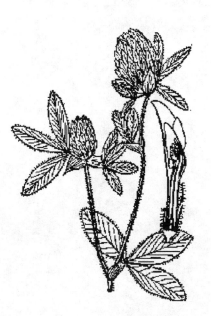

RED CLOVER:

Grows to 18" high. Pink to red, rounded heads (May to September). Found everywhere — fields, lawns, parks, roadsides, etc.

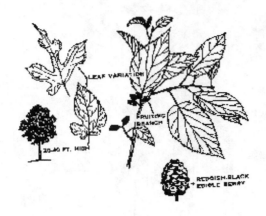

RED MULBERRY:

Grows to 60' tall. Heart-shaped leaves, sandpapery on top. Fruit similar to thin blackberry (June to July). Flowers in tight, drooping clusters (April to May). Found in rich soil, open woods, fence rows, thickets, etc.

RED RASPBERRY:

Grows to 5' tall. Upright, prickly shrub. White-powdered stems. Small clusters of tiny white flowers (June to October). Red, many sectioned fruits follow flowers. Found everywhere — roadsides, edges of woods, clearings, etc.

RHUBARB:

Large and compact. Thick stems. Leaves 2' to 4' long. Peculiar, aromatic odor. Crimson fruit. Flowers greenish-white in clusters (May to August). Found easily everywhere — fields, parks, roadsides, etc.

ROSE HIPS:

Look for fruit — fleshy red hips containing hair and seeds. Check wherever roses are seen growing. Found on edges of woods, pastures, yards, roadsides, etc.

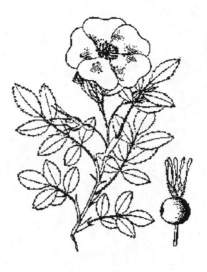

ROSEMARY:

Grows to 4' high. Extremely aromatic evergreen shrub. Narrow, dark green, leathery leaves are spiky. Clusters of pale blue flowers near branch ends (April to May). Found in light, dry, chalky soil, sheltered sunny places, etc.

SAGE:

Grows to 30" high. Stems covered with white down. Strong smell. Thick velvety leaves with hairy underside. Whorls of violet-blue flowers near stem ends (May to July). Found in dumps, gardens, wastelands, etc. Widely cultivated.

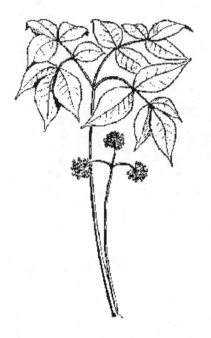

SARSAPARILLA:
Grows to 2' high. Climbing or trailing vine. Woody stems. Small, greenish-white flowers (May to July). Found in swamps, on riverbanks, in moist woods, etc.

SASSAFRAS:
Grows to 30' tall. Rough gray fissured bark. Large leaves have spicy odor when crushed. Pale yellow-green flowers in clusters (April to May). Blue-black, pea-size berries follow flowers. Found in poor soil, sandy soil, edges of woods, thickets, etc.

SENNA, WILD:
Grows to 6' tall. Slightly hairy, pale green stem. Yellow flowers in loose clusters (July to August). Found in dry thickets, open woods, beside streams, etc.

SHEPHERD'S PURSE:
Grows to 2' high. Dandelion-like, hairy leaves. Stems covered with hairs. Triangular fruit. Small erect spikes of tiny white flowers (January to December). Found in backyards, roadsides, gardens, fields, etc. Quite common.

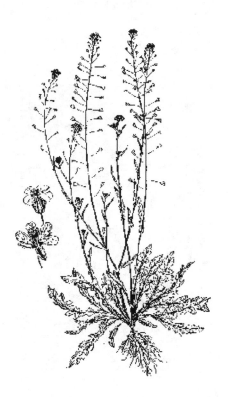

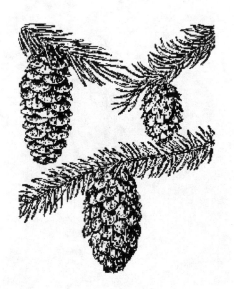

SILVER SPRUCE:
Grows to 100' tall. Evergreen. Silvery, pale gray-blue. Needles give off disagreeable odor when broken. Found in mountainous areas. Also on private land, parks, etc.

SKULLCAP (MAD-DOG-SKULLCAP):

Grows to 3' high. Widely branched. Square, slender stem. Leaves thin, pointed, coarsely serrated, lance-shaped. Bright violet-blue flowers bloom on spikes (June to September). Found in moist, shaded areas, meadows, woods, near ponds, etc.

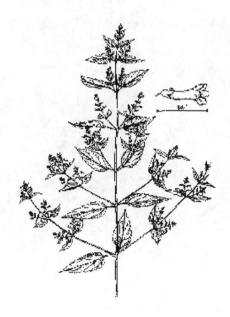

SLIPPERY ELM:

Grows to 50' tall. Rough bark. Mildly aromatic. Oval leaves like sandpaper on top. 3" to 6" long. Fruits nearly round and winged. Ripen (March to May). Flowers bloom (March to April). Found in moist woods, thickets, rich soil, rocky hillsides, etc.

SMARTWEED:

Grows to 2' high. Reddish stems. Lance-shaped leaves. Silver sheath at base of each leaf. Greenish flowers in arching clusters on spikes (July to October). Found in damp soil, near shores, pond edges, yards, ditches, etc.

444

SMOOTH SUMAC:

Grows to 15' tall. Straggly branched shrub. Hairless. Dark red fuzzy berries in dense clusters (August to October). Whitish flowers in cone-shaped clusters (June to July). Found in old fields, fringe of woods, road embankments, etc.

SNAKE ROOT, WHITE:

Grows to 5' tall. Slender stalks. Toothed leaves. Clusters of bright white flowers (July to October). Found in clearings, damp and shady pastures, fields, thickets, etc.

SOAPWORT (BOUNCING BET):

Grows to 2' high. Thick joints on stem. Leaves and roots make lather when crushed. Clusters of pale pink flowers (June to September). Found everywhere — roadsides, railway embankments, fields, etc.

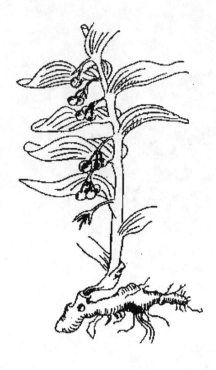

SOLOMON'S SEAL:
Grows to 3' high. Dark green leaves, numerous parallel veins. Drooping clusters greenish-white flowers (May to August). Found in woods, thickets, fields, etc. Cultivated in gardens.

SOURWOOD (SORREL TREE):
Grows to 60' tall. Deep green leaves to 6" long. Sour taste. Egg-shaped fruits about 1/2" long. Flowers white, urn-shaped in drooping clusters (June to July). Found in rich woods, thickets, mountain slopes, etc.

SPEARMINT:

Grows to 3' high. Oblong, bright green leaves. Distinct smell of spearmint. Slender, elongated spikes of pale pink flowers (June to October). Found in fairly shady, moist places. Extremely common.

SPIKENARD, AMERICAN:

Grows to 5' tall. Dark green or reddish stem. Aromatic root — smells like spices. Whitish flowers (June to August). Found in rich woods, old pastures, fields, etc.

SQUAW WEED:

Grows to 4' high. Smooth stem. Two leaf types — lower heart-shaped, upper lance shaped. Golden yellow flowers in flat-topped clusters (March to July). Found on stream and creek banks, in swamps, etc.

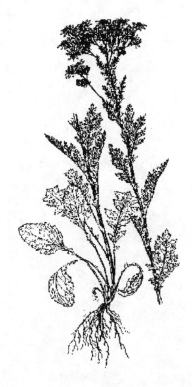

STINGING NETTLE:

Grows to 6' tall. Dark green leaves heart-shaped to egg-shaped. Surface of leaves covered with stiff stinging bristles (hairs). Greenish flower clusters on many spikes (June to September). Found in vacant lots, gardens, ditches, edge of damp woods, etc.

SWEET (BLACK) BIRCH:

Grows to 60' tall. Sweet, aromatic black or gray bark. Broken twigs and leaves have strong wintergreen smell. Found in rich woods, river banks, moist fertile grounds, etc.

SWEET EVERLASTING:

Grows to 2' high. Soft, hairy, lance-shaped leaves. No stalk. Flowers dirty white in spreading clusters. Found in dry soil, fields, pastures, clearings, etc.

SWEETFLAG (CALAMUS):

Grows to 4' high. Cattail-like leaves. Bruised leaves smell like tangerines. Stalks covered with tiny green-yellow flowers (May to August). Found in shallow water swamps, wet fields, stream edges, etc.

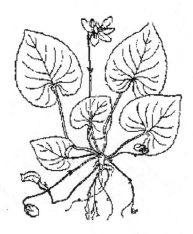

SWEET VIOLET:

Grows to 6" high. Green heart-shaped leaves. Scalloped edges. Sweet smelling, drooping, blue-violet flowers (April to June). Found on playgrounds, in yards, etc. Cultivated commercially.

TAMARACK:

Grows to 100' tall. Evergreen. Bears 3/4" long, oval cones. Needles in circular clusters. Thick foliage. Found in cold swamps, wet soils, lowlands, etc.

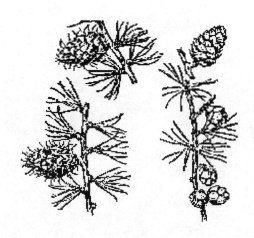

TANSY (COMMON):

Grows to 3' high. Smooth, dark green, fern-like leaves. Strong camphor smell. Tiny prickles on leaves. Loose clusters golden-yellow flowers (June to September). Found in fields, edges of woods, hedge rows, along roads, etc.

THYME, WILD:

Grows to 6" high. Bushy, short stalk, tiny leaves. Strongly pungent — spicy taste and odor. Tiny purple flower clusters at branch ends (July to August). Found in well-drained soil, rocky areas. etc. Widely cultivated.

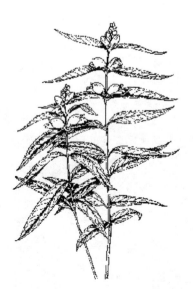

TURTLEHEAD:

Grows to 3' high. Lance-shaped, dark, shiny green leaves. White/pink flower clusters on spikes (August to September). Found near streams, wet woods, thickets, etc.

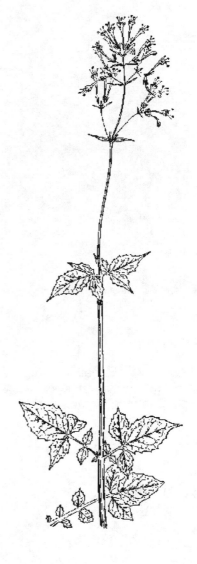

VALERIAN:

Grows to 5' tall. Almost hairless shrub. Root tastes warm, slightly bitter, nauseous. Mild fetid smell. Pink to white, mildly fragrant, flower clusters (June to July). Found in ditches, fields, stream banks, etc. Quite commonplace.

WAHOO TREE:

Grows to 25' tall. Slender shrub or small tree. Lusterless blue-green leaves, hairy underside. Crimson fruit (September). Deep purple flowers (June to July). Found in parks, gardens, rich woods, on river banks, etc.

WHITE OAK:

Grows to 110' tall. Bright olive-green leaves to 7" long. Light brown, sweet, edible acorns grow in pairs. Flowers (May and June). Found in dry woods, sandy places, gravelly ridges, etc.

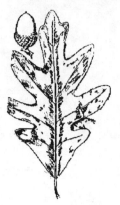

WHITE PINE:

Grows to 150' tall. Soft, light and dark blue-green needles. Dark brown cylindrical pine cones, 4" to 8" long. Found in light sandy soil. Common in eastern U.S.

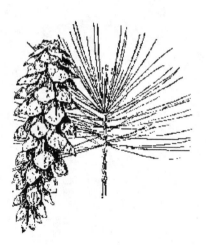

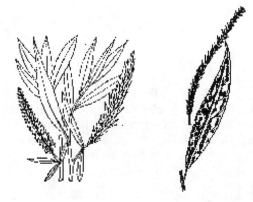

WHITE WILLOW:

Grows to 95' tall. Pliable, silky feeling branches. Bitter bark. Long, pointed, lance-shaped leaves. Hairy above and beneath. Found in Moist or wet woods, along river, stream edges, etc.

WILD HYDRANGEA:

Grows to 9' tall. Seven different bark layers on stem. Almost heart- shaped leaves, 3" to 6" long. Flat clusters of tiny white flowers (June to July). Found in rich woods, shaded places, banks of streams, etc.

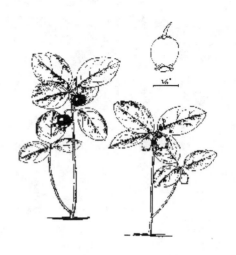

WINTERGREEN:

Grows to 6" high. Low creeping evergreen. Aromatic. Glossy, oval leaves. Dry red berries (Fall and Winter). Waxy, single white flowers like drooping bells (June to August). Found in dry, wooded areas, clearings, base of trees, etc.

WITCH HAZEL:

Grows to 15' tall. Hardy ornamental. Foliage turns bright orange, yellow and purple in Fall. Small clusters bright yellow flowers (September to December). Found in woods, moist and sandy soil, etc. Widely cultivated.

WOOD-SORREL, CREEPING:

Grows to 10" high. Clover-like leaves. Hairy stalks. White flowers, yellow at the base (April to June). Found on dry, open soil, pastures, lawns, gardens, etc.

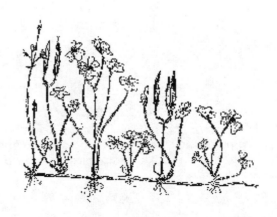

WORMWOOD:

Grows to 4' high. Extremely aromatic. Dark green leaves covered with silky hairs. Tiny drooping, greenish-yellow flowers (June to September). Found all over in waste grounds — roadsides, fields, etc.

YARROW COMMON:

Grows to 3' high. Lacy leaves. Attractive feathery foliage. Offensive smell and bitter taste. Daisy-like flower clusters on arching branches (May to October). Found throughout U.S. in pastures, roadsides, meadows, fields, etc.

YELLOW DOCK:

Grows to 5' tall. Lance-shaped leaves. Winged, heart-shaped seeds (June to September). Small green flowers in clusters at stem ends (May to September). Found in pastures, meadows, gardens, roadsides, etc.

YELLOWROOT:

Grows to 3' high. Small shrub. Bright yellow root. Small brown-purple flowers on drooping spikes (April to May). Found on moist stream banks, damp thickets and woods, etc.